Noise Theory and Application to Physics

Springer
New York
Berlin
Heidelberg
Hong Kong
London
Milan
Paris
Tokyo

Physics and Astronomy ONLINE LIBRARY

springeronline.com

Advanced Texts in Physics

This program of advanced texts covers a broad spectrum of topics that are of current and emerging interest in physics. Each book provides a comprehensive and yet accessible introduction to a field at the forefront of modern research. As such, these texts are intended for senior undergraduate and graduate students at the M.S. and Ph.D. levels; however, research scientists seeking an introduction to particular areas of physics will also benefit from the titles in this collection.

Philippe Réfrégier

Noise Theory and Application to Physics

From Fluctuations to Information

With 80 Figures

 Springer

Philippe Réfrégier
Institut Fresnel
D.U. St Jérôme
13397 Marseille cedex 20
France
philippe.refregier@fresnel.fr

Library of Congress Cataloging-in-Publication Data
Réfrégier, Philippe.
  Noise theory and application to physics : from fluctuations to information / Philippe Réfrégier.
    p.   cm. — (Advanced texts in physics)
  Includes bibliographical references and index.
  ISBN 0-387-20154-8 (alk. paper)
   1. Fluctuations (Physics)   2. Entropy (Information theory)   I. Title.   II. Series.
  QC6.4.F58R44   2003
  530.15′92—dc22   2003060458

ISBN 0-387-20154-8   Printed on acid-free paper.

Printed in the United States of America.   (AL/MVY)

9 8 7 6 5 4 3 2 1   SPIN 10949207

Springer-Verlag is a part of *Springer Science+Business Media*

springeronline.com

To the memory of my Father

Foreword

I had great pleasure in reading Philippe Réfrégier's book on the theory of noise and its applications in physics. The main aim of the book is to present the basic ideas used to characterize these unwanted random signals that obscure information content. To this end, the author devotes a sigificant part of his book to a detailed study of the probabilistic foundations of fluctuation theory.

Following a concise and accurate account of the basics of probability theory, the author includes a detailed study of stochastic processes, emphasizing the idea of the correlation function, which plays a key role in many areas of physics.

Physicists often assume that the noise perturbing a signal is Gaussian. This hypothesis is justified if one can consider that the noise results from the superposition of a great many independent random perturbations. It is this fact that brings the author to discuss the theory underlying the addition of random variables, accompanied by a wide range of illustrative examples.

Since noise affects information, the author is naturally led to consider Shannon's information theory, which in turn brings him to the altogether fundamental idea of entropy. This chapter is completed with a study of complexity according to Kolmogorov. This idea is not commonly discussed in physics and the reader will certainly appreciate the clear presentation within these pages.

In order to explain the nature of noise from thermal sources, Philippe Réfrégier then presents the essential features of statistical physics. This allows him to give a precise explanation of temperature. The chapter is very complete and omits none of the key ideas.

To conclude the work, the author devotes an important chapter to problems of estimation, followed by a detailed discussion of the examples presented throughout the book.

I am quite certain that this book will be highly acclaimed by physicists concerned with the problems raised by information transmission. It is well presented, rigorous without excess, and richly illustrated with examples which bring out the significance of the ideas under discussion.

February 2003   *Nino Boccara*

Preface

This book results from work carried out over the past few years as a member of the Physics and Image Processing team at the Fresnel Institute in the Ecole Nationale Supérieure de Physique de Marseille and the University of Aix Marseille III. In particular, it relates to the MSc programme in Optics, Image and Signal.

I would like to thank all my colleagues for many stimulating scientific discussions over the past years. Naturally, this involves for the main part the permanent academic staff of the Physics and Image Processing team, who are too numerous to list here. However, I wish to extend particular thanks to François Goudail for his invaluable help, both on the scientific level and in terms of the technical presentation of this book, and to Pierre Chavel for his judicious remarks and advice.

Nino Boccara, whose teachings have been extremely useful to me, accepted to write the foreword to this book, for which I am sincerely grateful.

Our understanding is fashioned and refined during scientific discussions with colleagues of the same or similar disciplines. I would therefore like to thank the GDR ISIS and the Société Française d'Optique who created such a favourable context for exchange.

Finally, I would like to acknowledge my debt towards Marie-Hélène and Nina who supported me when I undertook the task of writing this book.

Marseille, France   *Philippe Réfrégier*
July 2003

Contents

1 Introduction 1

2 Random Variables 5
   2.1 Random Events and Probability 6
   2.2 Random Variables 7
   2.3 Means and Moments 10
   2.4 Median and Mode of a Probability Distribution 12
   2.5 Joint Random Variables 13
   2.6 Covariance 16
   2.7 Change of Variables 18
   2.8 Stochastic Vectors 19
   Exercises 22

3 Fluctuations and Covariance 25
   3.1 Stochastic Processes 25
   3.2 Stationarity and Ergodicity 28
   3.3 Ergodicity in Statistical Physics 32
   3.4 Generalization to Stochastic Fields 34
   3.5 Random Sequences and Cyclostationarity 35
   3.6 Ergodic and Stationary Cases 40
   3.7 Application to Optical Coherence 41
   3.8 Fields and Partial Differential Equations 42
   3.9 Power Spectral Density 44
   3.10 Filters and Fluctuations 46
   3.11 Application to Optical Imaging 50
   3.12 Green Functions and Fluctuations 52
   3.13 Stochastic Vector Fields 56
   3.14 Application to the Polarization of Light 57
   3.15 Ergodicity and Polarization of Light 61
   3.16 Appendix: Wiener–Khinchine Theorem 64
   Exercises 66

4   Limit Theorems and Fluctuations 71
   4.1   Sum of Random Variables 71
   4.2   Characteristic Function 74
   4.3   Central Limit Theorem 76
   4.4   Gaussian Noise and Stable Probability Laws 80
   4.5   A Simple Model of Speckle 81
   4.6   Random Walks 89
   4.7   Application to Diffusion 92
   4.8   Random Walks and Space Dimensions 97
   4.9   Rare Events and Particle Noise 100
   4.10   Low Flux Speckle 102
   Exercises 104

5   Information and Fluctuations 109
   5.1   Shannon Information 109
   5.2   Entropy 111
   5.3   Kolmogorov Complexity 114
   5.4   Information and Stochastic Processes 117
   5.5   Maximum Entropy Principle 119
   5.6   Entropy of Continuous Distributions 122
   5.7   Entropy, Propagation and Diffusion 124
   5.8   Multidimensional Gaussian Case 128
   5.9   Kullback–Leibler Measure 130
   5.10   Appendix: Lagrange Multipliers 133
   Exercises 134

6   Thermodynamic Fluctuations 137
   6.1   Gibbs Statistics 137
   6.2   Free Energy 141
   6.3   Connection with Thermodynamics 142
   6.4   Covariance of Fluctuations 143
   6.5   A Simple Example 146
   6.6   Fluctuation–Dissipation Theorem 149
   6.7   Noise at the Terminals of an RC Circuit 153
   6.8   Phase Transitions 158
   6.9   Critical Fluctuations 161
   Exercises 163

7   Statistical Estimation 167
   7.1   The Example of Poisson Noise 167
   7.2   The Language of Statistics 169
   7.3   Characterizing an Estimator 169
   7.4   Maximum Likelihood Estimator 174
   7.5   Cramer–Rao Bound in the Scalar Case 177
   7.6   Exponential Family 179

7.7   Example Applications 181
7.8   Cramer–Rao Bound in the Vectorial Case 182
7.9   Likelihood and the Exponential Family 183
7.10   Examples in the Exponential Family 186
        7.10.1 Estimating the Parameter in the
               Poisson Distribution 187
        7.10.2 Estimating the Mean of the Gamma Distribution 187
        7.10.3 Estimating the Mean of the Gaussian Distribution 188
        7.10.4 Estimating the Variance of the Gaussian Distribution . . 189
        7.10.5 Estimating the Mean of the Weibull Distribution 190
7.11   Robustness of Estimators 192
7.12   Appendix: Scalar Cramer–Rao Bound 196
7.13   Appendix: Efficient Statistics 199
7.14   Appendix: Vectorial Cramer–Rao Bound 200
Exercises 205

8   Examples of Estimation in Physics 209
8.1   Measurement of Optical Flux 209
8.2   Measurement Accuracy in the Presence of Gaussian Noise 212
8.3   Estimating a Detection Efficiency 217
8.4   Estimating the Covariance Matrix 219
8.5   Application to Coherency Matrices 221
8.6   Making Estimates in the Presence of Speckle 224
8.7   Fluctuation–Dissipation and Estimation 225
Exercises 227

9   Solutions to Exercises 231
9.1   Chapter Two. Random Variables 231
9.2   Chapter Three. Fluctuations and Covariance 235
9.3   Chapter Four. Limit Theorems and Fluctuations 243
9.4   Chapter Five. Information and Fluctuations 250
9.5   Chapter Six. Statistical Physics 259
9.6   Chapter Seven. Statistical Estimation 266
9.7   Chapter Eight. Examples of Estimation in Physics 271

References 285

Index 287

1

Introduction

The aim of this book is to present the statistical basis for theories of noise in physics. More precisely, the intention is to cover the essential elements required to characterize noise (also referred to as fluctuations) and to describe optimization techniques for measurements carried out in the presence of such perturbations. Although this is one of the main concerns of any engineer or physicist, these ideas tend not to be tackled in a global manner. The approach developed here thus aims to provide the reader with a consistent view of this concept, in such a way that the various physical interpretations are not obscured by mathematical difficulties.

In the book, fluctuations are placed at the center of attention. In order to analyze them, our approach is based upon probability theory, the physics of linear systems, statistical physics, information theory and statistics.

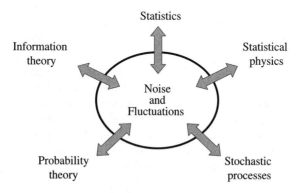

Fig. 1.1. The main themes of the book

Probability Theory

The principal mathematical tool used to describe noise and fluctuations is the theory of probability. Chapter 2 provides a brief presentation of the main ideas concerning random variables. Probability is not an easy subject and we strongly recommend the reader to check that he or she has a good grasp of the ideas in Chapter 2 before proceeding to the following chapters. There is perhaps a certain ambiguity in speaking of a mathematical tool and it might be better to speak rather of a language when the interaction between the physical situation under investigation and the mathematical concept chosen to describe it is so vast.

Stochastic Processes and Stochastic Fields

By far the most common way of characterizing noise consists in studying its second order properties as we discuss in chapter 3. The most general context for doing so is the framework of stochastic processes and stochastic fields, whose second order characterization is often a prerequisite for any quantitative study. We have sought to present a simplified version, common to physics and engineering science. Indeed, ideas and properties traditionally presented as different concepts often correspond to the same reality. In particular, we analyze problems related to the propagation of stochastic fields, filtering of stochastic processes and stochastic fields, and also stochastic vector fields, illustrating the latter by the study of light polarization. These ideas are applied to optical coherence and optical imaging.

Limit Behavior and Physical Applications

The random variables describing the phenomenon we happen to be observing may reflect a probability distribution that is encountered in a wide range of otherwise contrasting physical situations. Examples are the distributions that follow from the Gaussian, Gamma and Poisson probability laws. In Chapter 4, we show that these recurring patterns may be the result of limit behavior. An example is the behavior of a sum of random variables which, under certain rather general conditions, is described by a Gaussian random variable. This property provides us with many applications in physics: random walks, speckle in coherent imaging, and particle diffusion, not to mention Gaussian noise which constitutes a widely used model in physics. This chapter is also the place to introduce the characteriztic function, essential for calculating probabilities and characterizing noise in physical systems.

Information Theory and Applications to Noise

Although familiar to us, chance and randomness are complex notions, and the explanations used in the physical and engineering sciences are not always intuitively convincing. However, information theory can throw some light on these matters. Chapter 5 presents the main features of information theory insofar as they have a bearing upon the problem of noise. We analyze the question from two complementary standpoints: Shannon's approach and Chaitin and Kolmogorov's approach. These considerations lead naturally to the notion of entropy, which is a measure of the stochastic complexity of noise. We analyze and illustrate these ideas with a range of examples.

Statistical Physics

Any quantity associated with a physical system at nonzero temperature must fluctuate. This noise of thermal origins is thus inherent in the measurement of any macroscopic physical quantity and often represents the ultimate limit to the accuracy of our measurement systems. Statistical physics allows us to characterize these fluctuations at equilibrium in the sense of the second order moments. In Chapter 6, we first analyze those aspects of statistical physics needed to obtain a good understanding of fluctuations at thermodynamic equilibrium. We then focus on the characterization of fluctuations at equilibrium, illustrating these ideas in the context of electronic circuits.

Statistics

In a typical practical context, the physicist or engineer may be faced with quite the opposite problem to the one posed in the probabilistic approach. For the problem is often one of inferring quantities not considered as random, such as a mean particle flux, the mean voltage across a resistor, the covariance functions of a process, and so on, from a finite number of observations or measurements. This is therefore no longer a simple problem of probability, for we must now appeal to statistics. It is worth noting in this context that statistical physics is more probabilistic than genuinely statistical.

Statistics is a vast field and we shall only consider the rather restricted aspect of statistical inference whose aim is precisely to determine efficient methods of estimation from fluctuating measurements, i.e., from measurements made in the presence of noise. An important feature of statistics is the possibility of characterizing the accuracy of our estimates of measured quantities. The ultimate attainable accuracy cannot be arbitrarily small for a finite number of measurements and various lower bounds have been known for some time now. In chapter 7 we shall only consider the best known amongst these, the Cramer–Rao bound, and we shall study the conditions under which it may be attained.

Applications

In case there should be any doubt about it, this book is mainly oriented toward applications. The last chapter reconsiders the various examples used as illustrations throughout the rest of the book, turning to the specific problem of estimation. Characterizing, predicting and optimizing measurements are permanent concerns of the physicist or engineer. However, these goals cannot be achieved efficiently by applying rules of thumb. A deeper understanding of noise is essential. We hope that this book will help the reader by achieving a good compromise between theory and application.

2

Random Variables

The idea of a random variable or random event involves no assumption about the intrinsic nature of the phenomenon under investigation. Indeed, it may be a perfectly deterministic phenomenon, and yet a description of the measured quantities in terms of random variables can be extremely productive. If a mathematical theory is judged by its results, there can be no doubt that this is a very useful approach. In this book, we shall discuss this theory in detail, concentrating mainly on results that are relevant to applications in physics.

To fix ideas, let us illustrate the notion of a random variable by an example from everyday life. Time is a physical quantity whose deterministic nature is not difficult to accept. Suppose, however, that you are in the countryside without a watch and that the time is 16h37. It would be very difficult to estimate the time so precisely. On the other hand, you might be able to say that there are 90 chances out of 100 that the time is between 16h00 and 17h00; and if it is winter, that there are 100 chances out of 100 that the time is between 8h00 and 20h00, as there is rarely daylight during the night time. In other words, although the exact time can be considered as deterministic, the quantity to which one has access experimentally will only be an estimate. Now any estimate can be tainted with uncertainty, or noise as the engineers call it, which one may seek to characterize using the theory of probabilities.

We see therefore that, in the phenomenological context in which we find ourselves, the aim is not to investigate the intrinsic nature of the objects concerned, but rather to build up techniques using only the information available to us. We thus adopt this standpoint with regard to the relevant events, relaxing our hypotheses about the specific character of the events themselves. We do not pay attention to the essence of the object, but concentrate on the measurements and predictions we may make, a practice that has led to progress in a great many areas of physics.

2.1 Random Events and Probability

We begin with the straightforward observation of events which we describe as random in order to express the fact that we do not know what will be observed. Consider first the simple case where the set Ω of possible random events is finite, i.e., it contains a finite number of elements. This is the case, for example, for the set of possible outcomes on a lottery wheel. Suppose that this set Ω contains N possible events λ_i, where the index i takes values from 1 to N, so that $\Omega = \{\lambda_1, \ldots, \lambda_N\}$. This set Ω can be made up of quite arbitrary elements, with no particular mathematical structure. (The population described by the set is then said to be amorphous.) We may assign a number p_i between 0 and 1 to each event λ_i. This set of N positive numbers will be called a probability law on Ω if $p_1 + \cdots + p_N = 1$, or written more succinctly,

$$\sum_{i=1}^{N} p_i = 1 \, .$$

We then say that p_i is the probability of λ_i and we shall write $p_i = P(\lambda_i)$.

In the case where the set is infinite but countable, in the sense that the elements can be numbered by the positive integers, these ideas are easily generalized. We write $\Omega = \{\lambda_1, \lambda_2, \ldots, \lambda_n, \ldots\}$ and

$$p_1 + p_2 + \cdots + p_n + \cdots = 1 \quad \text{or} \quad \sum_{i=1}^{\infty} p_i = 1 \, .$$

With this definition of probability, it is sometimes possible to identify the probability with the frequency of occurrence of the relevant event. Consider the trivial example of tossing a coin. In this case, the two possible random events are "'heads" or "'tails." Therefore $\Omega = \{\text{tails}, \text{heads}\}$ with $\lambda_1 = \text{tails}$ and $\lambda_2 = \text{heads}$. Moreover, if the coin is not weighted, it is reasonable to set $p_1 = 1/2 = p_2$. Indeed, if the experiment is repeated a great many times, the coin will just as often give tails as heads. A six-sided die, or any other game, can be treated in the same manner.

The idea of identifying the probability with the frequency of occurrence of a random event, which one might call the frequency interpretation of probability, is the one most commonly adopted by physicists. It is nevertheless interesting to consider the possibility that the notion of probability might not be identified with the frequency of occurrence. Indeed, as explained above, one may be led to consider as random a quantity of perfectly deterministic nature. In this case, it will not be possible to carry out independent experiments and the probability will not be identifiable with a quantity resulting from an experiment. Much work has been devoted to this question, but we are not concerned here with such theoretical discussions. Let us simply note that everyday life is far from contradicting the former standpoint. Indeed it is common to hear such statements as: "This horse has a three in four chance of beating that one." The race itself is a single event and the probability of 3/4 mentioned here can in no way correspond to a frequency of occurrence. But this quantity may nevertheless prove useful to a gambler.

The set in question may be infinite. Consider a monkey typing on the keyboard of a computer. We may choose as random events the various possible words, i.e., sequences of letters that are not separated by a space. The set Ω of possible words is clearly infinite. To see this, we may imagine that the monkey typing on the computer keyboard never once presses on the space bar. One might object, quite rightly, that the animal has a finite lifespan, so that the set Ω must also be finite. Rather than trying to refine the example by finding a way around this objection, let us just say that it may be simpler to choose an infinite size for Ω than to estimate the maximal possible size. What matters in the end is the quality of the results obtained with the chosen model and the simplicity of that model.

Generalizing a little further, it should be noted that the set Ω of possible events may not only be infinite; it may actually be uncountable. In other words, it may be that the elements of the set Ω cannot be put into a one-to-one correspondence with the positive integers. To see this, suppose that we choose at random a real number between 0 and 1. In this case, we may identify Ω with an interval $[0, 1]$, and this is indeed uncountable in the above sense. This is a classic problem in mathematics. Let us outline our approach when the given set is uncountable. We consider the set of all subsets of Ω and we associate with every subset $\omega \subseteq \Omega$ a positive number $P(\omega)$. We then apply Kolmogorov's axioms to equip Ω with a probability law. To do so, the following conditions must be satisfied:

- $P(\Omega) = 1$ and $P(\emptyset) = 0$ (where $\emptyset$ is the empty set),
- if the subsets $A_1, A_2, \ldots, A_n, \ldots$ are pairwise disjoint, so that no pair of sets contains common elements, we must have[1]

$$P(A_1 \cup A_2 \cup \ldots \cup A_n \cup \ldots) = P(A_1) + P(A_2) + \cdots + P(A_n) + \cdots .$$

We see in this framework that we no longer speak of the probability of an event, but rather the probability of a set of events. Since a set may comprise a single element, the Kolmogorov axiom includes the definition of the probability of a single event. However, in the case where Ω is uncountably infinite, the probability of any single event will generally be zero. We shall return to this point when studying random variables, for the practical consequences here are very important.

2.2 Random Variables

A random variable is defined as a variable whose value is determined by a random experiment. More precisely, we consider a set Ω of random events λ

---

[1] $A \cup B$ denotes the set theoretic union of the two sets A and B.

and we associate with each of these events λ a value X_λ. If the possible values of X_λ are real numbers, we speak of a real random variable, whereas if they are complex numbers, we have a complex random variable. In the rest of this chapter we shall be concerned mainly with real- or integer-valued random variables. In the latter case, X_λ will be a whole number. In order to define the probability of a random variable, we proceed in two stages. We consider first the case where Ω is countable. The uncountable case will then lead to the idea of probability density.

If Ω is countable, we can define $p_i = P(\lambda_i)$ with $\sum_{i=1}^{\infty} p_i = 1$. The latter is also written $\sum_{\lambda \in \Omega} P(\lambda) = 1$, which simply means that the sum of the probabilities $P(\lambda)$ of each element of Ω must equal 1. Let x be a possible value of X_λ. Then $P(x)$ denotes the probability that X_λ is equal to x. We obtain this value by summing the probabilities of all random events in Ω such that $X_\lambda = x$. In the game of heads or tails, we may associate the value 0 to tails and 1 to heads. We thereby construct an integer-valued random variable. The probability $P(0)$ is thus $1/2$, as is $P(1)$. For a game with a six-sided die, we would have $P(1) = P(2) = \ldots = P(6) = 1/6$. If our die is such that the number 1 appears on one side, the number 2 on two sides, and the number 3 on three sides, we then set $P(1) = 1/6$, $P(2) = 1/3$, and $P(3) = 1/2$.

Letting D_x denote the set of possible values of X_λ, we see that we must have

$$\sum_{x \in D_x} P(x) = 1 \ .$$

Note in passing that, although X_λ is indeed a random variable, x itself is a parameter and hence a known quantity. This comment may appear a subtle theoretical distinction. However, a lack of understanding of this point could lead the reader into great difficulties later on.

We are now in a position to state some examples of well known and extremely useful probability laws. Bernoulli's law is perhaps one of the simplest. The random variable, also known as a Bernoulli variable, can take only the values 0 and 1. The probability that X_λ equals 1 is denoted q and the probability that X_λ equals 0 is thus $1 - q$. Hence, q is the only parameter of the Bernoulli law.

Poisson's law is also widely used. In this case the random variable X_λ can take any positive integer values. If $P(n)$ is the probability that X_λ is equal to n, Poisson's law is defined by

$$P(n) = e^{-\mu} \frac{\mu^n}{n!} \ ,$$

where μ is the single parameter determining the distribution and

$$n! = n \cdot (n-1) \cdot (n-2) \cdot \ldots \cdot 2 \cdot 1 \ .$$

As we shall discover later, the Poisson law is a simple model which allows us to describe a great many physical phenomena.

The situation is less simple when Ω is uncountable. It leads to further mathematical complexity in probability theory, requiring the use of measure theory. However, for our present purposes, it would not be useful to go into a detailed presentation of this subject. We shall therefore sidestep this difficulty by working directly with random variables. When we need to refer to random events to illustrate some physical concept, it will suffice to restrict ourselves to the countable case. Let us consider a random variable X_λ defined on an uncountable set Ω. In this case, the range of values of X_λ usually constitutes a continuous set. We then speak of a continuous random variable, as opposed to a discrete random variable which takes values in a countable set. The probability that X_λ is equal to some given precise value x is generally zero, rendering this notion somewhat irrelevant. It is basically for this reason that it is useful to introduce a distribution function $F_X(x)$ which gives the probability that X_λ is smaller than x. Letting ω_x be the subset of Ω containing those elements λ such that $X_\lambda \leq x$, we then have $P(\omega_x) = F_X(x)$. We define the probability density function of the variable X_λ as the derivative[2] of $F_X(x)$:

$$P_X(x) = \frac{\mathrm{d}F_X(x)}{\mathrm{d}x} .$$

As $F_X(\infty) = 1$ and $F_X(x) = \int_{-\infty}^{x} P_X(y)\mathrm{d}y$, we deduce that

$$\int_{-\infty}^{\infty} P_X(x)\mathrm{d}x = 1 .$$

We can give a simple interpretation of the probability density function. Suppose that the probability density is continuous at point x. The probability that X_λ lies between $x - \mathrm{d}x/2$ and $x + \mathrm{d}x/2$ is then of the order of $P_X(x)\mathrm{d}x$ for small $\mathrm{d}x$ and this relation is an equality in the limit as $\mathrm{d}x$ tends to 0. We should therefore bear in mind the fact that only $F_X(x)$ actually represents a probability, and that the same cannot be said of $P_X(x)$. Table 2.1 gives some of the more commonly encountered probability density functions in physics.

The appearance of the derivative should be treated with some caution from the mathematical standpoint, and precise mathematical conditions must be formulated in order to apply it correctly. One practical solution for simplifying the formalism consists in treating the quantities $F_X(x)$ and $P_X(x)$ as mathematical distributions. Here again we choose not to labor the details on this technical aspect. It is nevertheless worth noting that the use of distributions provides a unified framework for both discrete and continuous variables. Indeed, consider the probability law p_n, where n is a natural number. This law can be written in the form of a probability density function:

$$P_X(x) = \sum_n p_n \delta(x - n) ,$$

---

[2] We shall often need to consider this derivative in the sense of distributions.

Table 2.1. A selection of probability density functions commonly occurring in physics. Note that the function $\Gamma(x)$ is defined for positive x by $\Gamma(x) = \int_0^\infty u^{x-1}e^{-u}du$

Name	Probability density	Parameters
Uniform	$1/(b-a)$ if $a \le x \le b$ and 0 otherwise	a and b
Gaussian	$\dfrac{1}{\sqrt{2\pi}\sigma}\exp\left[-\dfrac{(x-m)^2}{2\sigma^2}\right]$	m and σ
Exponential	$a\exp(-ax)$ if $x \ge 0$ and 0 otherwise	a
Gamma	$\dfrac{\beta^\alpha x^{\alpha-1}}{\Gamma(\alpha)}\exp(-\beta x)$ if $x \ge 0$ and 0 otherwise	α and β
Cauchy	$\dfrac{a}{\pi(a^2+x^2)}$ with $a > 0$	a

where $\delta(x-n)$ is the Dirac distribution centered on n. The latter is defined by its action on continuous functions $f(x)$, viz.,

$$\int f(x)\delta(x-n)\,\mathrm{d}x = f(n) \ .$$

We thus find

$$\int f(x)P_X(x)\,\mathrm{d}x = \sum_n p_n f(n) \ .$$

2.3 Means and Moments

The moments of a probability law play an important role in physics. They are defined by

$$\langle X_\lambda^r \rangle = \int x^r P_X(x)\,\mathrm{d}x \ .$$

Moments are generally only considered for positive integer values of r. However, we may equally well choose any positive real value of r. It should be emphasized that there is *a priori* no guarantee of the existence of the moment of order r. For example, although all moments with $r \ge 0$ exist for the Gaussian law, only those moments $\langle X_\lambda^r \rangle$ with $r \in [0,1)$ exist for the Cauchy law. Indeed, in the latter case, the moment $\langle X_\lambda \rangle$ is not defined because

$$\int_{-\infty}^{\infty} \frac{ax}{\pi(a^2+x^2)}\,\mathrm{d}x$$

is not a convergent integral.

For integer-valued discrete random variables, we have

$$\langle X_\lambda^r \rangle = \sum_n n^r p_n .$$

The first moments play a special role, because they are very often considered in physical problems. For $r = 1$, we obtain the mean value of the random variable:

$$m_X = \langle X_\lambda \rangle = \int x P_X(x) \, dx .$$

We also consider the variance, which is the second central moment

$$\sigma_X^2 = \langle (X_\lambda - m_X)^2 \rangle = \int (x - m_X)^2 P_X(x) \, dx .$$

The physical interpretation is simple. σ_X, also known as the standard deviation, represents the width of the probability density function, whilst m_X is the value on which the probability density is centered (see Fig. 2.2). Quite generally, we define the central moment of order r by $\langle (x - \langle x \rangle)^r \rangle$. Table 2.2 shows the mean and variance of various probability laws.

Table **2.2.** Mean and variance of various probability laws

Name	Mean	Variance
Bernoulli	q	$q(1-q)$
Poisson	μ	μ
Uniform	$(a+b)/2$	$(a-b)^2/12$
Gaussian	m	σ^2
Exponential	$1/a$	$1/a^2$
Gamma	α/β	α/β^2
Cauchy	Undefined	Undefined

The moments just defined correspond to mean values involving the probability density function. They are referred to as expected values or expectation values, or simply expectations. These quantities are defined from deterministic variables x rather than random variables X_λ. They are clearly deterministic quantities themselves. Although they characterize the probability density function, they cannot be directly ascertained by experiment. In fact, they can only be estimated. We shall return to the idea of estimation in Chapter 7. Let us content ourselves here with a property which illustrates the practical significance of moments. This property follows from the weak law of large numbers, which tells us that if we consider independent realizations of random variables $Y_{\lambda(j)}$ distributed according to the same law with mean m_Y, then $S_\lambda(n)$ defined by

$$S_\lambda(n) = \frac{Y_{\lambda(1)} + Y_{\lambda(2)} + \cdots + Y_{\lambda(n)}}{n}$$

converges in probability to m_Y. Convergence in probability means that, as n tends to infinity ($n \to \infty$), the probability that $S_\lambda(n)$ takes a value different from m_Y tends to zero. We see that, if $\langle f(X_\lambda) \rangle$ exists [with $\langle f(X_\lambda) \rangle = \int f(x) P_X(x) \, \mathrm{d}x$], and if we make n measurements of X_λ, then

$$\frac{f(X_{\lambda(1)}) + f(X_{\lambda(2)}) + \cdots + f(X_{\lambda(n)})}{n}$$

will be an approximation for $\langle f(X_\lambda) \rangle$ which improves as the number of measurements n increases. It is this basic property which confers a genuine practical meaning upon the definitions of expectation values. We may thus envisage them as the result of calculating the mean over an infinite number of random experiments which are independent but arise from the same statistical ensemble with the same probability law. Indeed it is this fact which justifies the name of expectation value. Once again, the reader is encouraged to reflect carefully on the idea of expectation value. It is often a misunderstanding of this basic notion which leads to serious errors of reasoning.

2.4 Median and Mode of a Probability Distribution

We have just observed that the Cauchy distribution has no mean because the integral

$$\lim_{a \to -\infty} \lim_{b \to \infty} \int_a^b x P_X(x) \mathrm{d}x$$

does not converge. Note, however, that it is symmetrical about the origin, so that the value $x = 0$ must play some special role. This is indeed what we conclude when we introduce the notions of mode and median for this probability law.

When it exists, the median x_M is defined as the value of x such that

$$\int_{-\infty}^{x_\mathrm{M}} P_X(x) \, \mathrm{d}x = \int_{x_\mathrm{M}}^{\infty} P_X(x) \, \mathrm{d}x \ .$$

It is easy to check that the median of the Cauchy distribution is 0. It is also a straightforward matter to show that the median of the Gaussian distribution is just its mean. This last result, whereby the median is equal to the mean, is true for all probability laws that possess a mean and that are symmetrical about it.

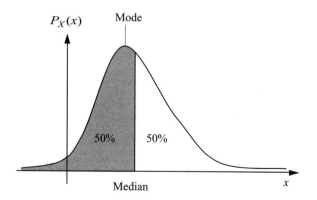

Fig. 2.1. Intuitive significance of the median and mode

The mode of a probability distribution corresponds to the most probable value in the case of discrete random variables and to the value at which the probability density function is maximal in the case of continuous random variables. The mode may not therefore consist of a unique value. Considering the case of continuous random variables, when the mode x_P exists and is unique, we must have

$$x_P = \operatorname*{argmax}_x \; [P_X(x)] \; ,$$

which simply means that it is the value of x which maximizes $P_X(x)$, see Fig. 2.2. It is then clear that, if $P_X(x)$ is differentiable, the mode satisfies the relation

$$\frac{\partial}{\partial x} P_X(x_P) = 0 \; ,$$

or equivalently,

$$\frac{\partial}{\partial x} \ln\left[P_X(x_P)\right] = 0 \; .$$

It is thus easy to see that the mode of a Gaussian distribution is equal to its mean.

2.5 Joint Random Variables

It is common in physics to measure several quantities during an experiment. These quantities are not necessarily independent and we are often led to characterize their dependence. The notion of correlation partly achieves this aim. Let us define the idea of noise in a rather pragmatic way for the time being as

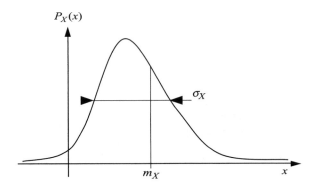

Fig. 2.2. Intuitive significance of mean and variance

something which impairs successive measurements of the same quantity with fluctuations. Imagine for example that we carry out a measurement using an experimental setup comprising a sensor and an amplifier. Suppose further that we wish to find out whether the noise observed during the recording of the measurements is noise arising in the sensor or noise arising in the amplifier. The notions of correlation and statistical dependence allow us to answer this practical question in a precise manner.

In a more mathematical context, suppose that for each random event λ we define two random variables X_λ and Y_λ. It may be interesting to consider the probability law for joint observation of certain values of X_λ and Y_λ. Likewise, knowing the value assumed by one of the random variables may provide information that will help us to determine the most likely values for the other random variable. Bayes' relation provides a precise formulation of this problem.

Consider two subsets A and B of Ω, where Ω represents the set of all possible random events. The probability of observing an event which belongs simultaneously to A and B is $P(A \cap B)$, where probabilities are measures of sets as illustrated in Fig. 2.3. $P(A \cup B)$ represents the probability that an element belongs to A or B. If we know *a priori* that the observed element belongs to B, the probability that it also belongs to A corresponds to the relative measure of A in B, that is, $P(A \cap B)/P(B)$, assuming of course that $P(B)$ is nonzero. If we use $P(A|B)$ to denote the probability that a random event belongs to A given that it belongs to B, we have Bayes' relation

$$P(A|B) = P(A, B)/P(B),$$

where we adopt the standard notation $P(A, B) = P(A \cap B)$. It follows immediately that

$$P(A, B) = P(A\,|\,B)P(B) = P(B\,|\,A)P(A) \ .$$

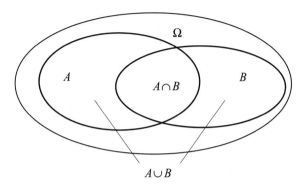

Fig. 2.3. Sets used in the definition of conditional probabilities

The joint distribution function $F_{X,Y}(x, y)$ relative to the two random variables X_λ and Y_λ is defined as the probability that simultaneously X_λ is less than x and Y_λ less than y. Let 1_x denote the set of random events in Ω such that X_λ is less than x. Likewise, let 1_y denote the set of random events in Ω such that Y_λ is less than y. Then it follows that $F_X(x) = P(1_x)$ and $F_Y(y) = P(1_y)$, and also that

$$F_{X,Y}(x, y) = P(1_x \cap 1_y) \ .$$

Since $P(1_x, 1_y) = P(1_x|1_y)P(1_y)$, we deduce that

$$F_{X,Y}(x, y) = F_{X|Y}(x|y)F_Y(y) \ .$$

We define the joint probability density function as

$$P_{X,Y}(x, y) = \frac{\partial^2 F_{X,Y}(x, y)}{\partial x \partial y} \ .$$

We have seen that $P_X(x) = \partial F_X(x)/\partial x$ and $P_Y(y) = \partial F_Y(y)/\partial y$, and we may now define the conditional probability density function as

$$P_{X|Y}(x|y) = \frac{P_{X,Y}(x, y)}{P_Y(y)} \ ,$$

and symmetrically,

$$P_{Y|X}(y|x) = \frac{P_{X,Y}(x, y)}{P_X(x)} \ .$$

To complete these relations, note that $F_{X,Y}(x,\infty) = F_X(x)$ and hence,

$$P_X(x) = \int\limits_{-\infty}^{\infty} P_{X,Y}(x,y)\mathrm{d}y \ ,$$

and likewise,

$$P_Y(y) = \int\limits_{-\infty}^{\infty} P_{X,Y}(x,y)\mathrm{d}x \ .$$

We say that $P_{X,Y}(x,y)$ is the joint probability law and that $P_X(x)$ and $P_Y(y)$ are the marginal probability laws.

2.6 Covariance

We are now in a position to give a precise definition of the independence of two random variables X_λ and Y_λ. These two random variables are independent if

$$P_{X,Y}(x,y) = P_X(x)P_Y(y) \ ,$$

which then implies that $P_{X|Y}(x|y) = P_X(x)$ and $P_{Y|X}(y|x) = P_Y(y)$. In other words, knowing the value of a realization of Y_λ tells us nothing about the value of X_λ since $P_{X|Y}(x|y) = P_X(x)$, and likewise, knowing the value of X_λ tells us nothing about the value of a realization of Y_λ.

The second extreme situation corresponds to the case where there is a perfectly deterministic relationship between X_λ and Y_λ, which we denote by $Y_\lambda = g(X_\lambda)$. Clearly, in this case, when the value of a realization of X_λ is known, only the value $g(X_\lambda)$ is possible for Y_λ, and we write

$$P_{Y|X}(y|x) = \delta\big(y - g(x)\big) \ .$$

Intermediate cases are interesting since they correspond to many practical situations. In order to measure the correlation between the two random variables X_λ and Y_λ, we might try to estimate the conditional probability density function $P_{X|Y}(x|y)$. However, the task is often impossible from a practical point of view and the notion of covariance is generally preferred. The covariance Γ_{XY} is defined by

$$\Gamma_{XY} = \langle X_\lambda Y_\lambda \rangle - \langle X_\lambda \rangle \langle Y_\lambda \rangle \ ,$$

or more explicitly,

$$\Gamma_{XY} = \iint (xy - m_X m_Y) P_{X,Y}(x,y)\mathrm{d}x\mathrm{d}y \ ,$$

where $m_X = \langle X_\lambda \rangle$ and $m_Y = \langle Y_\lambda \rangle$.

It can be shown that $|\Gamma_{XY}|^2 \leq \sigma_X^2 \sigma_Y^2$. Indeed, consider the quadratic form $(\alpha \delta X_\lambda - \delta Y_\lambda)^2$, where $\delta X_\lambda = X_\lambda - \langle X_\lambda \rangle$ and $\delta Y_\lambda = Y_\lambda - \langle Y_\lambda \rangle$. Since this form it positive definite, its expectation value must also be positive. Expanding out this expression, we obtain

$$\alpha^2 \langle (\delta X_\lambda)^2 \rangle - 2\alpha \langle \delta X_\lambda \delta Y_\lambda \rangle + \langle (\delta Y_\lambda)^2 \rangle \geq 0 \ .$$

The discriminant of this quadratic form in α must be negative, since it has at most one root. This implies that

$$\langle \delta X_\lambda \delta Y_\lambda \rangle^2 - \langle (\delta X_\lambda)^2 \rangle \langle (\delta Y_\lambda)^2 \rangle \leq 0 \ ,$$

which proves the claim.

It is therefore common to introduce the correlation coefficient, defined as $\rho_{XY} = \Gamma_{XY}/\sigma_X \sigma_Y$, which takes values between -1 and $+1$. From a practical standpoint, if the absolute value of ρ_{XY} is equal to 1, the two random variables are perfectly correlated. (To be precise, they must be proportional almost everywhere.) However, if ρ_{XY} is equal to 0, they are not correlated. This is the case, for example, if the two random variables X_λ and Y_λ are independent. It should nevertheless be borne in mind that, although the independence of two random variables does indeed imply that they are uncorrelated,[3] i.e., that $\rho_{XY} = 0$, the converse is false. This property is easy to demonstrate and it is a straightforward matter to construct examples of dependent random variables for which $\rho_{XY} = 0$. Consider, for example, the random variable Φ_λ uniformly distributed between 0 and 2π. Then set $X_\lambda = \sin \Phi_\lambda$ and $Y_\lambda = \cos \Phi_\lambda$. It follows that $\langle X_\lambda Y_\lambda \rangle = \langle \sin \Phi_\lambda \cos \Phi_\lambda \rangle$ or

$$\langle X_\lambda Y_\lambda \rangle = \int \sin \phi \cos \phi P(\phi) d\phi = \frac{1}{2\pi} \int_0^{2\pi} \sin \phi \cos \phi \, d\phi = 0 \ .$$

The random variables X_λ and Y_λ are therefore uncorrelated. However, they are not independent, since $(X_\lambda)^2 + (Y_\lambda)^2 = 1$.

---

[3] Introducing once again the centered variables $\delta X_\lambda = X_\lambda - \langle X_\lambda \rangle$ and $\delta Y_\lambda = Y_\lambda - \langle Y_\lambda \rangle$, it is easy to see that $\langle \delta X_\lambda \rangle = 0 = \langle \delta Y_\lambda \rangle$ and that $\Gamma_{XY} = \langle \delta X_\lambda \delta Y_\lambda \rangle$, i.e.,

$$\Gamma_{XY} = \iint xy P_{\delta X, \delta Y}(x, y) dx dy \ ,$$

noting that we are considering the probability density functions of the centered variables δX_λ and δY_λ. Clearly, we have $P_{\delta X}(x) = P_X(x - \langle X_\lambda \rangle)$ and $P_{\delta Y}(y) = P_Y(y - \langle Y_\lambda \rangle)$. Since by hypothesis $P_{X,Y}(x, y) = P_X(x)P_Y(y)$, we can deduce from the above that $P_{\delta X, \delta Y}(x, y) = P_{\delta X}(x)P_{\delta Y}(y)$. It thus follows that $\Gamma_{XY} = \iint xy P_{\delta X}(x)P_{\delta Y}(y) dx dy$ and hence

$$\Gamma_{XY} = \int x P_{\delta X}(x) dx \int y P_{\delta Y}(y) dy \ .$$

This in turn means that $\Gamma_{XY} = \langle \delta X_\lambda \rangle \langle \delta Y_\lambda \rangle$ and thus $\Gamma_{XY} = 0$.

2.7 Change of Variables

Given the probability density function $P_X(x)$ of a random variable X_λ, one often seeks in physics to determine the density of a related random variable $Y_\lambda = g(X_\lambda)$, where g is a function, assumed continuous. For example, in electromagnetism or optics, given the probability density function of the amplitude A_λ of the field, one may need to know the probability density function of the intensity $I_\lambda = |A_\lambda|^2$. In electronics, the output voltage V_λ of a component may depend on the applied voltage U_λ according to a relation of the form $V_\lambda = a \exp[\alpha(U_\lambda - U_0)]$. In order to determine the probability density function of fluctuations in the output physical quantity in terms of the probability density function of the input to the component, a change of variables calculation is required. This is the subject of the present section.

Suppose to begin with that the function $y = g(x)$ is increasing and differentiable, hence bijective. Let $F_X(x)$ and $F_Y(y)$ denote the distribution functions of X_λ and Y_λ. The probability that Y_λ is less than $g(x)$ is equal to the probability that X_λ is less than x. Hence, $F_Y[g(x)] = F_X(x)$. Differentiating, we obtain

$$\frac{\mathrm{d}F_Y[g(x)]}{\mathrm{d}x} = \frac{\mathrm{d}F_Y[g(x)]}{\mathrm{d}g(x)}\frac{\mathrm{d}g(x)}{\mathrm{d}x} = \frac{\mathrm{d}F_X(x)}{\mathrm{d}x} .$$

Moreover, since $y = g(x)$, writing

$$g'(x) = \frac{\mathrm{d}g(x)}{\mathrm{d}x} ,$$

we obtain

$$P_Y(y) = \frac{1}{g'(x)}P_X(x) .$$

Noting that $g'(x) = \mathrm{d}y/\mathrm{d}x$, the above expression can also be written in the more memorable form (see Fig. 2.4)

$$P_Y(y)\mathrm{d}y = P_X(x)\mathrm{d}x .$$

If the relation $y = g(x)$ is not bijective, the above argument can be applied to intervals where it is bijective, adding the contributions from the various intervals for each value of y.

Considering the case where the probability density function $P_A(a)$ of the amplitude A_λ (assumed to be real-valued) of the electric field is Gaussian with zero mean and variance σ^2, let us determine the probability density function $P_I(I)$ of the intensity $I_\lambda = |A_\lambda|^2$. To do so, we begin with the positive values of a. Hence,

$$P_I^+(I)\mathrm{d}I = \frac{1}{\sqrt{2\pi}\sigma} \exp\left(-\frac{a^2}{2\sigma^2}\right)\mathrm{d}a .$$

Now $\mathrm{d}I = 2a\,\mathrm{d}a$ and $a = \sqrt{I}$, which implies

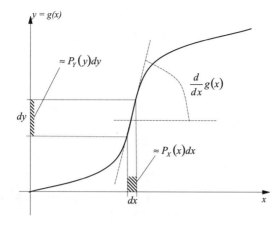

Fig. 2.4. Transformation of probability density upon change of variable

$$P_I^+(I) = \frac{1}{2\sqrt{2\pi I}\sigma} \exp\left(-\frac{I}{2\sigma^2}\right) \; .$$

In the same manner we obtain for negative values

$$P_I^-(I) = \frac{1}{2\sqrt{2\pi I}\sigma} \exp\left(-\frac{I}{2\sigma^2}\right) \; .$$

For each value of I, we may have $a = \sqrt{I}$ or $a = -\sqrt{I}$, and we thus deduce that $P_I(I) = P_I^+(I) + P_I^-(I)$. Hence,

$$P_I(I) = \frac{1}{\sqrt{2\pi I}\sigma} \exp\left(-\frac{I}{2\sigma^2}\right) \; .$$

2.8 Stochastic Vectors

A stochastic vector $\boldsymbol{X}_\lambda$ is a vector whose value is determined from the outcome of a random experiment. As for random variables, we consider a set Ω of random events λ and associate a random vector $\boldsymbol{X}_\lambda$ with λ. If the possible values of the components of $\boldsymbol{X}_\lambda$ are real numbers, we shall speak of a real stochastic vector. If they are complex numbers, we have a complex stochastic vector.

For the moment, we discuss the case of real N-dimensional stochastic vectors. The stochastic vector can be described by its components, viz.,

$$\boldsymbol{X}_\lambda = \big(X_\lambda(1), X_\lambda(2), \ldots, X_\lambda(N)\big)^{\mathrm{T}} \, ,$$

where the symbol T indicates that we consider the transposed vector. We thus see that a stochastic vector is simply equivalent to a system of N random variables. The distribution function $F_X(\boldsymbol{x})$ is the joint probability that $X_\lambda(j)$ is less than or equal to x_j for all j in the range from 1 to N, with $\boldsymbol{x} = (x_1, x_2, \ldots, x_N)^{\mathrm{T}}$. In other words,

$$F_X(\boldsymbol{x}) = \mathrm{Prob}\left[X_\lambda(1) \leqslant x_1,\ X_\lambda(2) \leqslant x_2,\ \ldots,\ X_\lambda(N) \leqslant x_N\right] .$$

In the case where the components are continuous variables, it is a simple matter to find the probability density function:

$$P_X(\boldsymbol{x}) = \frac{\partial^N}{\partial x_1 \partial x_2 \ldots \partial x_N} F_X(\boldsymbol{x}) .$$

In the complex case, let $X_\lambda(j) = X_\lambda^{\mathrm{R}}(j) + \mathrm{i} X_\lambda^{\mathrm{I}}(j)$, where $X_\lambda^{\mathrm{R}}(j)$ and $X_\lambda^{\mathrm{I}}(j)$ are the real and imaginary parts of the component $X_\lambda(j)$. The distribution function is then

$$F_X(\boldsymbol{x}) = \mathrm{Prob}\big[X_\lambda^{\mathrm{R}}(1) \leqslant x_1^{\mathrm{R}},\ X_\lambda^{\mathrm{I}}(1) \leqslant x_1^{\mathrm{I}},\ X_\lambda^{\mathrm{R}}(2) \leqslant x_2^{\mathrm{R}},\ X_\lambda^{\mathrm{I}}(2) \leqslant x_2^{\mathrm{I}},$$
$$\ldots,\ X_\lambda^{\mathrm{R}}(N) \leqslant x_N^{\mathrm{R}},\ X_\lambda^{\mathrm{I}}(N) \leqslant x_N^{\mathrm{I}}\big] ,$$

and the probability density function is

$$P_X(\boldsymbol{x}) = \frac{\partial^{2N}}{\partial x_1^{\mathrm{R}} \partial x_1^{\mathrm{I}} \partial x_2^{\mathrm{R}} \ldots \partial x_N^{\mathrm{R}} \partial x_N^{\mathrm{I}}} F_X(\boldsymbol{x}) .$$

An N-dimensional complex stochastic vector is thus equivalent to a $2N$-dimensional real stochastic vector.

The covariance matrix $\overline{\overline{\varGamma}}$ plays a central role in many situations. It is defined by its components:

$$\varGamma_{ij} = \big\langle X_\lambda(i)\left[X_\lambda(j)\right]^*\big\rangle - \langle X_\lambda(i)\rangle\langle[X_\lambda(j)]^*\rangle ,$$

where a^* is the complex conjugate of a. If the stochastic vector is real-valued, the above formula simplifies to

$$\varGamma_{ij} = \langle X_\lambda(i) X_\lambda(j)\rangle - \langle X_\lambda(i)\rangle\langle X_\lambda(j)\rangle .$$

The covariance matrix can be directly formulated in terms of the stochastic vector using

$$\overline{\overline{\varGamma}} = \Big\langle \boldsymbol{X}_\lambda\left[\boldsymbol{X}_\lambda\right]^\dagger\Big\rangle - \langle\boldsymbol{X}_\lambda\rangle\Big\langle\left[\boldsymbol{X}_\lambda\right]^\dagger\Big\rangle ,$$

where $\boldsymbol{a}^\dagger$ is the transposed conjugate of $\boldsymbol{a}$. Indeed, it is easy to see that

$$\varGamma_{ij} = \Big\langle X_\lambda(i)\left[X_\lambda(j)\right]^*\Big\rangle - \langle X_\lambda(i)\rangle\Big\langle\left[X_\lambda(j)\right]^*\Big\rangle$$

is equivalent to

$$\overline{\overline{\varGamma}} = \Big\langle \boldsymbol{X}_\lambda\left[\boldsymbol{X}_\lambda\right]^\dagger\Big\rangle - \langle\boldsymbol{X}_\lambda\rangle\Big\langle\left[\boldsymbol{X}_\lambda\right]^\dagger\Big\rangle .$$

Note that if $\boldsymbol{a}$ and $\boldsymbol{b}$ are two N-component vectors, $\boldsymbol{a}^\dagger \boldsymbol{b}$ is a scalar, since it is in fact the scalar product of $\boldsymbol{a}$ and $\boldsymbol{b}$, whilst $\boldsymbol{b} \boldsymbol{a}^\dagger$ is an $N \times N$ tensor with ij th component $b_i a_j^*$. This formulation is sometimes useful for simplifying certain proofs. For example, we can show that the covariance matrices are positive. For simplicity, we assume here that the mean value of $\boldsymbol{X}_\lambda$ is zero. If it is not, we can consider $\boldsymbol{Y}_\lambda = \delta \boldsymbol{X}_\lambda = \boldsymbol{X}_\lambda - \langle \boldsymbol{X}_\lambda \rangle$. For any vector $\boldsymbol{a}$, the modulus squared of the scalar product $\boldsymbol{a}^\dagger \boldsymbol{X}_\lambda$ is positive or zero, i.e.,

$$\left| \boldsymbol{a}^\dagger \boldsymbol{X}_\lambda \right|^2 \geq 0 .$$

This expression can be written

$$\left(\boldsymbol{a}^\dagger \boldsymbol{X}_\lambda \right) \left[\left(\boldsymbol{X}_\lambda \right)^\dagger \boldsymbol{a} \right] \geqslant 0 ,$$

or

$$\boldsymbol{a}^\dagger \boldsymbol{X}_\lambda \left(\boldsymbol{X}_\lambda \right)^\dagger \boldsymbol{a} \geqslant 0 .$$

Taking the expectation value of this expression, viz.,

$$\left\langle \boldsymbol{a}^\dagger \boldsymbol{X}_\lambda \left(\boldsymbol{X}_\lambda \right)^\dagger \boldsymbol{a} \right\rangle = \boldsymbol{a}^\dagger \left\langle \boldsymbol{X}_\lambda \left(\boldsymbol{X}_\lambda \right)^\dagger \right\rangle \boldsymbol{a} ,$$

and using the fact that

$$\overline{\overline{\varGamma}} = \left\langle \boldsymbol{X}_\lambda \left[\boldsymbol{X}_\lambda \right]^\dagger \right\rangle ,$$

we obtain for any $\boldsymbol{a}$ the relation

$$\boldsymbol{a}^\dagger \overline{\overline{\varGamma}} \boldsymbol{a} \geqslant 0 ,$$

which shows that any covariance matrix is positive. From

$$\varGamma_{ij} = \left\langle X_\lambda(i) \left[X_\lambda(j) \right]^* \right\rangle ,$$

we see immediately that we have a Hermitian matrix, i.e., $\overline{\overline{\varGamma}}^\dagger = \overline{\overline{\varGamma}}$, since

$$\left\{ X_\lambda(j) \left[X_\lambda(i) \right]^* \right\}^* = X_\lambda(i) \left[X_\lambda(j) \right]^* .$$

Now it is well known that any Hermitian matrix can be diagonalized by a unitary transition matrix and that it has real eigenvalues. The covariance matrix is thus diagonalizable with zero or positive real eigenvalues and mutually orthogonal eigenvectors.

Now consider the example of real Gaussian N-dimensional stochastic vectors with mean $\boldsymbol{m}$ and covariance matrix $\overline{\overline{\varGamma}}$. Let $\overline{\overline{K}}$ be the inverse matrix of $\overline{\overline{\varGamma}}$. Then the probability density function is

$$P_X(\boldsymbol{x}) = \frac{1}{\left(\sqrt{2\pi} \right)^N \sqrt{\left| \overline{\overline{\varGamma}} \right|}} \exp \left[-\frac{1}{2} (\boldsymbol{x} - \boldsymbol{m})^\dagger \overline{\overline{K}} (\boldsymbol{x} - \boldsymbol{m}) \right] ,$$

where $\left|\overline{\overline{\Gamma}}\right|$ is the determinant of $\overline{\overline{\Gamma}}$. This expression can be written in the form

$$P_X(\boldsymbol{x}) = \frac{1}{\left(\sqrt{2\pi}\right)^N \sqrt{\left|\overline{\overline{\Gamma}}\right|}} \exp\left[-\frac{1}{2}\sum_{i=1}^{N}\sum_{j=1}^{N}(x_i - m_i)K_{ij}(x_j - m_j)\right] .$$

This simply means that

$$P_X(x_1, x_2, \ldots, x_N) = \frac{1}{(\sqrt{2\pi})^N \sqrt{\left|\overline{\overline{\Gamma}}\right|}} \exp\left[-\frac{1}{2}Q(x_1, x_2, \ldots, x_N)\right] ,$$

where

$$Q(x_1, x_2, \ldots, x_N) = \sum_{i=1}^{N}\sum_{j=1}^{N}(x_i - m_i)K_{ij}(x_j - m_j) ,$$

and

$$\Gamma_{ij} = \int \ldots \int (x_i x_j - m_i m_j)P_X(x_1, x_2, \ldots, x_N)\mathrm{d}x_1\mathrm{d}x_2\ldots\mathrm{d}x_N .$$

Exercises

Exercise 2.1. Probability and Probability Density Function

Let X_λ be a random variable uniformly distributed between $-a$ and a, where $a > 0$. Consider the new variable Y_λ obtained from X_λ in the following way:

$$Y_\lambda = \begin{cases} -a/2 & \text{if } -a \leq X_\lambda \leq -a/2 , \\ X_\lambda & \text{if } -a/2 < X_\lambda < a/2 , \\ a/2 & \text{if } a/2 \leq X_\lambda \leq a . \end{cases}$$

Determine the probability density $P_Y(y)$ of Y_λ.

Exercise 2.2. Histogram Equalization

Let X_λ be a random variable with probability density function $P_X(x)$. Consider the new variable Y_λ obtained from X_λ in the following manner:

$$Y_\lambda = \int_{-\infty}^{X_\lambda} P_X(\eta)\mathrm{d}\eta .$$

Determine the probability density function $P_Y(y)$ of Y_λ.

Exercise 2.3. Moments of the Gaussian Distribution

Calculate the central moments of the Gaussian probability law.

Exercise 2.4. Stochastic Vector

Consider a central Gaussian stochastic vector in two real dimensions. Show that we can write

$$P_{X_1,X_2}(x_1, x_2) = \frac{1}{2\pi\sigma_1\sigma_2\sqrt{(1-\rho^2)}} \exp\left[-\frac{x_1^2/\sigma_1^2 + x_2^2/\sigma_2^2 - 2x_1x_2\rho/\sigma_1\sigma_2}{2(1-\rho^2)} \right].$$

Exercise 2.5

Let $G(x, y)$ be the probability that the random variable X_λ lies between x and y. Determine the probability density of X_λ as a function of $G(x, y)$.

Exercise 2.6. Distribution of a Mixture

A gas contains a mixture of two types of atom A_1 and A_2 with respective concentrations c_1 and c_2. The probability of photon emission by atoms A_1 is p_1, whilst that for atoms A_2 is p_2. What is the photon emission probability p for the mixed gas? Generalize to the case of an arbitrary mixture.

Exercise 2.7. Complex Gaussian Random Variable

Consider the complex random variable defined by $Z_\lambda = X_\lambda + iY_\lambda$ where $i^2 = -1$, and X_λ and Y_λ are independent Gaussian random variables with the same variance. Give an expression for the probability density of Z_λ.

Exercise 2.8. Weibull Variable

Determine the probability density function of Y_λ obtained from X_λ by the transformation $Y_\lambda = (X_\lambda)^\beta$, where $\beta > 0$ and X_λ is a random variable distributed according to the Gamma probability law. Analyze the special case where the Gamma distribution is exponential.

Exercise 2.9. Average of Noisy Measurements

A device measures a physical quantity g which is assumed to be constant in time. Several measurements with values F_i are made at N successive times. Each measurement is perturbed by noise B_i in such a way that

$$F_i = g + B_i.$$

B_i is a random variable, assumed to have a Gaussian distribution with mean 0 and standard deviation σ. For simplicity, assume that the dependence of the random variables on the random events is not noted. Assume also that the variables B_i are statistically independent of one another. The sum of all the measurements is evaluated, thereby producing a new random variable

$$Y = \frac{1}{N} \sum_{i=1}^{N} F_i \ .$$

(1) Calculate the probability density function of the random variable Y, assuming it obeys a Gaussian distribution.
(2) Why can we say that measurement of g using Y is more 'precise' than measurement using a single value F_i ?

Exercise 2.10. Change of Variable

Consider two independent random variables X_λ and Y_λ, identically distributed according to a Gaussian probability law with zero mean. Determine the probability density function of the quotient random variable $Z_{\lambda_T} = X_{\lambda_1}/Y_{\lambda_2}$.

3

Fluctuations and Covariance

In this chapter, we shall discuss random functions and fields, generally known as stochastic processes and stochastic fields, respectively, which can simply be understood as random variables depending on a parameter such as time or space. We may then consider new means with respect to this parameter and hence study new properties. We shall concentrate mainly on second order properties, i.e., properties of the first two moments of these random functions and fields.

3.1 Stochastic Processes

The idea of covariance is very productive in physics. In Chapter 1, we mentioned the case where Z_λ represents noise in the output of a measurement system. In that case, X_λ represented noise in the sensor and Y_λ noise in the amplifier (see Fig. 3.1). If G is the gain of the amplifier, assumed perfectly linear, the noise in the measurement can be written $Z_\lambda = GX_\lambda + Y_\lambda$. If Y_λ is much smaller than GX_λ, one would expect the absolute value of ρ_{ZX} to be close to 1 and ρ_{ZY} almost zero. On the other hand, if Y_λ is much bigger than GX_λ, then the absolute value of ρ_{ZY} will be of order 1 whilst ρ_{ZX} will be close to 0.

This notion is easily extended to a wide range of situations. Imagine for example that we are interested in temporal fluctuations in an electric field measured using an antenna. Assuming that we do not know *a priori* the field that we are going to measure, it may be useful to represent it by a time-dependent random variable $E_\lambda(t)$. We then define the covariance function[1] of the two random variables that represent the field at times t_1 and t_2:

---

[1] Many authors use the term correlation function. However, this generates an ambiguity as we shall see later. Indeed, the correlation function generally corresponds to another type of mean and the two quantities (covariance function and correlation function) are only equal under certain conditions to be examined below.

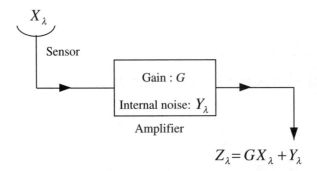

Fig. 3.1. A situation in which noise is additive

$$\Gamma(t_1, t_2) = \left\langle \left[E_\lambda(t_1)\right]^* E_\lambda(t_2) \right\rangle - \langle E_\lambda(t_1) \rangle^* \langle E_\lambda(t_2) \rangle \ ,$$

where z^* is the complex conjugate of z. (The notation has been simplified to make it more readable.) If the electric field is represented by a real variable, we then define

$$\Gamma(t_1, t_2) = \langle E_\lambda(t_1) E_\lambda(t_2) \rangle - \langle E_\lambda(t_1) \rangle \langle E_\lambda(t_2) \rangle \ .$$

In the last example, $E_\lambda(t)$ is a stochastic process, for it is a random variable which depends on time. In other words, it is a function whose value is determined by a random experiment. More precisely, in order to define the notion of a stochastic process, we consider a set Ω of random events and associate a function $X_\lambda(t)$ with each one of these random events λ (see Fig. 3.2). (Note that t does not have to be time.) If the possible values of $X_\lambda(t)$ are real numbers, we speak of a real stochastic process, whilst if the values are complex numbers, we refer to a complex stochastic process.

Note then that, for a fixed value of λ, i.e., when we consider some given realization of the stochastic process, there is nothing to distinguish $X_\lambda(t)$ from a deterministic function. However, when t is fixed and λ is undetermined, $X_\lambda(t)$ is simply a random variable.

The notion of covariance function plays such an important role that it is worth going into a little further detail. For a given realization $X_\lambda(t)$, i.e., for a given λ, let us consider the variation of this function between times t_1 and t_2. A measure of this variation can be obtained from $|X_\lambda(t_1) - X_\lambda(t_2)|^2$. Taking the expected value over the various possible realizations, we may thus consider

$$\langle |X_\lambda(t_1) - X_\lambda(t_2)|^2 \rangle \ .$$

Expanding out this expression, we obtain

$$\langle |X_\lambda(t_1) - X_\lambda(t_2)|^2 \rangle = \Gamma(t_1, t_1) + \Gamma(t_2, t_2) - \Gamma(t_1, t_2) - \Gamma(t_2, t_1) \ .$$

This result shows that the covariance function is a quantity which characterises the mean variation of realizations of the stochastic process between times t_1 and t_2.

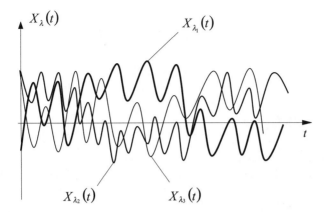

Fig. 3.2. Illustration of a stochastic process in which each curve corresponds to a single realization of the stochastic process and hence to a single realization of a random function

There are many situations in physics where one is interested in fluctuations at different points of space. The latter are represented by vectors. In the following, we shall identify the vector r having components $(x, y, z)^{\mathrm{T}}$ with the point having coordinates (x, y, z). (Then by abuse of language, we shall often speak of the point r.) When the dimension of the space is strictly greater than 1, there is obviously no ordering relation for different r, as there was for the time t. $X_\lambda(r)$ is then more readily defined as a stochastic field.

Let us consider a classic situation in physics where we wish to measure a microscopic quantity at different points in space. For example, we may be dealing with local fluctuations in magnetization in magnetism, polarization in electricity, density in acoustics, strain in mechanics, or velocity in hydrodynamics. If the quantities are complex-valued, the covariance of the stochastic field is written

$$\Gamma(r_1, r_2) = \langle E_\lambda^*(r_1) E_\lambda(r_2) \rangle - \langle E_\lambda^*(r_1) \rangle \langle E_\lambda(r_2) \rangle .$$

The most general case occurs when the fields depend on both space and time, as happens, for example, when we measure an electric field at a point r in space at a given time t. The stochastic field is then written $E_\lambda(r, t)$. We can define the covariance function for the two random variables which represent the field at point r_1 at time t_1 and the field at point r_2 at time t_2:

$$\Gamma(r_1, t_1, r_2, t_2) = \langle E_\lambda^*(r_1, t_1) E_\lambda(r_2, t_2) \rangle - \langle E_\lambda^*(r_1, t_1) \rangle \langle E_\lambda(r_2, t_2) \rangle .$$

This simple definition underpins the notion of coherence in classical optics. We then consider the electric field of light $\boldsymbol{E}_\lambda(r, t)$, which is a 3-component vector. In order to simplify the notation, we assume that it is parallel to the Oz axis, which has direction defined by the unit vector $\boldsymbol{e}_z$. We thus write $\boldsymbol{E}_\lambda(r, t) = E_\lambda^{(z)}(r, t) \boldsymbol{e}_z$ and the covariance becomes

$$\Gamma(\boldsymbol{r}_1, t_1, \boldsymbol{r}_2, t_2) =$$

$$\left\langle \left[E_\lambda^{(z)}(\boldsymbol{r}_1, t_1) \right]^* E_\lambda^{(z)}(\boldsymbol{r}_2, t_2) \right\rangle - \left\langle \left[E_\lambda^{(z)}(\boldsymbol{r}_1, t_1) \right]^* \right\rangle \left\langle E_\lambda^{(z)}(\boldsymbol{r}_2, t_2) \right\rangle .$$

The general case is described in Section 3.13.

We also introduce the normalized covariance function:

$$\rho(\boldsymbol{r}_1, t_1, \boldsymbol{r}_2, t_2) = \frac{\Gamma(\boldsymbol{r}_1, t_1, \boldsymbol{r}_2, t_2)}{\sqrt{\Gamma(\boldsymbol{r}_1, t_1, \boldsymbol{r}_1, t_1) \Gamma(\boldsymbol{r}_2, t_2, \boldsymbol{r}_2, t_2)}} .$$

The field is said to be coherent at points $\boldsymbol{r}_1$, $\boldsymbol{r}_2$ and times t_1, t_2 if the modulus of $\rho(\boldsymbol{r}_1, t_1, \boldsymbol{r}_2, t_2)$, written $|\rho(\boldsymbol{r}_1, t_1, \boldsymbol{r}_2, t_2)|$, is close to 1. On the other hand, if this value is close to 0, the field is incoherent. There are two special cases, when $\boldsymbol{r}_1 = \boldsymbol{r}_2$ and when $t_1 = t_2$:

- when $\boldsymbol{r}_1 = \boldsymbol{r}_2 = \boldsymbol{r}$, we speak of temporal coherence at the point $\boldsymbol{r}$ at times t_1 and t_2 if $|\rho(\boldsymbol{r}, t_1, \boldsymbol{r}, t_2)|$ is close to 1 and temporal incoherence when it is close to 0;
- when $t_1 = t_2 = t$, we speak of spatial coherence at points $\boldsymbol{r}_1$ and $\boldsymbol{r}_2$ at time t if $|\rho(\boldsymbol{r}_1, t, \boldsymbol{r}_2, t)|$ is close to 1 and spatial incoherence when it is close to 0.

The practical consequences of coherence are observed in light interference experiments, for example. The fact that the electric field oscillates in time complicates the situation and we shall return to the idea of coherence in more detail later. However, we must first define the key ideas of stationarity and ergodicity. Indeed, these will allow us to make the connection between the above theoretical quantities, defined in terms of expectation values and not directly measurable, and quantities that are more easily estimated by experiment.

3.2 Stationarity and Ergodicity

To begin with, we consider the case where t represents time. We have seen that a stochastic process is a function whose value is determined by a random experiment, since we associate a function $X_\lambda(t)$ with each random event λ in Ω. To simplify the analysis, we assume that the possible values of $X_\lambda(t)$ are real numbers. We have already seen that a stochastic process can be understood as a family of functions indexed by λ or as a random variable depending on a parameter t. This last approach allows us to apply all our definitions concerning random variables to stochastic processes.

At a given instant of time t, since the stochastic process $X_\lambda(t)$ is a simple random variable, we may associate a probability density function $P_{X,t}(x)$ with it. For example, if this probability density function is Gaussian, we speak of a Gaussian stochastic process. We may also be interested in the probability density function $P_{X,t_1,t_2}(x_1, x_2)$, which is the joint probability density function for the random variables $X_\lambda(t_1)$ and $X_\lambda(t_2)$. We could of course generalize this definition to an arbitrary number N of times to consider

$$P_{X,t_1,t_2,...,t_N}(x_1, x_2, \ldots, x_N) .$$

It is generally difficult to estimate these joint probability density functions and we are more often interested in the various moments. Indeed, we define the instantaneous moments $\langle [X_\lambda(t)]^n \rangle$, or more generally,

$$\langle [X_\lambda(t_1)]^{n_1} [X_\lambda(t_2)]^{n_2} \ldots [X_\lambda(t_P)]^{n_P} \rangle .$$

Note that the same random event occurs throughout the latter expression.

As a special case, we recover the first two moments arising in the expression for the covariance introduced in Section 3.2, viz.,

$$\Gamma_{XX}(t_1, t_2) = \langle X_\lambda(t_1)X_\lambda(t_2) \rangle - \langle X_\lambda(t_1) \rangle \langle X_\lambda(t_2) \rangle .$$

For simplicity, we shall restrict ourselves in the following to expressions involving the first two moments $m_X(t) = \langle X_\lambda(t) \rangle$ and $\Gamma_{XX}(t_1, t_2)$, which are moreover the most used in practice.

A stochastic process is said to be weakly stationary or stationary in the wide sense if it is stationary up to its second order moments, i.e., if $m_X(t)$ is independent of t and if $\Gamma_{XX}(t_1, t_2)$ only depends on $(t_1 - t_2)$. In this case, we write

$$m_X = \langle X_\lambda(t) \rangle \quad \text{and} \quad \Gamma_{XX}(t_2 - t_1) = \Gamma_{XX}(t_1, t_2) .$$

Note the abuse of notation in the second equation, since we use the same symbol to denote a function of two variables and a function of a single variable.

If a stochastic process is weakly stationary, its first two moments are unaffected by the choice of time origin. Indeed they are invariant under time translation. This is made even clearer if we set $t_1 = t$ and $t_2 = t + \tau$, for stationarity then implies that $\langle X_\lambda(t) \rangle$ and $\Gamma_{XX}(t, t+\tau)$ are independent of t.

In the stationary case, we considered expectation values, that is, averages taken with respect to λ. However, we can now introduce a new mean, obtained by integrating over t for fixed λ. Over an interval $[T_1, T_2]$, the mean of $X_\lambda(t)$ is written

$$\frac{1}{T_2 - T_1} \int_{T_1}^{T_2} X_\lambda(t)\mathrm{d}t .$$

The obvious problem with this definition is that it depends on the choice of T_1 and T_2. To get round this difficulty, we take the limit of the above mean when T_2 tends to infinity and T_1 tends to minus infinity. Note, however, that there is no guarantee that such a quantity actually exists, i.e., that the limit exists. When it does, it will be called the time average, written

$$\overline{X_\lambda(t)} = \lim_{\substack{T_1 \to -\infty \\ T_2 \to \infty}} \left[\frac{1}{T_2 - T_1} \int_{T_1}^{T_2} X_\lambda(t)\mathrm{d}t \right] .$$

In the same way, we can introduce a kind of "second temporal moment," viz.,

$$\overline{X_\lambda(t)X_\lambda(t+\tau)} = \lim_{\substack{T_1 \to -\infty \\ T_2 \to \infty}} \left[\frac{1}{T_2 - T_1} \int_{T_1}^{T_2} X_\lambda(t)X_\lambda(t+\tau)dt \right] .$$

In this way we define the correlation function by

$$C_{X_\lambda X_\lambda}(\tau) = \overline{X_\lambda(t)X_\lambda(t+\tau)} ,$$

and the centered correlation function by

$$C_{X_\lambda X_\lambda}^{\mathrm{c}}(\tau) = \overline{X_\lambda(t)X_\lambda(t+\tau)} - \overline{X_\lambda(t)}\ \overline{X_\lambda(t+\tau)} .$$

Clearly, $\overline{X_\lambda(t)}$ and $\overline{X_\lambda(t)X_\lambda(t+\tau)}$ cannot depend on t. However, they may depend on λ and τ. A stochastic process is said to be weakly ergodic or ergodic in the wide sense if it is ergodic up to second order moments, i.e., if $\overline{X_\lambda(t)}$ and $\overline{X_\lambda(t)X_\lambda(t+\tau)}$ do not depend on λ.

Note that this definition is a kind of dual to the definition of stationarity. A stochastic process is (weakly) stationary if the expectation relative to λ removes the dependence on t. A stochastic process is (weakly) ergodic if the average with respect to t removes the dependence on λ. It is common in physics books to define ergodicity only in the case of stationary stochastic processes. However, this approach tends to hide the symmetry between the definitions. Let us now illustrate these ideas with two simple examples.

Consider first the case where $X_\lambda(t) = A_\lambda$ and A_λ is a real random variable. This random variable is clearly stationary. (When there is no risk of ambiguity, although we speak simply of stationarity and ergodicity, it should be understood that we are referring to weak stationarity and weak ergodicity, up to second order moments.) Indeed, it is easy to see that $\langle X_\lambda(t)\rangle$ and $\langle X_\lambda(t)X_\lambda(t+\tau)\rangle$ are independent of t, since we have $\langle X_\lambda(t)\rangle = \langle A_\lambda\rangle$ and $\langle X_\lambda(t)X_\lambda(t+\tau)\rangle = \langle (A_\lambda)^2\rangle$. On the other hand, it is not ergodic because the time average does not eliminate the dependence on λ. Indeed, $\overline{X_\lambda(t)} = A_\lambda$ and $\overline{X_\lambda(t)X_\lambda(t+\tau)} = (A_\lambda)^2$. These results are easy to interpret. As the process is time-independent, it is invariant under time translations and hence obviously stationary. Now ergodicity means that time averages should "rub out" any dependence on the particular realization of the stochastic process that we are analyzing. In other words, each realization should be representative (up to the second order moment) of the family of functions defining the stochastic process. It is clear that this cannot be the case when $X_\lambda(t) = A_\lambda$ (unless of course we have the trivial situation in which A_λ is constant as a function of λ, i.e., A_λ is a fixed value).

We now discuss the case where $X_\lambda(t) = A\cos(\omega t + \varphi_\lambda)$, with A a real parameter and φ_λ a real random variable taking values between 0 and 2π with probability density $P(\varphi)$.

Let us begin by studying the more general case of a stochastic process $X_\lambda(t)$ constructed from a non-random function $f(t)$ by introducing a dependence on a random variable q_λ. Expectation values are easy to determine. Indeed, we have $X_\lambda(t) = f(t, q_\lambda)$ and hence $\langle X_\lambda(t) \rangle = \langle f(t, q_\lambda) \rangle$. At a given time t, $f(t, q_\lambda)$ is a random variable which can be understood as resulting from a change of variable from q_λ. When the relation between $f(t, q)$ and q is bijective, we thus find $P_{f(t,q)}(f)\mathrm{d}f = P_q(q)\mathrm{d}q$ and hence,

$$\langle f(t, q_\lambda) \rangle = \int f P_{f(t,q)}(f)\mathrm{d}f = \int f(t, q) P_q(q)\mathrm{d}q \ .$$

In the present example, $\langle X_\lambda(t) \rangle = \langle A\cos(\omega t + \varphi_\lambda) \rangle$, i.e.,

$$\langle X_\lambda(t) \rangle = \int\limits_0^{2\pi} A\cos(\omega t + \varphi)P(\varphi)\mathrm{d}\varphi \ ,$$

and this integral is not generally independent of t. The stochastic process is not stationary to first order, and so it is not stationary up to second order. (A stochastic process is weakly stationary if its first two moments are independent of t. If the first moment is not, this is enough to assert that the stochastic process is not weakly stationary.) Note that concerning the second order moment we have

$$\langle X_\lambda(t)X_\lambda(t + \tau) \rangle = \int\limits_0^{2\pi} A^2 \cos(\omega t + \varphi)\cos\big[\omega(t + \tau) + \varphi\big]P(\varphi)\mathrm{d}\varphi \ ,$$

hence,

$$\langle X_\lambda(t)X_\lambda(t + \tau) \rangle = \frac{1}{2}A^2 \int\limits_0^{2\pi} \big[\cos(2\omega t + \omega\tau + 2\varphi) + \cos(\omega\tau)\big]P(\varphi)\mathrm{d}\varphi \ ,$$

or again,

$$\langle X_\lambda(t)X_\lambda(t + \tau) \rangle = \frac{1}{2}A^2 \cos(\omega\tau) + \frac{1}{2}A^2 \int\limits_0^{2\pi} \cos(2\omega t + \omega\tau + \varphi)P(\varphi)\mathrm{d}\varphi \ .$$

The second term is not generally independent of time t.

Regarding the question of ergodicity, let us examine the time averages. We have

$$\overline{X_\lambda(t)} = \lim_{\substack{T_1 \to -\infty \\ T_2 \to \infty}} \frac{1}{T_2 - T_1} \int\limits_{T_1}^{T_2} A\cos(\omega t + \varphi_\lambda)\mathrm{d}t \ ,$$

which is independent of λ because the integral is actually zero. The process $X_\lambda(t) = A\cos(\omega t + \varphi_\lambda)$ is thus ergodic to first order. To second order,

$$\overline{X_\lambda(t)X_\lambda(t+\tau)} = \lim_{\substack{T_1 \to -\infty \\ T_2 \to \infty}} \frac{1}{T_2 - T_1} \int_{T_1}^{T_2} A^2 \cos(\omega t + \varphi) \cos\left[\omega(t+\tau) + \varphi\right] dt \, ,$$

or

$$\overline{X_\lambda(t)X_\lambda(t+\tau)} = \lim_{\substack{T_1 \to -\infty \\ T_2 \to \infty}} \frac{1}{T_2 - T_1} \int_{T_1}^{T_2} \frac{A^2}{2} \left[\cos(2\omega t + \omega\tau + \varphi) + \cos(\omega\tau) \right] dt \, ,$$

and hence finally,

$$\overline{X_\lambda(t)X_\lambda(t+\tau)} = \frac{A^2}{2} \cos(\omega\tau) \, ,$$

which is independent of λ. The function is thus weakly ergodic.

The two simple examples we have just examined illustrate how a stochastic process may be either stationary or ergodic. Let us emphasize once again that we have here two quite separate notions. Of course, there are many examples of stochastic processes which are both stationary and ergodic. This is true for example in the second example if $P(\varphi)$ is constant between 0 and 2π.

3.3 Ergodicity in Statistical Physics

In statistical physics, ergodicity is often tackled in a different way. We begin by defining the state Q_t of a physical system at time t as the set of variables allowing us to integrate its dynamical equation. It is thus generally a very high-dimensional vector, since it includes the generalized coordinates of each particle making up the system. In the case of a monoatomic gas, the state Q_t corresponds to the set of positions and velocities of each molecule at the relevant time t.

In the simple case of an isolated system (microcanonical case), we are interested in the evolution operator of the system, denoted $\chi^\tau[\,]$, which relates the state Q_t of the system at time t with the state $Q_{t+\tau}$ it will have at time $t + \tau$. (A priori, this evolution operator may itself depend on t and on τ. However, if the system is isolated, as we are assuming here, it only depends on τ.) We write

$$Q_{t+\tau} = \chi^\tau[Q_t] \, .$$

The state of the physical system can thus be represented by a point in a space, generally of very high dimension, known as the phase space. In reality, the variables for each particle can only take values within a bounded set and the state Q_t of the system belongs to a bounded subset of phase space which we shall call S. A physical system is then said to be ergodic if the only invariant subspaces of S under action of the operator $\chi^\tau[\,]$ are the set S itself and the empty set (see Fig. 3.3). In other words, during its evolution, an ergodic system visits all its possible states. This would not be the case if, during its evolution, the system could remain trapped in some subspace S_E of S. Certain states, such as those in the complement of S_E in S, would not then be visited, and this would mean that the system was not ergodic.

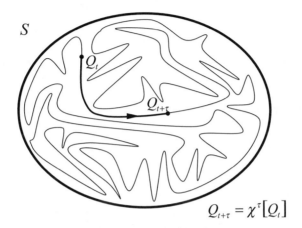

$$Q_{t+\tau} = \chi^\tau[Q_t]$$

Fig. 3.3. In physics, a system is said to be ergodic if the only subspaces of the phase space that remain invariant under the action of the evolution operator χ^τ are the whole space and the empty set

As an example, let us consider a ferromagnetic system. This type of material spontaneously acquires a nonzero magnetization when cooled below a certain temperature. As for magnets, this macroscopic magnetization can change direction under the action of an intense magnetic field. However, in the absence of any magnetic field, these changes of direction are very rare and we shall ignore them. The magnetization of the physical system we are considering is the sum of the magnetic moments of all the atoms making up the material. As we just pointed out, as time goes by, the system evolves, whilst maintaining its magnetization along a fixed direction. The states of the material that would lead to a macroscopic magnetization oriented along the opposite direction thus remain unvisited. This system is not therefore ergodic in the sense of statistical physics. Suppose, however, that we are interested in the fluctuations of the magnetization along a direction perpendicular to the observed macroscopic magnetization. There is no *a priori* reason why this particular stochastic process should not be ergodic in the sense of moments.

The approach adopted in the last section, defining ergodicity in terms of second order moments, is clearly more phenomenological than the one adopted in statistical physics. It is the first approach that is generally adopted in signal theory. It can be shown that, if a system is stationary and ergodic in the sense of statistical physics, it is also ergodic in the sense of signal theory. The converse is false, however. In the following, we limit our discussion to the notions of stationarity and ergodicity in the wide sense, since this approach has the advantage of being less restrictive.

3.4 Generalization to Stochastic Fields

The above definitions can be generalized to random functions that depend on the coordinates of points in space rather than on time. As mentioned before, space points are represented by the vector r. In this case, we speak of homogeneity rather than stationarity. The stochastic field $X_\lambda(r)$ is said to be homogeneous (in the wide sense) if its first two moments do not depend on the choice of spatial origin. We must therefore have translational invariance. More precisely, homogeneity implies that $\langle X_\lambda(r) \rangle$ and $\Gamma_{XX}(r, r+d)$ do not depend on r, where the covariance function is in this case

$$\Gamma_{XX}(r, r+d) = \langle X_\lambda^*(r) X_\lambda(r+d) \rangle - \langle X_\lambda^*(r) \rangle \langle X_\lambda(r+d) \rangle .$$

Translations are not the only transformations of space. In particular, we could consider the rotations. Let R_ω denote the rotation operator effecting a rotation through angle $|\omega|$ about an axis through O parallel to the vector ω. To any point r, we can associate a new point u defined by $u = R_\omega[r]$. The stochastic field is said to be isotropic (in the wide sense) if its first two moments are unaffected by such rotations. More precisely, isotropy in the wide sense implies that, for any rotation, i.e., for any vector ω, we must have

$$\langle X_\lambda(r) \rangle = \langle X_\lambda(R_\omega[r]) \rangle$$

and

$$\langle X_\lambda^*(r_1) X_\lambda(r_2) \rangle = \langle X_\lambda^*(R_\omega[r_1]) X_\lambda(R_\omega[r_2]) \rangle ,$$

or

$$\Gamma_{XX}(r_1, r_2) = \Gamma_{XX}(R_\omega[r_1], R_\omega[r_2]) .$$

The question of ergodicity is a little more delicate and we shall limit our discussion to the effects of translation. We define spatial averages by

$$\overline{X_\lambda(x,y,z)} = \lim_{V \to \infty} \left[\frac{1}{|V|} \iiint\limits_V X_\lambda(x,y,z) \, \mathrm{d}x \, \mathrm{d}y \, \mathrm{d}z \right] ,$$

which we shall write more simply in the form

$$\overline{X_\lambda(r)} = \lim_{V \to \infty} \left[\frac{1}{|V|} \iiint\limits_V X_\lambda(r) \mathrm{d}r \right] ,$$

where $|V|$ is the volume of the subspace V and $V \to \infty$ means that the volume increases to cover the whole space. Note that (x,y,z) represents the components of r.

For the second spatial moment, we then have

$$\overline{X_\lambda^*(\boldsymbol{r})X_\lambda(\boldsymbol{r}+\boldsymbol{d})} = \lim_{V\to\infty}\left[\frac{1}{|V|}\iiint\limits_V X_\lambda^*(\boldsymbol{r})X_\lambda(\boldsymbol{r}+\boldsymbol{d})\mathrm{d}\boldsymbol{r}\right] ,$$

where $\overline{X_\lambda^*(\boldsymbol{r})X_\lambda(\boldsymbol{r}+\boldsymbol{d})} = \overline{X_\lambda^*(r_1,r_2,r_3)X_\lambda(r_1+d_1,r_2+d_2,r_3+d_3)}$.

A stochastic field can then be described as ergodic in the wide sense or weakly ergodic if $\overline{X_\lambda(\boldsymbol{r})}$ and $\overline{X_\lambda^*(\boldsymbol{r})X_\lambda(\boldsymbol{r}+\boldsymbol{d})}$ do not depend on λ. Note that this definition is a kind of dual to the definition of homogeneity.

3.5 Random Sequences and Cyclostationarity

It is interesting to look more closely at the consequences of stationarity when the stochastic processes in question are sampled at a finite number of times. In this case, the stochastic processes are simply random sequences, which we shall denote by $X_\lambda(n)$, where $n \in [1, N]$. As before, and without loss of generality, we assume that the sequences are real with zero mean. However, it is difficult to define stationarity for a random sequence of finite length. In fact, it is easier if we construct a periodic sequence of infinite length:

$$X_\lambda^{\mathrm{P}}(n) = X_\lambda(\mathrm{mod}_N[n]) ,$$

where the function $\mathrm{mod}_N[n]$ is defined by $\mathrm{mod}_N[n] = n - pN$ and p is a whole number chosen so that $n - pN \in [1, N]$ (see Fig. 3.4).

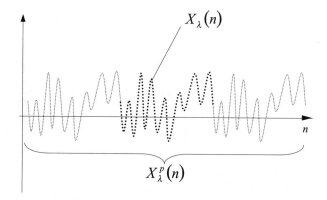

Fig. 3.4. Periodic sequence constructed from a finite sequence

The sequence $X_\lambda(n)$ is said to be weakly cyclostationary or cyclostationary in the wide sense if the two expectation values $\langle X_\lambda^{\mathrm{P}}(n)X_\lambda^{\mathrm{P}}(n+m)\rangle$ and $\langle X_\lambda^{\mathrm{P}}(n)\rangle$ are independent of n. In this case, the covariance matrix $\overline{\overline{\varGamma}}$ has a special mathematical structure. (In fact, it is said to be a circulant Toeplitz matrix.) Suppose first that $\langle X_\lambda(n) \rangle = 0$. We then find that

$$\Gamma_{nm} = \langle X_\lambda^{\mathrm{P}}(n) X_\lambda^{\mathrm{P}}(m) \rangle = \Gamma(m-n) \ . $$

We can calculate the discrete Fourier transform of $X_\lambda(n)$, since it is a finite sequence. (There is thus no problem of non-convergence as might happen with a continuous signal, or a signal with unbounded temporal support.) The result is

$$X_\lambda(n) = \frac{1}{N} \sum_{\nu=0}^{N-1} \hat{X}_\lambda(\nu) \exp\left(\mathrm{i}\frac{2\pi\nu n}{N}\right) \ ,$$

where $\nu \in [0, N-1]$ and

$$\hat{X}_\lambda(\nu) = \sum_{n=1}^{N} X_\lambda(n) \exp\left(-\mathrm{i}\frac{2\pi\nu n}{N}\right) \ .$$

Hence,

$$\hat{X}_\lambda(\nu_1)\hat{X}_\lambda^*(\nu_2) = \sum_{m=1}^{N}\sum_{n=1}^{N} X_\lambda(n) X_\lambda(m) \exp\left(-\mathrm{i}\frac{2\pi\nu_1 n}{N}\right) \exp\left(\mathrm{i}\frac{2\pi\nu_2 m}{N}\right) \ .$$

We now set $m' = m - n$ and calculate the expectation value

$$\langle \hat{X}_\lambda(\nu_1)\hat{X}_\lambda^*(\nu_2) \rangle = \sum_{m'=0}^{N-1}\sum_{n=1}^{N} \langle X_\lambda^{\mathrm{P}}(n) X_\lambda^{\mathrm{P}}(n+m') \rangle$$

$$\times \exp\left(-\mathrm{i}\frac{2\pi\nu_1 n}{N}\right) \exp\left[\mathrm{i}\frac{2\pi\nu_2(n+m')}{N}\right] \ ,$$

or

$$\langle \hat{X}_\lambda(\nu_1)\hat{X}_\lambda^*(\nu_2) \rangle = \sum_{n=1}^{N}\sum_{m=0}^{N-1} \Gamma(m) \exp\left(\mathrm{i}\frac{2\pi\nu_2 m}{N}\right) \exp\left[\mathrm{i}\frac{2\pi n(\nu_2 - \nu_1)}{N}\right] \ .$$

Since $\nu_1, \nu_2 \in [0, N-1]$, it is a straightforward matter to show that,

$$\sum_{n=1}^{N} \exp\left[\mathrm{i}\frac{2\pi n(\nu_2 - \nu_1)}{N}\right] = N\delta_{\nu_1 - \nu_2} \ ,$$

where δ_ν is the Kronecker delta defined by

$$\delta_\nu = \begin{cases} 1 & \text{if } \nu = 0 \ , \\ 0 & \text{otherwise} \ . \end{cases}$$

Finally, we obtain

$$\langle \hat{X}_\lambda(\nu_1)\hat{X}_\lambda^*(\nu_2)\rangle = N^2 \hat{\Gamma}(\nu_2)\delta_{\nu_1-\nu_2} \, , \tag{3.1}$$

where we have put

$$\hat{\Gamma}(\nu) = \frac{1}{N}\sum_{m=0}^{N-1} \Gamma(m)\exp\left(\mathrm{i}\frac{2\pi\nu m}{N}\right) \, .$$

In fact, this defines the spectral density $\hat{\Gamma}(\nu)$ of $X_\lambda(n)$, since $\hat{\Gamma}(\nu) = \langle \hat{X}_\lambda(\nu)\hat{X}_\lambda^*(\nu)\rangle/N^2$. The factor $1/N^2$ has been introduced so that $\Gamma(0) = \sum_{\nu=0}^{N-1}\hat{\Gamma}(\nu)$, which represents the power of $X_\lambda(n)$.

Equation (3.1) shows that, if $\nu_1 \neq \nu_2$, then $\hat{X}_\lambda(\nu_1)$ and $\hat{X}_\lambda(\nu_2)$ are uncorrelated random variables. This property follows directly from the stationarity of $X_\lambda^\mathrm{P}(n)$. But the stationarity of $X_\lambda^\mathrm{P}(n)$ implies another interesting consequence. We have

$$\hat{X}_\lambda(\nu) = \sum_{n=1}^{N} X_\lambda(n)\exp\left(-\mathrm{i}\frac{2\pi\nu n}{N}\right) \, ,$$

and hence,

$$\langle \hat{X}_\lambda(\nu)\rangle = 0 \, .$$

Indeed,

$$\langle \hat{X}_\lambda(\nu)\rangle = \sum_{n=1}^{N}\langle X_\lambda(n)\rangle\exp\left(-\mathrm{i}\frac{2\pi\nu n}{N}\right) \, ,$$

and

$$\langle X_\lambda(n)\rangle = \langle X_\lambda(1)\rangle = m_X = 0,$$

so that

$$\langle \hat{X}_\lambda(\nu)\rangle = m_X \sum_{n=1}^{N}\exp\left(-\mathrm{i}\frac{2\pi\nu n}{N}\right) \, .$$

We thus find that for any frequency ν, we have $\langle \hat{X}_\lambda(\nu)\rangle = 0$.[2]

We shall now determine the probability density of $\hat{X}_\lambda(\nu)$ when $X_\lambda(n)$ is distributed according to a zero mean Gaussian probability law. In this case, $\hat{X}_\lambda(\nu)$ will also be Gaussian. [In fact, even if $X_\lambda(n)$ is not Gaussian, $\hat{X}_\lambda(\nu)$ will nevertheless generally be approximately Gaussian as a consequence of the

---

[2] If $\langle X_\lambda(n)\rangle = m_X \neq 0$, we would simply have

$$\langle \hat{X}_\lambda(\nu)\rangle = m_X N\delta_\nu \, ,$$

and $\langle \hat{X}_\lambda(\nu)\rangle = 0$ would hold for any frequency $\nu \in [1, N-1]$.

central limit theorem, as we shall see in the next chapter.] We shall show that $\hat{X}_\lambda(\nu)$ is then a complex random variable with probability density

$$P_{\hat{X}}(\alpha, \beta) = \frac{1}{\pi N^2 \hat{\Gamma}(\nu)} \exp\left[-\frac{1}{N^2 \hat{\Gamma}(\nu)}(\alpha^2 + \beta^2)\right] ,$$

where α and β are real and represent the real and imaginary parts of the complex variable $x = \alpha + i\beta$. In particular, note that the phase of $\hat{X}_\lambda(\nu)$ is random and uniformly distributed over the interval from 0 to 2π.

Given that the probability density of $\hat{X}_\lambda(\nu)$ is Gaussian with zero mean, we need only calculate the second order moments. Considering nonzero frequencies, we set $\hat{X}_\lambda(\nu) = \hat{X}_\lambda^R(\nu) + i\hat{X}_\lambda^I(\nu)$, where $\hat{X}_\lambda^R(\nu)$ and $\hat{X}_\lambda^I(\nu)$ are the real and imaginary parts of $\hat{X}_\lambda(\nu)$, respectively. Hence,

$$\hat{X}_\lambda^R(\nu) = \sum_{n=1}^{N} X_\lambda(n) \cos\left(\frac{2\pi\nu n}{N}\right) ,$$

and

$$\hat{X}_\lambda^I(\nu) = -\sum_{n=1}^{N} X_\lambda(n) \sin\left(\frac{2\pi\nu n}{N}\right) .$$

We thus deduce that

$$\langle|\hat{X}_\lambda^R(\nu)|^2\rangle = \sum_{m=1}^{N} \sum_{n=1}^{N} \langle X_\lambda(m) X_\lambda(n)\rangle \cos\left(\frac{2\pi\nu m}{N}\right) \cos\left(\frac{2\pi\nu n}{N}\right) .$$

We have $\langle X_\lambda(m) X_\lambda(n)\rangle = \Gamma(m - n)$ and set $m - n = m'$. We then obtain

$$\langle|\hat{X}_\lambda^R(\nu)|^2\rangle = \sum_{m'=1}^{N} \Gamma(m') \sum_{n=1}^{N} \cos\left[\frac{2\pi\nu(n + m')}{N}\right] \cos\left(\frac{2\pi\nu n}{N}\right) .$$

Now

$$\cos\left[\frac{2\pi\nu(n + m')}{N}\right] \cos\left(\frac{2\pi\nu n}{N}\right)$$
$$= \frac{1}{2}\left\{\cos\left[\frac{2\pi\nu(2n + m')}{N}\right] + \cos\left(\frac{2\pi\nu m'}{N}\right)\right\} .$$

Since

$$\sum_{n=1}^{N} \cos\left[\frac{2\pi\nu(2n + m')}{N}\right] = 0 ,$$

we obtain

$$\langle|\hat{X}_\lambda^R(\nu)|^2\rangle = \frac{N}{2} \sum_{m'=1}^{N} \Gamma(m') \cos\left(\frac{2\pi\nu m'}{N}\right) .$$

Since $\Gamma(m)$ is an even function,[3] we have

$$\sum_{m'=1}^{N} \Gamma(m') \cos\left(\frac{2\pi\nu m'}{N}\right) = N\hat{\Gamma}(\nu) ,$$

and hence,

$$\langle |\hat{X}_\lambda^{\mathrm{R}}(\nu)|^2 \rangle = \frac{N^2}{2} \hat{\Gamma}(\nu) .$$

In the same way, we find that

$$\langle |\hat{X}_\lambda^{\mathrm{I}}(\nu)|^2 \rangle = \frac{N}{2} \sum_{m'=1}^{N} \Gamma(m') \cos\left(\frac{2\pi\nu m'}{N}\right) .$$

Let us now calculate $\langle \hat{X}_\lambda^{\mathrm{R}}(\nu)\hat{X}_\lambda^{\mathrm{I}}(\nu) \rangle$:

$$\langle \hat{X}_\lambda^{\mathrm{R}}(\nu)\hat{X}_\lambda^{\mathrm{I}}(\nu) \rangle = -\sum_{m=1}^{N}\sum_{n=1}^{N} \langle X_\lambda(m)X_\lambda(n) \rangle \cos\left(\frac{2\pi\nu m}{N}\right) \sin\left(\frac{2\pi\nu n}{N}\right) ,$$

or, proceeding as before,

$$\langle \hat{X}_\lambda^{\mathrm{R}}(\nu)\hat{X}_\lambda^{\mathrm{I}}(\nu) \rangle$$

$$= -\frac{N}{2} \sum_{m'=1}^{N} \Gamma(m') \sum_{n=1}^{N} \left\{ \sin\left[\frac{2\pi\nu(2n+m')}{N}\right] + \sin\left(\frac{2\pi\nu m'}{N}\right) \right\} .$$

Now

$$\sum_{n=1}^{N} \sin\left[2\pi\nu(2n+m')/N\right] = 0$$

and since $\Gamma(m)$ is an even function, we have

$$\sum_{m'=1}^{N} \Gamma(m') \sin\left(\frac{2\pi\nu m'}{N}\right) = 0 .$$

To sum up, we thus obtain

$$\langle |\hat{X}_\lambda^{\mathrm{R}}(\nu)|^2 \rangle = \langle |\hat{X}_\lambda^{\mathrm{I}}(\nu)|^2 \rangle = \frac{N^2}{2} \hat{\Gamma}(\nu) ,$$

and

$$\langle \hat{X}_\lambda^{\mathrm{R}}(\nu)\hat{X}_\lambda^{\mathrm{I}}(\nu) \rangle = 0 .$$

This explains the form of the probability density function mentioned above.

---

[3] For real cyclostationary sequences, we have

$$\langle X_\lambda(m-n)X_\lambda(0) \rangle = \langle X_\lambda(0)X_\lambda(n-m) \rangle .$$

3.6 Ergodic and Stationary Cases

To begin with, we consider the case where t represents time. If the stochastic process is real, stationary and ergodic, $\langle X_\lambda(t) \rangle$ and $\langle X_\lambda(t) X_\lambda(t + \tau) \rangle$ do not depend on t, and $\overline{X_\lambda(t)}$ and $\overline{X_\lambda(t) X_\lambda(t + \tau)}$ do not depend on λ. Hence,

$$\left\langle \overline{X_\lambda(t)} \right\rangle = \overline{X_\lambda(t)} \, ,$$

$$\left\langle \overline{X_\lambda(t) X_\lambda(t + \tau)} \right\rangle = \overline{X_\lambda(t) X_\lambda(t + \tau)} \, ,$$

and also

$$\overline{\langle X_\lambda(t) \rangle} = \langle X_\lambda(t) \rangle \, ,$$

$$\overline{\langle X_\lambda(t) X_\lambda(t + \tau) \rangle} = \langle X_\lambda(t) X_\lambda(t + \tau) \rangle \, .$$

If we assume that we can change the order of the integrals in the expectation values and the time averages,[4] we obtain the fundamental relations,

$$\overline{X_\lambda(t)} = \langle X_\lambda(t) \rangle \, ,$$

and

$$\overline{X_\lambda(t) X_\lambda(t + \tau)} = \langle X_\lambda(t) X_\lambda(t + \tau) \rangle \, .$$

We thus obtain the fundamental result that the ensemble average $\langle X_\lambda(t) \rangle$ and the covariance function $\Gamma_{XX}(\tau)$ can be obtained by calculating the time averages $\overline{X_\lambda(t)}$ and $\overline{X_\lambda(t) X_\lambda(t + \tau)}$. The latter are more easily estimated than the expectation values, which require us to carry out independent experiments. [Note that the covariance function $\Gamma_{XX}(t, t + \tau)$ is then equal to the centered correlation function $C^c_{X_\lambda X_\lambda}(t + \tau) = \overline{X_\lambda(t) X_\lambda(t + \tau)} - \overline{X_\lambda(t)}\ \overline{X_\lambda(t + \tau)}$.]

In the case of homogeneous and ergodic real stochastic fields,

$$\langle X_\lambda(\boldsymbol{r}) \rangle = \overline{X_\lambda(\boldsymbol{r})} \, ,$$

and

$$\langle X_\lambda(\boldsymbol{r}) X_\lambda(\boldsymbol{r} + \boldsymbol{d}) \rangle = \overline{X_\lambda(\boldsymbol{r}) X_\lambda(\boldsymbol{r} + \boldsymbol{d})} \, ,$$

whilst the spatial averages are

$$\overline{X_\lambda(\boldsymbol{r})} = \lim_{V \to \infty} \left[\frac{1}{|V|} \iiint_V X_\lambda(\boldsymbol{r}) \mathrm{d}\boldsymbol{r} \right] \, ,$$

and

$$\overline{X_\lambda(\boldsymbol{r}) X_\lambda(\boldsymbol{r} + \boldsymbol{d})} = \lim_{V \to \infty} \left[\frac{1}{|V|} \iiint_V X_\lambda(\boldsymbol{r}) X_\lambda(\boldsymbol{r} + \boldsymbol{d}) \mathrm{d}\boldsymbol{r} \right] \, .$$

Summing up and simplifying somewhat, we can say that the expectation values (that is the statistical averages) can be determined theoretically, whilst the time and space averages can be more easily measured. Stationarity (or homogeneity) and ergodicity thus serve to relate what can be calculated and what can be measured.

---

[4] This amounts to assuming that we can apply Fubini's theorem.

3.7 Application to Optical Coherence

Let us now reconsider the concept of optical coherence. To avoid overcomplicating the notation, we shall once again assume that the electric field $\boldsymbol{E}_\lambda(\boldsymbol{r},t)$ lies parallel to the Oz axis, whose direction is defined by the unit vector $\boldsymbol{e}_z$. Hence, $\boldsymbol{E}_\lambda(\boldsymbol{r},t) = E_\lambda^{(z)}(\boldsymbol{r},t)\boldsymbol{e}_z$. Recall that the covariance is defined by

$$\Gamma(\boldsymbol{r}_1,t_1,\boldsymbol{r}_2,t_2)$$
$$= \left\langle \left[E_\lambda^{(z)}(\boldsymbol{r}_1,t_1)\right]^* E_\lambda^{(z)}(\boldsymbol{r}_2,t_2)\right\rangle - \left\langle \left[E_\lambda^{(z)}(\boldsymbol{r}_1,t_1)\right]^*\right\rangle \left\langle E_\lambda^{(z)}(\boldsymbol{r}_2,t_2)\right\rangle .$$

If the stochastic field is stationary, we have $m_E(\boldsymbol{r}) = \langle E_\lambda^{(z)}(\boldsymbol{r},t)\rangle$ and also

$$\left\langle \left[E_\lambda^{(z)}(\boldsymbol{r}_1,t)\right]^* E_\lambda^{(z)}(\boldsymbol{r}_2,t+\tau)\right\rangle = \left\langle \left[E_\lambda^{(z)}(\boldsymbol{r}_1,0)\right]^* E_\lambda^{(z)}(\boldsymbol{r}_2,\tau)\right\rangle .$$

We then write simply, with the usual abuse of notation,

$$\Gamma(\boldsymbol{r}_1,t_1,\boldsymbol{r}_2,t_2) = \Gamma(\boldsymbol{r}_1,\boldsymbol{r}_2,t_2 - t_1) .$$

Moreover, if the stochastic field is ergodic, we can estimate $m_E(\boldsymbol{r})$ by integrating over a long enough time interval $[T_1,T_2]$, i.e.,

$$m_E(\boldsymbol{r}) \approx \frac{1}{T_2 - T_1} \int_{T_1}^{T_2} E_\lambda^{(z)}(\boldsymbol{r},t)\mathrm{d}t .$$

The electric field of an optical wave oscillates about a zero value and we thus see that $m_E(\boldsymbol{r})$ is zero. We therefore obtain

$$\Gamma(\boldsymbol{r}_1,\boldsymbol{r}_2,\tau) \approx \frac{1}{T_2 - T_1} \int_{T_1}^{T_2} \left[E_\lambda^{(z)}(\boldsymbol{r}_1,t)\right]^* E_\lambda^{(z)}(\boldsymbol{r}_2,t+\tau)\mathrm{d}t .$$

To begin with, consider the case of a point light source and assume that the light is able to follow two different paths, as happens in the Mach–Zehnder interferometer shown in Fig. 3.5.

The dependence on the space variable $\boldsymbol{r}$ is irrelevant so, to simplify the notation, we write the field before the beam splitter in the form $E_\lambda(t) = E_\lambda^{(z)}(\boldsymbol{r},t)$. We also assume that $\langle E_\lambda(t)\rangle = 0$. The effect of the two arms of the interferometer is to introduce different delays, denoted τ_1 in the first arm and τ_2 in the second arm. The electric field at the detector is proportional to $E_\lambda(t-\tau_1) + E_\lambda(t-\tau_2)$ and the intensity is thus proportional to $|E_\lambda(t-\tau_1) + E_\lambda(t-\tau_2)|$, which can be written

$$|E_\lambda(t-\tau_1)|^2 + |E_\lambda(t-\tau_2)|^2$$
$$+ \left[E_\lambda(t-\tau_1)\right]^* E_\lambda(t-\tau_2) + E_\lambda(t-\tau_1)\left[E_\lambda(t-\tau_2)\right]^* .$$

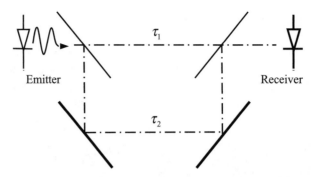

Fig. 3.5. Schematic illustration of an interferometry experiment using the Mach–Zehnder interferometer

If we assume that the electric field is stationary and ergodic, a good approximation for the intensity can be found by integrating over a sufficiently long time interval to obtain

$$2\sigma_E^2 + \Gamma_{EE}(\tau_2 - \tau_1) + \Gamma_{EE}(\tau_1 - \tau_2) \,,$$

where $\sigma_E^2 = \langle |E_\lambda(t)|^2 \rangle$ and $\Gamma_{E,E}(\tau_2 - \tau_1) = \langle [E_\lambda(t - \tau_1)]^* E_\lambda(t - \tau_2) \rangle$. We thus find that, if the field is temporally incoherent for large differences $|\tau_1 - \tau_2|$, the intensity at the detector will be independent of τ_1 and τ_2. On the other hand, if the field is coherent for certain values of $|\tau_1 - \tau_2|$, it will vary as a function of $\tau_1 - \tau_2$. It is sometimes possible to define a coherence time τ_c. This may be taken as the time τ_c for which $\Gamma_{EE}(\tau_c) = \Gamma_{EE}(0)/e$. For example, if we have $\Gamma_{EE}(\tau) = \Gamma_0 \exp(-|\tau|/\alpha) \cos(\omega\tau)$, the coherence time will be defined by $\tau_c = \alpha$.

In an analogous manner, if we consider the electric field at the same time but at two different points, we can sometimes define a coherence length. For example, if we have

$$\Gamma_{EE}(\boldsymbol{r}_1, \boldsymbol{r}_2) = \Gamma_0 \exp\left[-\frac{|\boldsymbol{r}_1 - \boldsymbol{r}_2|}{\xi}\right] \cos\left[\boldsymbol{k}\cdot(\boldsymbol{r}_1 - \boldsymbol{r}_2)\right] \,,$$

the coherence length will be defined by $\ell_c = \xi$.

More generally, in the case of a stationary and homogeneous optical field, we write

$$\Gamma_{EE}(\boldsymbol{r}_1, t_1, \boldsymbol{r}_2, t_2) = \Gamma_{EE}(\boldsymbol{r}_2 - \boldsymbol{r}_1, t_2 - t_1) \,.$$

The coherence domain is then defined by the set of coordinates $(\boldsymbol{r}, t)$ such that $\Gamma_{EE}(\boldsymbol{r}, t)$ is not negligible.

3.8 Fields and Partial Differential Equations

In vacuum, the electric field of an electromagnetic wave propagates according to the partial differential equation

$$\Delta E(\boldsymbol{r},t) - \frac{1}{c^2}\frac{\partial^2 E(\boldsymbol{r},t)}{\partial t^2} = 0 \ ,$$

where $\Delta E(\boldsymbol{r},t) = \partial^2 E(\boldsymbol{r},t)/\partial x^2 + \partial^2 E(\boldsymbol{r},t)/\partial y^2 + \partial^2 E(\boldsymbol{r},t)/\partial z^2$, c is the speed of light in vacuum, and x, y, z are the coordinates of the point $\boldsymbol{r}$.

The evolution of a physical quantity $X(\boldsymbol{r},t)$ is often described by a partial differential equation. This equation will of course depend on the problem under investigation, and we shall express it generically in the form

$$H\left[X(\boldsymbol{r},t)\right] = 0 \ .$$

Let us assume that the evolution of the relevant field $X(\boldsymbol{r},\ t)$ is described by a linear partial differential equation. Recall that a partial differential equation is said to be linear if, for all fields $X_1(\boldsymbol{r},t)$ and $X_2(\boldsymbol{r},t)$ satisfying $H\left[X_1(\boldsymbol{r},t)\right] = 0$ and $H\left[X_2(\boldsymbol{r},t)\right] = 0$, and for all scalars a and b, we have

$$H\left[aX_1(\boldsymbol{r},t) + bX_2(\boldsymbol{r},t)\right] = 0 \ .$$

We also assume that the field is described by a real-valued stochastic field which we shall denote by $X_\lambda(\boldsymbol{r},t)$. If the partial differential equation refers to variables $\boldsymbol{r}_1$, t_1, we shall write $H_1\left[X_\lambda(\boldsymbol{r}_1,t_1)\right] = 0$, and if it refers to variables $\boldsymbol{r}_2$, t_2, we shall write $H_2\left[X_\lambda(\boldsymbol{r}_2,t_2)\right] = 0$. We thus have

$$H_1\left[\langle X_\lambda(\boldsymbol{r}_1,t_1)X_\lambda(\boldsymbol{r}_2,t_2)\rangle\right] = \langle H_1\left[X_\lambda(\boldsymbol{r}_1,t_1)\right]X_\lambda(\boldsymbol{r}_2,t_2)\rangle \ ,$$

$$H_2\left[\langle X_\lambda(\boldsymbol{r}_1,t_1)X_\lambda(\boldsymbol{r}_2,t_2)\rangle\right] = \langle X_\lambda(\boldsymbol{r}_1,t_1)H_2\left[X_\lambda(\boldsymbol{r}_2,t_2)\right]\rangle \ ,$$

and hence,

$$H_1\left[\Gamma_{XX}(\boldsymbol{r}_1,t_1,\boldsymbol{r}_2,t_2)\right] = H_2\left[\Gamma_{XX}(\boldsymbol{r}_1,t_1,\boldsymbol{r}_2,t_2)\right] = 0 \ .$$

The field and its covariance evolve according to the same partial differential equation.

If we also assume that the stochastic field is stationary and homogeneous, we have

$$\Gamma_{XX}(\boldsymbol{r},t)$$
$$= \langle X_\lambda(\boldsymbol{r}_1,t_1)X_\lambda(\boldsymbol{r}_1+\boldsymbol{r},t_1+t)\rangle - \langle X_\lambda(\boldsymbol{r}_1,t_1)\rangle\langle X_\lambda(\boldsymbol{r}_1+\boldsymbol{r},t_1+t)\rangle \ ,$$

and hence,

$$H\left[\Gamma_{XX}(\boldsymbol{r},t)\right] = 0 \ ,$$

where $H[\ ]$ applies to variables $\boldsymbol{r}$ and t.

Let us consider the particular case where an electromagnetic wave propagates in vacuum. The covariance function of the electric field thus satisfies the equation

$$\Delta_1\Gamma_{XX}(\boldsymbol{r}_1,t_1,\boldsymbol{r}_2,t_2) - \frac{1}{c^2}\frac{\partial^2\Gamma_{XX}(\boldsymbol{r}_1,t_1,\boldsymbol{r}_2,t_2)}{\partial t_1^2} = 0 \ ,$$

where

$$\Delta_1 = \frac{\partial^2}{\partial x_1^2} + \frac{\partial^2}{\partial y_1^2} + \frac{\partial^2}{\partial z_1^2}$$

and $r_1 = (x_1, y_1, z_1)^{\mathrm{T}}$. In the particular case where the electric field of an electromagnetic wave is stationary and homogeneous, its covariance function evolves according to the equation

$$\Delta\Gamma_{EE}(r,t) - \frac{1}{c^2}\frac{\partial^2\Gamma_{EE}(r,t)}{\partial t^2} = 0 \,,$$

where

$$\Delta = \frac{\partial^2}{\partial x^2} + \frac{\partial^2}{\partial y^2} + \frac{\partial^2}{\partial z^2}$$

and $r = (x, y, z)^{\mathrm{T}}$. In this way, we can describe the evolution of coherence in optics.

3.9 Power Spectral Density

We consider a real stationary stochastic field $X_\lambda(t)$. To simplify the analysis, we assume in this section that $\langle X_\lambda(t)\rangle = 0$, so that the covariance is simply $\Gamma_{XX}(t) = \langle X_\lambda(t_1)X_\lambda(t_1 + t)\rangle$. Suppose that we can define the Fourier transform of the restriction of $X_\lambda(t)$ to the interval $[T_1, T_2]$, viz., $\int_{T_1}^{T_2} X_\lambda(t)\mathrm{e}^{-\mathrm{i}2\pi\nu t}\mathrm{d}t$. Clearly, this Fourier transform depends on the choice of T_1 and T_2. We could let these values tend to plus and minus infinity, respectively, i.e., $T_1 \to -\infty$ and $T_2 \to \infty$, but a fundamental problem arises in this case. Indeed, the limit

$$\lim_{\substack{T_1 \to -\infty \\ T_2 \to \infty}}\left[\frac{1}{\sqrt{(T_2 - T_1)}}\int_{T_1}^{T_2} X_\lambda(t)\exp(-\mathrm{i}2\pi\nu t)\mathrm{d}t\right] ,$$

does not generally exist, because the phase of $\int_{T_1}^{T_2} X_\lambda(t)\mathrm{e}^{-\mathrm{i}2\pi\nu t}\mathrm{d}t$ may not converge. We thus define the power spectral density of $X_\lambda(t)$, also called the spectrum of the signal, by

$$\hat{S}_{XX}(\nu) = \lim_{\substack{T_1 \to -\infty \\ T_2 \to \infty}}\left[\frac{1}{T_2 - T_1}\left\langle\left|\int_{T_1}^{T_2} X_\lambda(t)\exp(-\mathrm{i}2\pi\nu t)\mathrm{d}t\right|^2\right\rangle\right] .$$

For stationary stochastic processes $X_\lambda(t)$, the Wiener–Khinchine theorem (see Section 3.16) shows that $\hat{S}_{XX}(\nu)$ and the covariance $\Gamma_{XX}(\tau)$ are related by a Fourier transformation:

$$\hat{S}_{XX}(\nu) = \int_{-\infty}^{\infty} \Gamma_{XX}(t)\exp(-\mathrm{i}2\pi\nu t)\mathrm{d}t \ .$$

The instantaneous power of the fluctuations is defined by

$$P_X(t) = \langle X_\lambda(t)X_\lambda(t)\rangle \ .$$

In the stationary case, it is independent of time t and thus equal to the mean power, denoted by P_X. Since $\Gamma_{XX}(t) = \langle X_\lambda(t_1)X_\lambda(t_1+t)\rangle$, it is easy to see that $P_X = \Gamma_{XX}(0)$. For stationary stochastic processes, the Wiener–Khinchine theorem allows one to express this mean power of the fluctuations in terms of the power spectral density:

$$P_X = \int_{-\infty}^{\infty} \hat{S}_{XX}(\nu)\mathrm{d}\nu \ .$$

There is a class of signals known as filtered white noise which plays an important role in physics. White noise is a stochastic signal with positive, constant power spectral density. Such signals raise significant problems in physics because they have infinite power, as implied by the relation $P_X = \int_{-\infty}^{\infty} \hat{S}_{XX}(\nu)\mathrm{d}\nu$. This difficulty is overcome in a simple manner by considering that the power spectral density is constant and nonzero only in a certain frequency band $[\nu_1, \nu_2]$. We thus define white noise with bounded spectrum. In the general case where the spectral density of the stochastic process is quite arbitrary, we speak of colored noise.

We can define the spectral density of a real homogeneous stochastic field $X_\lambda(x, y, z)$. We also assume that $\langle X_\lambda(x, y, z)\rangle = 0$ so that the covariance is

$$\Gamma_{XX}(x, y, z) = \langle X_\lambda(x_1, y_1, z_1)X_\lambda(x_1+x, y_1+y, z_1+z)\rangle \ .$$

The power spectral density of $X_\lambda(x, y, z)$ is then defined by

$$\hat{S}_{XX}(k_x, k_y, k_z)$$

$$= \lim_{V\to\infty} \frac{1}{|V|}\left\langle \left|\iiint_V X_\lambda(x, y, z)\mathrm{e}^{-\mathrm{i}2\pi[k_x x + k_y y + k_z z]}\mathrm{d}x\mathrm{d}y\mathrm{d}z\right|^2 \right\rangle \ ,$$

where $|V|$ is the measure of the volume of V and we have assumed that we can define the Fourier transform of a restriction of $X_\lambda(x, y, z)$ to arbitrary finite regions V. For stationary stochastic fields, the Wiener–Khinchine theorem then establishes the result

$$\hat{S}_{XX}(k_x, k_y, k_z) = \iiint \Gamma_{XX}(x, y, z)\mathrm{e}^{-\mathrm{i}2\pi[k_x x + k_y y + k_z z]}\mathrm{d}x\mathrm{d}y\mathrm{d}z \ .$$

The mean power of fluctuations is then

$$P_X = \iiint \hat{S}_{XX}(k_x, k_y, k_z) \mathrm{d}k_x \mathrm{d}k_y \mathrm{d}k_z \ .$$

We sometimes use the vector notation

$$(x, y, z) \to \boldsymbol{r} \quad \text{and} \quad (k_x, k_y, k_z) \to \boldsymbol{k} \ ,$$

using which we may write

$$\hat{S}_{XX}(\boldsymbol{k}) = \int \Gamma_{XX}(\boldsymbol{r}) \mathrm{e}^{-\mathrm{i}2\pi \boldsymbol{k} \cdot \boldsymbol{r}} \mathrm{d}\boldsymbol{r} \ ,$$

and also

$$P_X = \int \hat{S}_{XX}(\boldsymbol{k}) \mathrm{d}\boldsymbol{k} \ .$$

3.10 Filters and Fluctuations

A concrete or abstract system will be defined by its action on the signals that can be applied to it. Among the concrete analog or digital systems that come to mind, one might mention electronic systems which modify an applied voltage, for example. A great many physical systems can indeed be viewed from this standpoint.

In macroscopic physics, the quantities considered as input signals are often intensive quantities, whilst the conjugate extensive quantities constitute the output signal, or more simply, the response of the system. These terms, conjugate extensive and intensive quantities, arise in thermodynamics. An extensive quantity has a value proportional to the number of particles making up the physical system under consideration, whilst an intensive quantity is independent of the number of particles. Two quantities are said to be conjugate if their product has units of energy and if they arise in different thermodynamical energy functions. Table 3.1 shows several examples of pairs of conjugate extensive and intensive quantities that are frequently encountered in physics.

It should be remembered, however, that the notion of a system does not require the existence of a material physical system. We may indeed consider abstract systems in order to represent mathematical operators such as the propagation operator, discussed further below.

We first consider functions depending on time. The relation between the output $s(t)$ and the input $e(t)$ will be denoted symbolically by $s(t) = S[e(t)]$. A system is said to be linear if, for any $e_1(t)$ and $e_2(t)$ and any numbers α and β (which may be real or complex depending on the situation), we have

$$\left. \begin{array}{l} s_1(t) = S[e_1(t)] \\ s_2(t) = S[e_2(t)] \end{array} \right\} \Rightarrow S[\alpha \, e_2(t) + \beta \, e_1(t)] = \alpha \, s_2(t) + \beta \, s_1(t) \ .$$

This linearity property is often an approximation for small applied signals. A system is said to be stationary if, for any $e(t)$ and real number τ, we have

Table 3.1. Pairs of conjugate quantities in thermodynamics

Intensive quantity	Extensive quantity
Electric field E	Electrical polarization P
Magnetic field H	Magnetization M
Electrical potential V	Electrical charge Q
Pressure P	Volume V
Chemical potential μ	Particle number N
Applied stress t	Strain u

$$s(t) = S\left[e(t)\right] \implies S\left[e(t-\tau)\right] = s(t-\tau) \, .$$

Here we find once again the property of time translation invariance used to define stationarity. It should be remembered, however, that we are now concerned with the idea of a stationary system rather than a stationary stochastic process. Stationarity of a system means that it possesses no internal clock and that it therefore reacts in the same way at whatever instant of time the input signal is applied. Most stationary linear systems possess a relation between input and output that can be written in the form of a convolution:

$$s(t) = \int_{-\infty}^{\infty} \chi(t-\tau)e(\tau)\mathrm{d}\tau \, ,$$

where $\chi(t)$ represents the convolution kernel (see Fig. 3.6). In this case we speak of a convolution filter, or more simply, a linear filter. It is a well known mathematical result that the Fourier transforms $\hat{e}(\nu)$ and $\hat{s}(\nu)$ of $e(t)$ and $s(t)$ are related by

$$\hat{s}(\nu) = \hat{\chi}(\nu)\hat{e}(\nu) \, ,$$

where $\hat{\chi}(\nu)$ is the Fourier transform of $\chi(t)$. In physics, $\chi(t)$ is called the susceptibility or impulse response since it is the response of the system if a Dirac impulse is applied as input, viz., $e(t) = \delta(t)$.

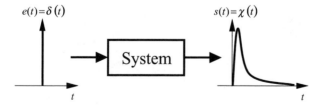

Fig. 3.6. Schematic illustration of impulse response

In physics, one usually measures the response function $\sigma(t)$ which is defined (see Fig. 3.7) as the response to an input of the form $e(t) = e_0 [1 - \theta(t)]$. Here the Heaviside step function is defined by $\theta(t) = 1$ if $t \geq 0$ and $\theta(t) = 0$ if $t < 0$.

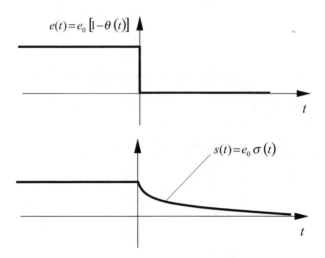

Fig. 3.7. Schematic illustration of the response function

As mentioned above, linearity is often an approximation for small applied signals. However, this is not the case when the input signal is a Dirac impulse. On the other hand, it is generally possible to measure the response $s(t)$ to a signal $e(t) = e_0[1-\theta(t)]$ applied with e_0 as small as necessary to obtain a linear response to the required accuracy. It then suffices to determine $\sigma(t)$ using the relation $\sigma(t) = s(t)/e_0$. The impulse response function is easily obtained from $\chi(t) = -d\sigma(t)/dt$.

We can now characterize the effect of a convolution filter with kernel $\chi(t)$ on fluctuations described by stationary stochastic processes $X_\lambda(t)$ with zero mean, i.e., $\langle X_\lambda(t) \rangle = 0$. Let $Y_\lambda(t)$ be the output stochastic process for the system, so that

$$Y_\lambda(\tau) = \int_{-\infty}^{\infty} \chi(\tau - t)X_\lambda(t)\mathrm{d}t ,$$

which can also be written

$$Y_\lambda(\tau) = \int_{-\infty}^{\infty} \chi(t)X_\lambda(\tau - t)\mathrm{d}t ,$$

leading to

$$Y_\lambda^*(\tau_1)Y_\lambda(\tau_2) = \int\limits_{-\infty}^{\infty}\int\limits_{-\infty}^{\infty} \chi^*(t_1)\chi(t_2)X_\lambda^*(\tau_1 - t_1)X_\lambda(\tau_2 - t_2)\mathrm{d}t_1\mathrm{d}t_2 \ .$$

If we assume that we can change the order of the various integrals, we then find that

$$\langle Y_\lambda^*(\tau_1)Y_\lambda(\tau_2)\rangle = \int\limits_{-\infty}^{\infty}\int\limits_{-\infty}^{\infty} \chi^*(t_1)\chi(t_2)\langle X_\lambda^*(\tau_1 - t_1)X_\lambda(\tau_2 - t_2)\rangle\mathrm{d}t_1\mathrm{d}t_2 \ .$$

Substituting $\Gamma_{XX}(\tau_2 - \tau_1)$ for $\langle X_\lambda^*(\tau_1)X_\lambda(\tau_2)\rangle$, the last equation becomes

$$\langle Y_\lambda^*(\tau_1)Y_\lambda(\tau_2)\rangle = \int\limits_{-\infty}^{\infty}\int\limits_{-\infty}^{\infty} \chi^*(t_1)\chi(t_2)\Gamma_{XX}(\tau_2 - t_2 - \tau_1 + t_1)\mathrm{d}t_2\mathrm{d}t_1 \ .$$

Note that $\langle Y_\lambda^*(\tau_1)Y_\lambda(\tau_2)\rangle$ only depends on $\tau_2 - \tau_1$, so that $Y_\lambda(t)$ is therefore stationary and we may write $\Gamma_{YY}(\tau_2 - \tau_1) = \langle Y_\lambda^*(\tau_1)Y_\lambda(\tau_2)\rangle$. Hence,

$$\Gamma_{YY}(\tau) = \int\limits_{-\infty}^{\infty}\int\limits_{-\infty}^{\infty} \chi(t_2)\chi^*(t_1)\Gamma_{XX}(\tau + t_1 - t_2)\mathrm{d}t_2\mathrm{d}t_1 \ .$$

By Fourier transform, we can then show that

$$\hat{S}_{YY}(\nu) = |\hat{\chi}(\nu)|^2\,\hat{S}_{XX}(\nu) \ .$$

The generalization to stochastic fields is very simple. A convolution filter is written

$$s(x,y,z) = \iiint \chi(x - \varsigma, y - \xi, z - \zeta)e(\varsigma,\xi,\zeta)\mathrm{d}\varsigma\mathrm{d}\xi\mathrm{d}\zeta \ .$$

For the three-dimensional Fourier transforms, we then have

$$\hat{s}(k_x, k_y, k_z) = \hat{\chi}(k_x, k_y, k_z)\hat{e}(k_x, k_y, k_z) \ .$$

It is clearly more convenient to use a vector notation and define the spatial Fourier transform by

$$\hat{a}(\boldsymbol{k}) = \int a(\boldsymbol{r})\exp\left(-\mathrm{i}2\pi\boldsymbol{k}\cdot\boldsymbol{r}\right)\mathrm{d}\boldsymbol{r} \ ,$$

which simply means

$$\hat{a}(\boldsymbol{k}) = \hat{a}(k_x, k_y, k_z) = \iiint a(x,y,z)\exp\left[-\mathrm{i}2\pi(k_xx + k_yy + k_zz)\right]\mathrm{d}x\mathrm{d}y\mathrm{d}z \ .$$

For homogeneous stochastic fields, we then obtain

$$\Gamma_{YY}(r) = \iint \chi(r_2)\chi^*(r_1)\Gamma_{XX}(r + r_1 - r_2)\mathrm{d}r_2\mathrm{d}r_1 \ ,$$

which reads

$$\Gamma_{YY}(x, y, z) = \iiint \iiint \chi(x_2, y_2, z_2)\chi^*(x_1, y_1, z_1)$$
$$\times \Gamma_{XX}(x + x_1 - x_2, y + y_1 - y_2, z + z_1 - z_2)\mathrm{d}x_2\mathrm{d}y_2\mathrm{d}z_2\mathrm{d}x_1\mathrm{d}y_1\mathrm{d}z_1 \ .$$

By Fourier transform, we can then deduce the relation

$$\hat{S}_{YY}(k) = |\hat{\chi}(k)|^2 \hat{S}_{XX}(k) \ ,$$

i.e.,

$$\hat{S}_{YY}(k_x, k_y, k_z) = |\hat{\chi}(k_x, k_y, k_z)|^2 \hat{S}_{XX}(k_x, k_y, k_z) \ .$$

3.11 Application to Optical Imaging

In this section, we shall illustrate the above ideas in the context of optical imaging. We therefore consider an imaging system between the plane P_1 of the object and the plane P_2 on which the image is formed (see Fig. 3.8).

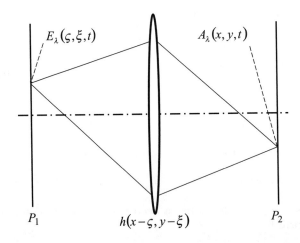

Fig. 3.8. Schematic illustration of an optical imaging system

We describe the electric field at a point r in plane P_1 and at time t by the scalar spatio-temporal field $E_\lambda(r, t)$. [The vector r has components x, y, i.e., $r = (x, y)^{\mathrm{T}}$.] The electric field has zero mean, i.e., $\langle E_\lambda(r, t) \rangle = 0$, and the covariance function of the field between point r_1 at time t_1 and point r_2 at time t_2 is

$$\Gamma_{EE}(\mathbf{r}_1, \mathbf{r}_2, t_1, t_2) = \langle E_\lambda^*(\mathbf{r}_1, t_1) E_\lambda(\mathbf{r}_2, t_2) \rangle .$$

At the detector in plane P_2, the detected field $A_\lambda(\mathbf{r}, t)$ is a linear transformation of the emitted field $E_\lambda(\mathbf{r}, t)$. Optical systems can be constructed in such a way that a convolution relation is a good approximation to the relation between emitted and detected fields:

$$A_\lambda(x, y, t) = \iint h(x - \varsigma, y - \xi) E_\lambda(\varsigma, \xi, t) \mathrm{d}\varsigma \mathrm{d}\xi ,$$

or more simply

$$A_\lambda(\mathbf{r}, t) = \int h(\mathbf{r} - \mathbf{u}) E_\lambda(\mathbf{u}, t) \mathrm{d}\mathbf{u} .$$

Note that, in this section, we are neglecting delays due to propagation of optical signals. It is easy to show that they have little effect on the results of our analysis in this context.

In order to keep the notation as simple as possible, we shall not take into account the magnification factors present in most optical systems. The received intensity $I_\lambda^{\mathrm{R}}(\mathbf{r}, t) = |A_\lambda(\mathbf{r}, t)|^2$ is thus

$$I_\lambda^{\mathrm{R}}(\mathbf{r}, t) = \iint h^*(\mathbf{r} - \mathbf{u}_1) h(\mathbf{r} - \mathbf{u}_2) E_\lambda^*(\mathbf{u}_1, t) E_\lambda(\mathbf{u}_2, t) \mathrm{d}\mathbf{u}_1 \mathrm{d}\mathbf{u}_2 .$$

Setting $I^{\mathrm{R}}(x, y, t) = \langle |A_\lambda(x, y, t)|^2 \rangle$ and assuming that we can change the order of integration, we then obtain

$$I^{\mathrm{R}}(\mathbf{r}, t) = \iint h^*(\mathbf{r} - \mathbf{u}_1) h(\mathbf{r} - \mathbf{u}_2) \Gamma_{EE}(\mathbf{u}_1, \mathbf{u}_2, t, t) \mathrm{d}\mathbf{u}_1 \mathrm{d}\mathbf{u}_2 . \qquad (3.2)$$

If the field $E_\lambda(\mathbf{r}, t)$ is homogeneous, we have by Fourier transform,

$$\hat{I}^{\mathrm{R}}(\mathbf{k}, t) = \left| \hat{h}(\mathbf{k}) \right|^2 S_{EE}(\mathbf{k}, t) ,$$

where the spatial Fourier transforms are defined by

$$\hat{a}(\mathbf{k}, t) = \hat{a}(k_x, k_y, t) = \iint a(x, y, t) \exp\left[-\mathrm{i}2\pi(k_x x + k_y y)\right] \mathrm{d}x \mathrm{d}y .$$

However, the interesting practical cases correspond to inhomogeneous fields. We must then consider the more general relation (3.2) between $I^{\mathrm{R}}(\mathbf{r}, t)$ and $\Gamma_{EE}(\mathbf{u}_1, \mathbf{u}_2, t, t)$.

We can now consider two extreme cases, namely, when the field is totally coherent or totally incoherent. In reality, there are two characteristic lengths in this problem. The first is defined by the size of the object in the plane P_1, which we denote by L. The second is related to the resolution δ introduced by the optical system. (The resolution is defined qualitatively as the smallest distance δ between two points which produces an image with two points that are distinct according to a certain mathematical criterion.) The field will then be described as spatially coherent if the coherence length is much larger than L and spatially incoherent if it is much smaller than δ.

We begin by considering the spatially incoherent case. We may then use the following approximation for the covariance of the emitted field: $\Gamma_{EE}(\boldsymbol{u_1}, \boldsymbol{u_2}, t, t) = I^{\mathrm{E}}(\boldsymbol{u_1}, t)\delta(\boldsymbol{u_1} - \boldsymbol{u_2})$, where $I^{\mathrm{E}}(\boldsymbol{r}, t) = \langle|E_\lambda(\boldsymbol{r}, t)|^2\rangle$. We thus obtain

$$I^{\mathrm{R}}(\boldsymbol{r}, t) = \int |h(\boldsymbol{r} - \boldsymbol{u})|^2\, I^{\mathrm{E}}(\boldsymbol{u}, t)\mathrm{d}\boldsymbol{u}\ ,$$

which implies that there is a convolution relation between the intensities in the planes of the object and the detector. In other words, the system is linear and stationary as far as the intensity is concerned.

In the spatially coherent case, we have

$$\Gamma_{EE}(\boldsymbol{r_1}, \boldsymbol{r_2}, t_1, t_2) = F^*(\boldsymbol{r_1})F(\boldsymbol{r_2})\Gamma_{00}(t_1, t_2)\ .$$

To see this, consider the simple situation in which the object has transparency $F(\boldsymbol{r})$ (possibly complex-valued, in order to describe the phenomena of absorption and phase shift), and we illuminate with a perfectly coherent and uniform field $E_\lambda(t)$. After the object, the field will be $E_\lambda(\boldsymbol{r}, t) = F(\boldsymbol{r})E_\lambda(t)$, and the covariance will be

$$\Gamma_{EE}(\boldsymbol{r_1}, \boldsymbol{r_2}, t_1, t_2) = F^*(\boldsymbol{r_1})F(\boldsymbol{r_2})\langle E_\lambda^*(t_1)E_\lambda(t_2)\rangle\ ,$$

or $\Gamma_{EE}(\boldsymbol{r_1}, \boldsymbol{r_2}, t_1, t_2) = F^*(\boldsymbol{r_1})F(\boldsymbol{r_2})\Gamma_{00}(t_1, t_2)$. The intensity at the detector is thus

$$I^{\mathrm{R}}(\boldsymbol{r}, t) = \iint h^*(\boldsymbol{r} - \boldsymbol{u_1})h(\boldsymbol{r} - \boldsymbol{u_2})F^*(\boldsymbol{u_1})F(\boldsymbol{u_2})\Gamma_{00}(t, t)\mathrm{d}\boldsymbol{u_1}\mathrm{d}\boldsymbol{u_2}\ .$$

Since the intensity $I_0(t)$ before the object is $\langle E_\lambda^*(t)E_\lambda(t)\rangle$, we have $\Gamma_{00}(t, t) = I_0(t)$ and we can thus write

$$I^{\mathrm{R}}(\boldsymbol{r}, t) = I_0(t)\left|\int h(\boldsymbol{r} - \boldsymbol{u})F(\boldsymbol{u})\mathrm{d}\boldsymbol{u}\right|^2\ .$$

The situation is therefore very different from the totally incoherent case. The relation is in fact linear and stationary in the field amplitude, and hence non-linear in the intensity, in contrast to the case of the totally incoherent field.

The result that we have just established in optics is encountered in many different areas of physics. A convolution relation in amplitude in the case of correlated fields becomes a convolution relation in intensity in the case of uncorrelated fields.

3.12 Green Functions and Fluctuations

As already mentioned in Section 3.8, the evolution of a physical quantity $X(\boldsymbol{r}, t)$ is often described by a linear partial differential equation whose specific form depends on the problem under investigation.

For example, when it propagates in vacuum, the electric field $E(\boldsymbol{r}, t)$ of an electromagnetic wave satisfies

$$\Delta E(x, y, z, t) - \frac{1}{c^2} \frac{\partial^2 E(x, y, z, t)}{\partial t^2} = 0 \,,$$

where

$$\Delta E(x, y, z, t) = \frac{\partial^2 E(x, y, z, t)}{\partial x^2} + \frac{\partial^2 E(x, y, z, t)}{\partial y^2} + \frac{\partial^2 E(x, y, z, t)}{\partial z^2} \,,$$

c is the speed of light in vacuum, and x, y, z are the coordinates of the point $\boldsymbol{r}$.

In the following chapter, we shall see that the diffusion equation is a partial differential equation describing macroscopic phenomena such as the diffusion of particles through a solvent. In a homogeneous medium, this equation is

$$\Delta N(x, y, z, t) - \frac{2}{\chi^2} \frac{\partial N(x, y, z, t)}{\partial t} = 0 \,,$$

where $N(x, y, z, t)$ represents the concentration of particles at time t at the point with coordinates x, y, z, and χ is the diffusion coefficient.

In this section we shall therefore assume that the field $X(\boldsymbol{r}, t)$ that interests us evolves according to a linear partial differential equation which we shall write in the form

$$H\left[X(\boldsymbol{r}, t)\right] = 0 \,.$$

We use the notation $\boldsymbol{r} = (x, y, z)^{\mathrm{T}}$ to shorten the equations. We can now introduce the Green function $G(\boldsymbol{r}, t, \boldsymbol{r}', t')$ (where $t > t'$) which solves the partial differential equation

$$H\left[G(\boldsymbol{r}, t, \boldsymbol{r}', t')\right] = 0 \,,$$

with initial conditions $X(\boldsymbol{r}, t') = \delta(\boldsymbol{r} - \boldsymbol{r}')$ at time t', where $H[\ ]$ acts on the coordinates $\boldsymbol{r}, t$. In other words, we have $H\left[G(\boldsymbol{r}, t, \boldsymbol{r}', t')\right] = 0$ for all $t > t'$ and

$$G(\boldsymbol{r}, t', \boldsymbol{r}', t') = \delta(\boldsymbol{r} - \boldsymbol{r}') \,.$$

$G(\boldsymbol{r}, t, \boldsymbol{r}', t')$ thus represents the field at time t and at the point $\boldsymbol{r}$ which results from the propagation according to the partial differential equation $H\left[G(\boldsymbol{r}, t, \boldsymbol{r}', t')\right] = 0$, with the initial condition that the field is concentrated at the point $\boldsymbol{r}'$ at time t'.

The solution $X(\boldsymbol{r}, t)$ to the partial differential equation with initial conditions $X(\boldsymbol{r}, t') = F(\boldsymbol{r})$ at time t' is then

$$X(\boldsymbol{r}, t) = \int G(\boldsymbol{r}, t, \boldsymbol{r}', t') F(\boldsymbol{r}') \mathrm{d}\boldsymbol{r}' \,.$$

Indeed, we have

$$H\left[X(\boldsymbol{r},t)\right] = \int H\left[G(\boldsymbol{r},t,\boldsymbol{r}',t')\right]F(\boldsymbol{r}')\mathrm{d}\boldsymbol{r}' \ ,$$

and $H\left[G(\boldsymbol{r},t,\boldsymbol{r}',t')\right] = 0$ thus implies that $H\left[X(\boldsymbol{r},t)\right] = 0$. Moreover, from the definition, $G(\boldsymbol{r},t',\boldsymbol{r}',t')$ is equal to $\delta(\boldsymbol{r}-\boldsymbol{r}')$ and hence $X(\boldsymbol{r},t') = F(\boldsymbol{r})$. We have thus found the solution to the partial differential equation which satisfies the initial conditions.

The situation is analogous to the one in Section 3.10. However, there are at least two important differences. We are considering a spatio-temporal field and we are not integrating with respect to the variable t, but rather with respect to $\boldsymbol{r}'$. The covariance function is defined by

$$\Gamma_{XX}(\boldsymbol{r}_1,t_1,\boldsymbol{r}_2,t_2) = \langle X_\lambda^*(\boldsymbol{r}_1,t_1)X_\lambda(\boldsymbol{r}_2,t_2)\rangle - \langle X_\lambda^*(\boldsymbol{r}_1,t_1)\rangle\langle X_\lambda(\boldsymbol{r}_2,t_2)\rangle \ .$$

Symbolically, we write

$$G(\boldsymbol{r},t,\boldsymbol{r}',t') \otimes F(\boldsymbol{r}') = \int G(\boldsymbol{r},t,\boldsymbol{r}',t')F(\boldsymbol{r}')\mathrm{d}\boldsymbol{r}' \ ,$$

or in long-hand

$$G(\boldsymbol{r},t,\boldsymbol{r}',t') \otimes F(\boldsymbol{r}') = \iiint G(x,y,z,t,x',y',z',t')F(x',y',z')\mathrm{d}x'\mathrm{d}y'\mathrm{d}z' \ .$$

We thus obtain $X_\lambda(\boldsymbol{r}_1,t_1) = G(\boldsymbol{r}_1,t_1,\boldsymbol{r},t)\otimes X_\lambda(\boldsymbol{r},t)$, where $t_1 \geq t$, and hence,

$$\begin{aligned}\langle X_\lambda^*(\boldsymbol{r}_1,t_1)&X_\lambda(\boldsymbol{r}_2,t_2)\rangle \\ &= G^*(\boldsymbol{r}_1,t_1,\boldsymbol{r}_1',t_1') \otimes G(\boldsymbol{r}_2,t_2,\boldsymbol{r}_2',t_2') \otimes \langle X_\lambda^*(\boldsymbol{r}_1',t_1')X_\lambda(\boldsymbol{r}_2',t_2')\rangle \ ,\end{aligned}$$

where $t_1 \geq t_1'$ and $t_2 \geq t_2'$. This can also be written

$$\Gamma_{XX}(\boldsymbol{r}_1,t_1,\boldsymbol{r}_2,t_2) = G^*(\boldsymbol{r}_1,t_1,\boldsymbol{r}_1',t_1') \otimes G(\boldsymbol{r}_2,t_2,\boldsymbol{r}_2',t_2') \otimes \Gamma_{XX}(\boldsymbol{r}_1',t_1',\boldsymbol{r}_2',t_2') \ ,$$

or more fully,

$$\begin{aligned}\Gamma_{XX}&(x_1,y_1,z_1,t_1,x_2,y_2,z_2,t_2) \\ &= \iiint \iiint G^*(x_1,y_1,z_1,t_1,x_1',y_1',z_1',t_1')G(x_2,y_2,z_2,t_2,x_2',y_2',z_2',t_2') \\ &\quad \times \Gamma_{XX}(x_1',y_1',z_1',t_1',x_2',y_2',z_2',t_2') \ \mathrm{d}x_1'\mathrm{d}y_1'\mathrm{d}z_1'\mathrm{d}x_2'\mathrm{d}y_2'\mathrm{d}z_2' \ .\end{aligned}$$

Let $X_F(\boldsymbol{r},t)$ be the solution to the partial differential equation

$$H\left[X(\boldsymbol{r},t)\right] = 0 \ ,$$

with initial conditions $X(\boldsymbol{r},t') = F(\boldsymbol{r})$. This partial differential equation $H\left[X(\boldsymbol{r},t)\right] = 0$ is said to be stationary if, for any initial conditions $X(\boldsymbol{r},t') = F(\boldsymbol{r})$ and any τ, the solution with initial conditions $X(\boldsymbol{r},t'+\tau) = F(\boldsymbol{r})$ is simply $X_F(\boldsymbol{r},t+\tau)$. In this case the Green function can be written in the form $G(\boldsymbol{r},\boldsymbol{r}',t-t')$.

Likewise, the partial differential equation $H\left[X(r,t)\right] = 0$ is said to be homogeneous if, for any initial conditions $X(r,t') = F(r)$ and for any ρ, the solution with initial conditions $X(r,t') = F(r + \rho)$ is simply $X_F(r + \rho, t)$. In this case, the Green function can be written in the form $G(r - r', t, t')$.

If the partial differential equation is stationary and homogeneous, the Green function can be written $G(r - r', t - t')$ and hence,

$$\Gamma_{XX}(r_1, t_1, r_2, t_2)$$
$$= G^*(r_1 - r'_1, t_1 - t'_1) \otimes G(r_2 - r'_2, t_2 - t'_2) \otimes \Gamma_{XX}(r'_1, t'_1, r'_2, t'_2) \, .$$

Moreover, if the field is stationary, we can write

$$\Gamma_{XX}(r_1, r_2, \tau) = \langle X^*_\lambda(r_1, t) X_\lambda(r_2, t + \tau) \rangle - \langle X^*_\lambda(r_1, t) \rangle \langle X_\lambda(r_2, t + \tau) \rangle \, ,$$

and hence,

$$\Gamma_{XX}(r_1, t_1, r_2, t_2)$$
$$= G^*(r_1 - r'_1, t_1 - t'_1) \otimes G(r_2 - r'_2, t_2 - t'_2) \otimes \Gamma_{XX}(r'_1, r'_2, t'_2 - t'_1) \, ,$$

or

$$\Gamma_{XX}(r_1, t_1, r_2, t_2)$$
$$= G^*(r_1 - r'_1, t_1 - t'_1) \otimes G(r_2 - r'_2, t_2 - t'_1 - \tau') \otimes \Gamma_{XX}(r'_1, r'_2, \tau') \, .$$

Setting $t_2 - t_1 = \tau$, it follows that

$$\Gamma_{XX}(r_1, t_1, r_2, t_1 + \tau)$$
$$= G^*(r_1 - r'_1, t_1 - t'_1) \otimes G(r_2 - r'_2, t_1 + \tau - t'_1 - \tau') \otimes \Gamma_{XX}(r'_1, r'_2, \tau') \, .$$

Since this relation is true for any $t'_1 \leqslant t_1$, we can choose $t'_1 = t_1$, which leads to

$$\Gamma_{XX}(r_1, t_1, r_2, t_1 + \tau)$$
$$= G^*(r_1 - r'_1, 0) \otimes G(r_2 - r'_2, \tau - \tau') \otimes \Gamma_{XX}(r'_1, r'_2, \tau') \, .$$

This shows that $\Gamma_{XX}(r_1, t_1, r_2, t_1 + \tau) = \Gamma_{XX}(r_1, 0, r_2, \tau)$, so that we may write it simply as $\Gamma_{XX}(r_1, r_2, \tau)$. We have already seen that $G(r, r', 0) = \delta(r - r')$, so that finally,

$$\Gamma_{XX}(r_1, r_2, \tau) = G(r_2 - r'_2, \tau - \tau') \otimes \Gamma_{XX}(r_1, r'_2, \tau') \, .$$

These results are summarized in Fig. 3.9

When written out in full, the equation is rather heavy, but more explicit:

$$\Gamma_{XX}(r_1, r_2, \tau) = \int G(r_2 - r'_2, \tau - \tau') \Gamma_{XX}(r_1, r'_2, \tau') \mathrm{d}r'_2 \, .$$

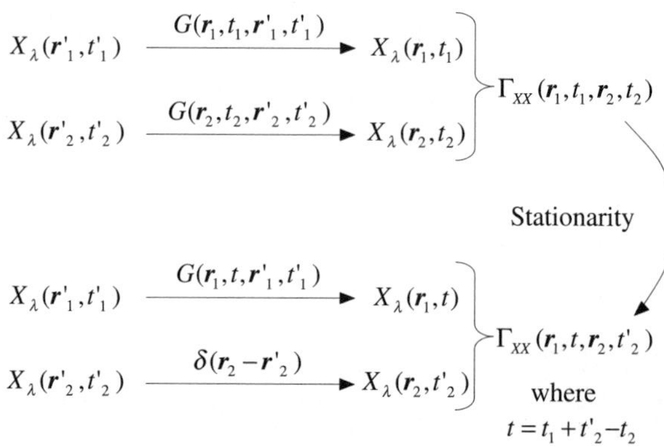

$$X_\lambda(r'_1,t'_1) \xrightarrow{\; G(r_1,t_1,r'_1,t'_1) \;} X_\lambda(r_1,t_1) \left.\right\}$$

$$X_\lambda(r'_2,t'_2) \xrightarrow{\; G(r_2,t_2,r'_2,t'_2) \;} X_\lambda(r_2,t_2) \left.\right\} \; \Gamma_{XX}(r_1,t_1,r_2,t_2)$$

Stationarity

$$X_\lambda(r'_1,t'_1) \xrightarrow{\; G(r_1,t,r'_1,t'_1) \;} X_\lambda(r_1,t) \left.\right\}$$

$$X_\lambda(r'_2,t'_2) \xrightarrow{\; \delta(r_2 - r'_2) \;} X_\lambda(r_2,t'_2) \left.\right\} \; \Gamma_{XX}(r_1,t,r_2,t'_2)$$

where
$$t = t_1 + t'_2 - t_2$$

Fig. 3.9. Illustration of results concerning the dynamical behavior of covariance functions using Green functions

Writing out the spatial coordinates, this gives

$$\Gamma_{XX}(x_1, y_1, z_1, x_2, y_2, z_2, \tau) = \iiint G(x_2 - x'_2, y_2 - y'_2, z_2 - z'_2, \tau - \tau')$$
$$\times \Gamma_{XX}(x_1, y_1, z_1, x'_2, y'_2, z'_2, \tau') \mathrm{d}x'_2 \mathrm{d}y'_2 \mathrm{d}z'_2 \; .$$

We thus note that, as time goes by, there is a spatial filtering of the covariance function $\Gamma_{XX}(r_1, r_2, \tau')$ by the convolution kernel $G(r_2, \tau - \tau')$ which represents the Green function.

It is no surprise that the covariance function and the field itself are filtered by the same kernel in the Green function formulation, since they obey the same partial differential equation (see Section 3.8).

3.13 Stochastic Vector Fields

In this section we investigate how we can generalize the above notions to a stochastic field $\boldsymbol{E}_\lambda(\boldsymbol{r}, t)$ which is a 3-component vector. It may represent the electric field of light, for example. The generalization to vector fields of arbitrary dimension is then immediate. We write

$$\boldsymbol{E}_\lambda(\boldsymbol{r}, t) = E_\lambda^{(x)}(\boldsymbol{r}, t)\boldsymbol{e}_x + E_\lambda^{(y)}(\boldsymbol{r}, t)\boldsymbol{e}_y + E_\lambda^{(z)}(\boldsymbol{r}, t)\boldsymbol{e}_z \; .$$

The analysis is simplified if we introduce the centered quantities

$$\delta E_\lambda^{(w)}(\boldsymbol{r}_1, t_1) = E_\lambda^{(w)}(\boldsymbol{r}_1, t_1) - \langle E_\lambda^{(w)}(\boldsymbol{r}_1, t_1) \rangle \; ,$$

where $w = x, y, z$, and the covariance matrix is then a 3×3 matrix

$$\overline{\overline{\Gamma}}(\boldsymbol{r}_1, t_1, \boldsymbol{r}_2, t_2) = \begin{pmatrix} \Gamma_{XX}^{(1,2)} & \Gamma_{YX}^{(1,2)} & \Gamma_{ZX}^{(1,2)} \\ \Gamma_{XY}^{(1,2)} & \Gamma_{YY}^{(1,2)} & \Gamma_{ZY}^{(1,2)} \\ \Gamma_{XZ}^{(1,2)} & \Gamma_{YZ}^{(1,2)} & \Gamma_{ZZ}^{(1,2)} \end{pmatrix} ,$$

with

$$\Gamma_{UV}^{(1,2)} = \Gamma_{UV}(\boldsymbol{r}_1, t_1, \boldsymbol{r}_2, t_2) = \left\langle \left[\delta E_{\lambda}^{(u)}(\boldsymbol{r}_1, t_1) \right]^* \delta E_{\lambda}^{(v)}(\boldsymbol{r}_2, t_2) \right\rangle .$$

For the field to be stationary, it is enough for it to be so to order one for each coordinate and to order two for each pair of coordinates (U, V). In other words, to order two, $\forall U \in \{X, Y, Z\}$ and $\forall V \in \{X, Y, Z\}$, $\Gamma_{UV}(\boldsymbol{r}_1, t_1, \boldsymbol{r}_2, t_2)$ must only depend on $t_2 - t_1$. We proceed in an analogous way for the properties of ergodicity and homogeneity. For example, in the case of homogeneous and stationary fields, for every coordinate pair (U, V), we have

$$\Gamma_{UV}(\boldsymbol{r}_1, t_1, \boldsymbol{r}_2, t_2) = \Gamma_{UV}(\boldsymbol{r}_2 - \boldsymbol{r}_1, t_2 - t_1) .$$

Several special cases can now be studied on the basis of this definition. For example, we may be concerned with only two coordinates, or we may wish to study the covariance matrix at a single point $\boldsymbol{r}$ or at the same times. This is what happens classically when we analyze the polarization properties of electromagnetic waves, as we shall see in Section 3.14.

3.14 Application to the Polarization of Light

We consider the electric field $\boldsymbol{E}_\lambda(\boldsymbol{r}, t)$ of light. (We could also consider the magnetic field.) In vacuum, if the light propagates along the Oz axis, the vector $\boldsymbol{E}_\lambda(\boldsymbol{r}, t)$ lies in the plane Ox, Oy. We may therefore write

$$\boldsymbol{E}_\lambda(\boldsymbol{r}, t) = E_\lambda^{(X)}(\boldsymbol{r}, t)\boldsymbol{e}_x + E_\lambda^{(Y)}(\boldsymbol{r}, t)\boldsymbol{e}_y ,$$

where $\boldsymbol{e}_x$ and $\boldsymbol{e}_y$ are unit vectors along the orthogonal axes Ox and Oy. We shall assume that the wave is stationary and homogeneous. We shall not be concerned with properties that depend on the space coordinates and we shall no longer include the dependence on $\boldsymbol{r}$. Letting ν_0 denote the central frequency of the optical wave and using the complex notation, we thus write

$$\boldsymbol{E}_\lambda(t) = \left[U_\lambda^{(X)}(t)\boldsymbol{e}_x + U_\lambda^{(Y)}(t)\boldsymbol{e}_y \right] e^{-i2\pi\nu_0 t} .$$

A perfectly monochromatic wave polarized along the Ox axis would be written $\boldsymbol{E}_\lambda(t) = U_\lambda^{(X)} e^{-i2\pi\nu_0 t}\boldsymbol{e}_x$. Of course, this is an ideal case that could never be achieved in reality. A purely monochromatic signal, for example, is incompatible with a signal of finite duration. We may define the covariance matrix by

$$\overline{\overline{J}}(\tau) = \begin{pmatrix} \left\langle \left[U_\lambda^{(X)}(t)\right]^* U_\lambda^{(X)}(t+\tau) \right\rangle & \left\langle \left[U_\lambda^{(Y)}(t)\right]^* U_\lambda^{(X)}(t+\tau) \right\rangle \\ \left\langle \left[U_\lambda^{(X)}(t)\right]^* U_\lambda^{(Y)}(t+\tau) \right\rangle & \left\langle \left[U_\lambda^{(Y)}(t)\right]^* U_\lambda^{(Y)}(t+\tau) \right\rangle \end{pmatrix} .$$

In practice, in the field of optics, one often defines the coherency matrix, which is the covariance matrix when $\tau = 0$. This matrix provides interesting information about the polarization state of the light. We shall illustrate this point using two concrete examples and in the two limiting cases of perfectly coherent and perfectly incoherent light.

We begin with the case of perfectly coherent light. If the light is linearly polarized along the Ox axis, we can write

$$\boldsymbol{E}_\lambda(t) = U_\lambda^{(X)} e^{-i2\pi\nu_0 t} \boldsymbol{e}_x .$$

For example, $U_\lambda^{(X)}$ may be a complex random variable of given modulus and phase uniformly distributed between 0 and 2π. This model would correspond to the fact that, when a perfectly coherent ideal source is switched on, the phase of the wave cannot generally be predicted with total certainty. The coherency matrix is then

$$\overline{\overline{J}} = \overline{\overline{J}}(0) = \left\langle \left| U_\lambda^{(X)} \right|^2 \right\rangle \begin{pmatrix} 1 & 0 \\ 0 & 0 \end{pmatrix} .$$

If the coherent light is linearly polarized along an axis at an angle θ with respect to the Ox axis, we can then write

$$\boldsymbol{E}_\lambda(t) = U_\lambda(\cos\theta\,\boldsymbol{e}_x + \sin\theta\,\boldsymbol{e}_y)e^{-i2\pi\nu_0 t} ,$$

and the coherency matrix becomes

$$\overline{\overline{J}} = \langle |U_\lambda|^2 \rangle \begin{pmatrix} \cos^2\theta & \cos\theta\sin\theta \\ \cos\theta\sin\theta & \sin^2\theta \end{pmatrix} .$$

For a given direction of rotation, circularly polarized coherent light is written

$$\boldsymbol{E}_\lambda(t) = U_\lambda(\boldsymbol{e}_x + e^{i\pi/2}\boldsymbol{e}_y)e^{-i2\pi\nu_0 t} .$$

The coherency matrix will then be

$$\overline{\overline{J}} = \langle |U_\lambda|^2 \rangle \begin{pmatrix} 1 & -i \\ i & 1 \end{pmatrix} .$$

Figure 3.10 summarizes the main polarization states of perfectly polarized light.

Consider now the case of incoherent light. This light can be linearly polarized along the Ox axis, in which case

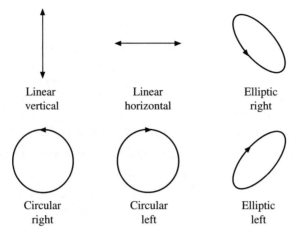

Linear vertical

Linear horizontal

Elliptic right

Circular right

Circular left

Elliptic left

Fig. 3.10. Schematic representation of the main polarization states of perfectly polarized light

$$\boldsymbol{E}_\lambda(t) = U_\lambda^{(X)}(t) e^{-\mathrm{i}2\pi\nu_0 t} \boldsymbol{e}_x \;,$$

where $U_\lambda^{(X)}(t)$ is a stochastic process. In this case, the coherency matrix is simply

$$\overline{\overline{J}} = \left\langle \left| U_\lambda^{(X)}(t) \right|^2 \right\rangle \begin{pmatrix} 1 & 0 \\ 0 & 0 \end{pmatrix} \;.$$

If the incoherent light is totally unpolarized, this means that we can write

$$\boldsymbol{E}_\lambda(t) = \left[U_\lambda^{(X)}(t) \boldsymbol{e}_x + U_\lambda^{(Y)}(t) \boldsymbol{e}_y \right] e^{-\mathrm{i}2\pi\nu_0 t} \;,$$

where $U_\lambda^{(X)}(t)$ and $U_\lambda^{(Y)}(t)$ are independent stochastic processes with zero mean and the same variance. The terms $\left\langle \left[U_\lambda^{(X)}(t) \right]^* U_\lambda^{(Y)}(t) \right\rangle$ are then zero and the coherency matrix assumes diagonal form:

$$\overline{\overline{J}} = \left\langle \left| U_\lambda^{(X)}(t) \right|^2 \right\rangle \begin{pmatrix} 1 & 0 \\ 0 & 1 \end{pmatrix} \;.$$

We can also define different polarization states for incoherent light which are intermediate between the two cases we have just described. The general coherency matrix is

$$\overline{\overline{J}} = \begin{pmatrix} I_X & \rho \\ \rho^* & I_Y \end{pmatrix} \;,$$

because $I_X = \left\langle \left| U_\lambda^{(X)}(t) \right|^2 \right\rangle$, $I_Y = \left\langle \left| U_\lambda^{(Y)}(t) \right|^2 \right\rangle$ and

$$\rho = \left\langle \left[U_\lambda^{(Y)}(t) \right]^* U_\lambda^{(X)}(t) \right\rangle .$$

Like any covariance matrix, the coherency matrix is Hermitian. It can therefore be diagonalized and has orthogonal eigenvectors. Like any covariance matrix (see Section 2.8), it is positive and its eigenvalues are therefore positive. We denote them by λ_1 and λ_2, where $\lambda_1 \geq \lambda_2$. (The eigenvalues of partially polarized light are represented schematically in Fig. 3.11.) As the eigenvectors are orthogonal, the change of basis matrix $\overline{\overline{M}}$ used to diagonalize the coherency matrix (i.e., such that $\overline{\overline{M}} \, \overline{\overline{J}} \, \overline{\overline{M}}^\dagger$ is diagonal, where $\overline{\overline{M}}^\dagger$ is the conjugate transpose of $\overline{\overline{M}}$) is therefore unitary, i.e., it satisfies the relation $\overline{\overline{M}} \, \overline{\overline{M}}^\dagger = \overline{\overline{M}}^\dagger \, \overline{\overline{M}} = \overline{\overline{\mathrm{Id}}}_2$ where $\overline{\overline{\mathrm{Id}}}_2$ is the 2×2 identity matrix. It is common practice to define the degree of polarization of light by

$$\mathcal{P} = \frac{\lambda_1 - \lambda_2}{\lambda_1 + \lambda_2} .$$

There are two invariants under orthonormal basis change (i.e., when the change of basis matrix is unitary), viz., the trace $\mathcal{T}$ and the determinant $\mathcal{D}$ of the matrix. Since $\mathcal{T} = \lambda_1 + \lambda_2$ and $\mathcal{D} = \lambda_1 \lambda_2$, it is a straightforward matter to deduce that

$$\mathcal{P} = \sqrt{1 - 4\frac{\mathcal{D}}{\mathcal{T}^2}} .$$

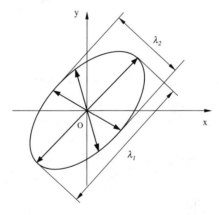

Fig. 3.11. Schematic representation of the eigenvalues of partially polarized light

The coherency matrix of light that is linearly polarized along an axis making an angle θ to the Ox axis is given by

$$\overline{\overline{J}} = \left\langle \left| U_\lambda^{(X)}(t) \right|^2 \right\rangle \begin{pmatrix} \cos^2\theta & \cos\theta\sin\theta \\ \cos\theta\sin\theta & \sin^2\theta \end{pmatrix}.$$

It is easy to check that this has zero determinant and hence that $\mathcal{P} = 1$. The same is true for circularly polarized light, whose coherency matrix is proportional to

$$\overline{\overline{J}} = \left\langle \left| U_\lambda(t) \right|^2 \right\rangle \begin{pmatrix} 1 & -\mathrm{i} \\ \mathrm{i} & 1 \end{pmatrix}.$$

In contrast, it is easy to see that, for totally unpolarized light, we have $\mathcal{P} = 0$.

3.15 Ergodicity and Polarization of Light

We shall now bring out another aspect which clearly illustrates the phenomenological nature of the idea of stochastic process. To begin with, we note that in the above discussion there is no conceptual difference on the mathematical level between perfectly coherent linearly polarized light and perfectly incoherent linearly polarized light. We have just seen that, in the first case, we have

$$\boldsymbol{E}_\lambda(t) = U_\lambda^{(X)} \mathrm{e}^{-\mathrm{i}2\pi\nu_0 t} \boldsymbol{e}_x ,$$

whilst in the second case,

$$\boldsymbol{E}_\lambda(t) = U_\lambda^{(X)}(t) \mathrm{e}^{-\mathrm{i}2\pi\nu_0 t} \boldsymbol{e}_x .$$

From a mathematical standpoint, we can define stochastic processes with constant value and thus set $U_\lambda^{(X)}(t) = U_\lambda^{(X)}$. The formulation $U_\lambda^{(X)}(t)$ is therefore the more general. We have already pointed out that the representation $\boldsymbol{E}_\lambda(t) = U_\lambda^{(X)} \mathrm{e}^{-\mathrm{i}2\pi\nu_0 t} \boldsymbol{e}_x$ for a coherent wave can only correspond to a limiting case. Indeed, emitted waves always have a nonzero natural spectral width, if only the one due to their finite temporal support. A better model of the coherent wave is therefore $\boldsymbol{E}_\lambda(t) = U_\lambda^{(X)}(t) \mathrm{e}^{-\mathrm{i}2\pi\nu_0 t} \boldsymbol{e}_x$, where $U_\lambda^{(X)}(t)$ is a function which varies little. To simplify the discussion, it is nevertheless common to write in the coherent case $\boldsymbol{E}_\lambda(t) = U_\lambda^{(X)} \mathrm{e}^{-\mathrm{i}2\pi\nu_0 t} \boldsymbol{e}_x$.

We shall now show that difficulties can arise if we do not appeal to the ideas we have defined for stochastic processes.

Consider the case of incoherent light that is linearly polarized along an axis making an angle θ_μ with the Ox axis. The quantity θ_μ is a random variable distributed uniformly between 0 and 2π and μ is a random event independent of λ. We then write

$$\boldsymbol{E}_\lambda(t) = U_\lambda(t)(\cos\theta_\mu \, \boldsymbol{e}_x + \sin\theta_\mu \, \boldsymbol{e}_y) \mathrm{e}^{-\mathrm{i}2\pi\nu_0 t}$$

and the coherency matrix becomes

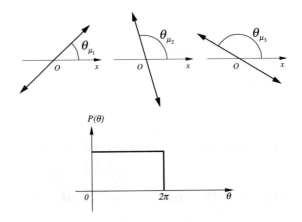

Fig. 3.12. Light with linear polarization along an axis making a random angle θ_μ with the Ox axis

$$\overline{\overline{J}}(0) = \left\langle |U_\lambda(t)|^2 \right\rangle \begin{pmatrix} \langle \cos^2 \theta_\mu \rangle & \langle \cos \theta_\mu \sin \theta_\mu \rangle \\ \langle \cos \theta_\mu \sin \theta_\mu \rangle & \langle \sin^2 \theta_\mu \rangle \end{pmatrix},$$

or

$$\overline{\overline{J}}(0) = \frac{1}{2} \left\langle |U_\lambda(t)|^2 \right\rangle \begin{pmatrix} 1 & 0 \\ 0 & 1 \end{pmatrix},$$

which corresponds to the coherency matrix of incoherent and totally unpolarized light. We thus see that, in the context of this model, no distinction is made between a completely unpolarized incoherent wave and a linearly polarized incoherent wave whose angle of polarization is a time-constant random variable, uniformly distributed between 0 and 2π. There is, however, a fundamental difference between the two physical situations. It can be brought out explicitly in the context of our model by analyzing the ergodicity properties of the relevant stochastic processes.

We write $\boldsymbol{E}_\lambda(t) = \boldsymbol{A}_\lambda(t)e^{-i2\pi\nu_0 t}$. In the case of completely unpolarized incoherent light, we have

$$\boldsymbol{A}_\lambda(t) = U_\lambda^{(X)}(t)\boldsymbol{e}_x + U_\lambda^{(Y)}(t)\boldsymbol{e}_y,$$

where $U_\lambda^{(X)}(t)$ and $U_\lambda^{(Y)}(t)$ are independent stochastic processes with zero mean and the same variance. To speak of totally incoherent light amounts to assuming that, between two distinct instants of time, the states of the electric field are independent of one another. The covariance function is then zero between these two times. It is reasonable to assume that, between different times, the states of the electric field are independent and explore the complete set of possible configurations of the polarization. The time average $\boldsymbol{A}_\lambda(t)$ is then easily determined:

$$\overline{\boldsymbol{A}_\lambda(t)} = \lim_{\substack{T_1 \to -\infty \\ T_2 \to \infty}} \left[\frac{1}{T_2 - T_1} \int_{T_1}^{T_2} \boldsymbol{A}_\lambda(t) dt \right] = 0 \, .$$

For the second temporal moment, we find that

$$\overline{\boldsymbol{A}_\lambda(t + \tau) \left[\boldsymbol{A}_\lambda(t) \right]^\dagger} = \lim_{\substack{T_1 \to -\infty \\ T_2 \to \infty}} \left[\frac{1}{T_2 - T_1} \int_{T_1}^{T_2} \boldsymbol{A}_\lambda(t + \tau) \left[\boldsymbol{A}_\lambda(t) \right]^\dagger dt \right] ,$$

where $\boldsymbol{A}^\dagger$ corresponds to the transposed complex conjugate of $\boldsymbol{A}$. Hence,

$$\boldsymbol{A}_\lambda(t + \tau) \left[\boldsymbol{A}_\lambda(t) \right]^\dagger = \begin{pmatrix} \left[U_\lambda^{(X)}(t) \right]^* U_\lambda^{(X)}(t + \tau) & \left[U_\lambda^{(Y)}(t) \right]^* U_\lambda^{(X)}(t + \tau) \\ \left[U_\lambda^{(X)}(t) \right]^* U_\lambda^{(Y)}(t + \tau) & \left[U_\lambda^{(Y)}(t) \right]^* U_\lambda^{(Y)}(t + \tau) \end{pmatrix} .$$

$\boldsymbol{A}_\lambda(t + \tau) \left[\boldsymbol{A}_\lambda(t) \right]^\dagger$ is therefore a matrix. Under the above hypotheses, it is reasonable to assume that

$$\lim_{\substack{T_1 \to -\infty \\ T_2 \to \infty}} \left[\frac{1}{T_2 - T_1} \int_{T_1}^{T_2} \boldsymbol{A}_\lambda(t + \tau) \left[\boldsymbol{A}_\lambda(t) \right]^\dagger dt \right] = \sigma_A^2 \overline{\overline{\mathrm{Id}}}_2 \delta_\tau \, ,$$

where σ_A^2 is independent of λ, δ_τ is the Kronecker delta and $\overline{\overline{\mathrm{Id}}}_2$ the 2×2 identity matrix. In other words, we assume that the stochastic process $\boldsymbol{A}_\lambda(t)$ is ergodic and that each component is white noise.

For coherent or incoherent light that is linearly polarized in such a way that the angle of polarization is a random variable uniformly distributed between 0 and 2π, we have $\boldsymbol{A}_{\lambda,\mu}(t) = U_\lambda(t) \left(\cos \theta_\mu \, \boldsymbol{e}_x + \sin \theta_\mu \, \boldsymbol{e}_y \right)$ and hence,

$$\overline{\boldsymbol{A}_{\lambda,\mu}(t)} = \overline{U_\lambda(t)} \left(\cos \theta_\mu \, \boldsymbol{e}_x + \sin \theta_\mu \, \boldsymbol{e}_y \right) ,$$

where

$$\overline{U_\lambda(t)} = \lim_{\substack{T_1 \to -\infty \\ T_2 \to \infty}} \left[\frac{1}{T_2 - T_1} \int_{T_1}^{T_2} U_\lambda(t) dt \right] .$$

To second order, we have

$$\overline{\boldsymbol{A}_{\lambda,\mu}(t) \left[\boldsymbol{A}_{\lambda,\mu}(t + \tau) \right]^\dagger} = \overline{[U_\lambda(t) U_\lambda^*(t + \tau)]} M(\theta_\mu) ,$$

where

$$M(\theta_\mu) = \begin{pmatrix} \cos^2 \theta_\mu & \cos \theta_\mu \sin \theta_\mu \\ \cos \theta_\mu \sin \theta_\mu & \sin^2 \theta_\mu \end{pmatrix} .$$

The above time averages do not therefore remove the dependence on the random event μ, and this shows clearly that $\boldsymbol{A}_{\lambda,\mu}(t)$ is not ergodic.

To conclude, we find the same coherency matrices for incoherent light that is completely unpolarized or incoherent light that is completely polarized along an axis at a uniformly distributed angle between 0 and 2π relative to some reference axis. The difference between these two cases is indeed related to their ergodicity property. Generally speaking, caution is required when dealing with covariance matrices of non-ergodic processes.

3.16 Appendix: Wiener–Khinchine Theorem

In this section, we shall demonstrate the Wiener–Khinchine theorem, which is very often used in physics. We shall assume throughout that the stochastic process is centered, i.e., $\langle X_\lambda(t) \rangle = 0$, so that the covariance is simply

$$\Gamma_{XX}(t_1, t_2) = \langle X_\lambda^*(t_1) X_\lambda(t_2) \rangle \ .$$

We define the spectral power density of $X_\lambda(t)$ by

$$\hat{S}_{XX}(\nu) = \lim_{\substack{T_1 \to -\infty \\ T_2 \to \infty}} \left\langle \left[\frac{1}{T_2 - T_1} \left| \int_{T_1}^{T_2} X_\lambda(t) \exp(-\mathrm{i}2\pi\nu t)\mathrm{d}t \right|^2 \right] \right\rangle \ .$$

This can be rewritten

$$\hat{S}_{XX}(\nu) = \lim_{\substack{T_1 \to -\infty \\ T_2 \to \infty}} \frac{1}{T_2 - T_1} \int_{T_1}^{T_2} \int_{T_1}^{T_2} \langle X_\lambda^*(t_2) X_\lambda(t_1) \rangle e^{-\mathrm{i}2\pi\nu(t_1 - t_2)} \mathrm{d}t_1 \mathrm{d}t_2 \ ,$$

or

$$\hat{S}_{XX}(\nu) = \lim_{\substack{T_1 \to -\infty \\ T_2 \to \infty}} \frac{1}{T_2 - T_1} \int_{T_1}^{T_2} \int_{T_1}^{T_2} \Gamma_{XX}(t_1, t_2) e^{-\mathrm{i}2\pi\nu(t_1 - t_2)} \mathrm{d}t_1 \mathrm{d}t_2 \ .$$

If the stochastic process is stationary, we have $\Gamma_{XX}(t_1, t_2) = \Gamma_{XX}(t_2 - t_1)$. Setting $t_1 - t_2 = \tau$ and $t_2 + t_1 = \mu$, we then obtain

$$\begin{pmatrix} \mu \\ \tau \end{pmatrix} = \begin{pmatrix} 1 & 1 \\ 1 & -1 \end{pmatrix} \begin{pmatrix} t_1 \\ t_2 \end{pmatrix} \ .$$

The Jacobian $\overline{\overline{J}}$ for the transformation is thus

$$\begin{bmatrix} 1 & 1 \\ 1 & -1 \end{bmatrix}$$

and the absolute value of its determinant $|\overline{\overline{J}}|$ is equal to 2.

More generally, when we carry out a change of variables

$$\mu = f_1(t_1, t_2) \quad \text{and} \quad \tau = f_2(t_1, t_2) \,,$$

with Jacobian $\overline{\overline{J}}(\mu, \tau)$, we have

$$\int \int \frac{1}{|\overline{\overline{J}}(\mu, \tau)|} F(\mu, \tau) \mathrm{d}\mu \mathrm{d}\tau = \int \int G(t_1, t_2) \mathrm{d}t_1 \mathrm{d}t_2 \,,$$

where $G(t_1, t_2) = F\big[f_1(t_1, t_2), f_2(t_1, t_2)\big]$ and $|\overline{\overline{J}}(\mu, \tau)|$ is the determinant of $\overline{\overline{J}}(\mu, \tau)$. In the present case, $G(t_1, t_2)$ is

$$G(t_1, t_2) = \Gamma_{XX}(t_1, t_2) \mathrm{Rect}_{T_1, T_2}(t_1) \mathrm{Rect}_{T_1, T_2}(t_2) \,,$$

where

$$\mathrm{Rect}_{T_1, T_2}(t) = \begin{cases} 1 & \text{if } t \in [T_1, T_2] \,, \\ 0 & \text{otherwise} \,. \end{cases}$$

We have $t_1 = (\mu + \tau)/2$ and $t_2 = (\mu - \tau)/2$, and hence,

$$\hat{S}_{XX}(\nu) = \lim_{\substack{T_1 \to -\infty \\ T_2 \to \infty}} \frac{1}{2(T_2 - T_1)}$$

$$\left[\int_{-\infty}^{\infty} \int_{-\infty}^{\infty} \mathrm{Rect}_{T_1, T_2}\left(\frac{\mu + \tau}{2}\right) \mathrm{Rect}_{T_1, T_2}\left(\frac{\mu - \tau}{2}\right) \Gamma_{XX}(\tau) e^{-\mathrm{i}2\pi\nu\tau} \mathrm{d}\mu \mathrm{d}\tau \right] \,,$$

so that

$$\hat{S}_{XX}(\nu) = \lim_{\substack{T_1 \to -\infty \\ T_2 \to \infty}} \frac{1}{2(T_2 - T_1)} \int_{T_1-T_2}^{T_2-T_1} \Gamma_{XX}(\tau) e^{-\mathrm{i}2\pi\nu\tau} \big[2(T_2 - T_1 - |\tau|)\big] \mathrm{d}\tau \,.$$

We now define

$$\Lambda_T(t) = \begin{cases} 1 - |t|/T & \text{if } |t| < T \,, \\ 0 & \text{otherwise} \,. \end{cases}$$

Provided we assume that we can change the order of the limit and the integral. With the above notation, we then have

$$\hat{S}_{XX}(\nu) = \int_{-\infty}^{\infty} \lim_{\substack{T_1 \to -\infty \\ T_2 \to \infty}} \Lambda_{T_2-T_1}(\tau) \Gamma_{XX}(\tau) e^{-\mathrm{i}2\pi\nu\tau} \mathrm{d}\tau \,,$$

and hence,

$$\hat{S}_{XX}(\nu) = \int_{-\infty}^{\infty} \Gamma_{XX}(\tau) e^{-\mathrm{i}2\pi\nu\tau} \mathrm{d}\tau \,.$$

This is precisely the Wiener–Khinchine theorem. It says that the power spectral density $\hat{S}_{XX}(\nu)$ of a stochastic process which is stationary to second order is equal to the Fourier transform of its covariance $\Gamma_{XX}(\tau)$.

Exercises

Exercise 3.1.

X_λ and Y_λ are two random variables with variances σ_X^2 and σ_Y^2, respectively. Denoting their correlation coefficient by Γ_{XY}, show that $\Gamma_{XY} \leq (\sigma_X^2 + \sigma_Y^2)/2$. Use the fact that

$$\langle (X_\lambda^c - Y_\lambda^c)^2 \rangle \geq 0 \,,$$

where $X_\lambda^c = X_\lambda - \langle X_\lambda \rangle$ and $Y_\lambda^c = Y_\lambda - \langle Y_\lambda \rangle$.

Exercise 3.2. Stochastic Process

A signal $x(t) > 0$ is perturbed by a multiplicative noise of speckle type, viz., $Y_\lambda(t) = x(t)B_\lambda(t)$, where we assume that $B_\lambda(t)$ is noise with probability density function described by the Gamma distribution:

$$P_B(b) = \begin{cases} \dfrac{b^{r-1}}{a^r \Gamma(r)} \exp\left(-\dfrac{b}{a}\right) & \text{if } b \geq 0 \,, \\ 0 & \text{otherwise} \,. \end{cases}$$

where $\Gamma(r)$ is the Gamma function. Setting $Z_\lambda(t) = \ln Y_\lambda(t)$, calculate the probability density function of the fluctuations in $Z_\lambda(t)$.

Exercise 3.3. Stochastic Process

Consider a stochastic process $X_\lambda(t)$ with Gaussian probability density function having zero mean and variance σ^2. This process is multiplied by a strictly positive function $g(t)$ to produce a new stochastic process $Y_\lambda(t) = g(t)X_\lambda(t)$. Calculate the probability density function for $Y_\lambda(t)$.

Exercise 3.4. Ergodicity and Stationarity

Let $f_T(t)$ be a periodic function with period T. Using $f_T(t)$, we construct the stochastic process

$$\Omega \to \mathbb{R}$$
$$\lambda \mapsto f_T(t - \tau_\lambda) \,,$$

where Ω is the space of random events λ, and $\mathbb{R}$ is the set of real numbers. We assume that the probability density function for τ_λ is constant in the interval $[0, T]$. We will be interested in the ergodicity and stationarity in the sense of first and second order moments.

(1) Determine whether or not this stochastic process is stationary.
(2) Determine whether or not it is ergodic.

Exercise 3.5. Stationarity

Let $f_\lambda(t)$ be a real stochastic process of finite power and infinite energy. This stochastic process is assumed weakly stationary and such that $\langle f_\lambda(t) \rangle = F \neq 0$. Let $g(t)$ be an arbitrary real function. We define $h_\lambda(t) = g(t) f_\lambda(t)$.

What are the conditions that $g(t)$ must satisfy if $h_\lambda(t)$ is to be weakly stationary?

Exercise 3.6. Stationarity and Ergodicity

Let $X_\lambda(t)$ be a real-valued stochastic process. Consider a linear system which transforms the noise according to

$$Y_\lambda(t) = \sum_{n=1}^{N} a_n X_\lambda(t - \tau_n) \,.$$

Suggest simple sufficient conditions on $X_\lambda(t)$ to ensure that $Y_\lambda(t)$ is weakly stationary and ergodic.

Exercise 3.7. Stationarity and Ergodicity

$X_\lambda(t)$ is a real-valued stochastic process. Consider a non-linear system which transforms the noise according to

$$Y_\lambda(t) = a_1 X_\lambda(t) + a_2 \left[X_\lambda(t) \right]^2 \,.$$

Suggest simple sufficient conditions on $X_\lambda(t)$ to ensure that $Y_\lambda(t)$ is weakly stationary and ergodic.

Exercise 3.8. Stationarity

Let $f_\lambda(t)$ be the periodic function of period T defined by

$$f_\lambda(t) = a \exp\left(2i\pi \frac{t}{T} - i\phi_\lambda \right) ,$$

where ϕ_λ is a random variable with values in the interval $[0, 2\pi]$ and $i^2 = -1$.

(1) Let $\hat{f}_\lambda(\nu)$ be the Fourier transform of $f_\lambda(t)$. Determine the phase of $\hat{f}_\lambda(\nu)$.
(2) Is $f_\lambda(t)$ weakly stationary?
(3) What can you deduce concerning $\langle \hat{f}_\lambda(\nu) \rangle$, where $\langle \ \rangle$ denotes the mean with respect to λ?
(4) Determine the Fourier transform of a signal of form

$$f_\lambda(t) = \sum_{n=-\infty}^{\infty} a_n \exp\left(2i\pi n \frac{t}{T} - i\phi_{n,\lambda} \right)$$

in the case where the $\phi_{n,\lambda}$ are independent random variables uniformly distributed over the interval $[0, 2\pi]$.
(5) What can you deduce concerning $\langle \hat{f}_\lambda^*(\nu_1) \hat{f}_\lambda(\nu_2) \rangle$?

Exercise 3.9. Power Spectral Density

Consider a system in which an emitted signal $r(t)$ is received after having followed two possible paths. In the absence of noise, the measured signal is modeled by $s(t) = (1 - a)r(t) + ar(t - \tau)$.

(1) Show that this is indeed a convolution system and determine the transfer function.
(2) The emitted signal is in fact white noise (hence weakly stationary) defined on a frequency band $[-\nu_B, \nu_B]$. Determine the spectral density of the measured noise.

Exercise 3.10. Power Spectral Density

Consider a stationary white noise signal defined in the frequency band $[-B, B]$, with power $2B\sigma_B^2$ and described by the stochastic process $X_\lambda(t)$. Assume that the power spectral density is

$$\hat{S}_{XX}(\nu) = \begin{cases} \sigma_B^2 & \text{if } \nu \in [-B, B] , \\ 0 & \text{otherwise} , \end{cases}$$

where B is positive and sufficiently large to replace by $+\infty$ in the calculations. This noise is filtered by a linear filter with impulse response

$$h(t) = \begin{cases} a \exp(-at) & \text{if } t \geq 0 , \\ 0 & \text{otherwise} . \end{cases}$$

(1) Calculate the autocorrelation function of the noise after filtering $X_\lambda(t)$ by $h(t)$ when $B \to +\infty$.
(2) Deduce the total power of the fluctuations after filtering.
(3) What happens if $a \to +\infty$?

Exercise 3.11. Power Spectral Density

Let $X_\lambda(t)$ be a weakly stationary real stochastic process such that $\langle X_\lambda(t) \rangle = 0$. We define

$$Y_\lambda(t) = \int_t^{t+T} X_\lambda(\xi) d\xi .$$

(1) Express the spectral density of $Y_\lambda(t)$ in terms of the spectral density of $X_\lambda(t)$.
(2) What happens if the spectral density $\hat{S}_{XX}(\nu)$ of $X_\lambda(t)$ is such that $\hat{S}_{XX}(\nu) = \sigma^2 \delta(\nu - n/T)$, where n is a nonzero natural number and $\delta(x)$ is the Dirac distribution?
(3) If now $\langle X_\lambda(t_1)X_\lambda(t_2) \rangle = \delta(t_1 - t_2)$, what happens to the spectral density of $Y_\lambda(t)$?
(4) How does the power of $Y_\lambda(t)$ vary?

Exercise 3.12. Noise and Impulse Response

Let $X_\lambda(t)$ and $Y_\lambda(t)$ be two real random signals, both stationary with finite power, where $Y_\lambda(t)$ is the result of filtering $X_\lambda(t)$ by a convolution
filter (hence linear and stationary) with impulse response $h(t)$. We write
$\Gamma_{XY}(\tau) = \langle X_\lambda(t)Y_\lambda(t+\tau)\rangle$ and $\Gamma_{XX}(\tau) = \langle X_\lambda(t)X_\lambda(t+\tau)\rangle$, where $\langle\ \rangle$ represents the mean with respect to outcomes of random events λ.
    We wish to estimate $h(t)$ from $\Gamma_{XY}(\tau)$ and $\Gamma_{XX}(\tau)$.

(1) Determine $h(t)$ in terms of $\Gamma_{XY}(\tau)$ and $\Gamma_{XX}(\tau)$.
(2) Write down the Fourier transform of this relation.
(3) What condition can you deduce on the spectral density of $X_\lambda(t)$ in order
    to determine $h(t)$?
(4) What happens if $X_\lambda(t)$ is white noise in the frequency band between $-B$
    and B?

4

Limit Theorems and Fluctuations

Sums of random variables are a fascinating subject for they lead to certain universal types of behavior. More precisely, if we add together independent random variables distributed according to the same probability density functions and not too widely scattered in the sense that they have a finite second moment, then the new random variable obtained in this way will be described to a good approximation by a Gaussian random variable. This property has a great many applications in physics. We shall describe a certain number of them: the random walk, speckle in coherent imaging, particle diffusion, and Gaussian noise, which is a widely used model in physics. The Gaussian distribution is not the only one to appear as a limiting case. The Poisson distribution can be introduced by analogous arguments and it is also very important because it provides simple models of fluctuations resulting from detection of low particle fluxes.

4.1 Sum of Random Variables

Consider a sequence of random variables $X_{\lambda(1)}, X_{\lambda(2)}, \ldots, X_{\lambda(n)}, \ldots$, with finite means and second moments. The mean and variance of $X_{\lambda(n)}$ will be denoted by m_n and σ_n^2, respectively. The sum random variable is defined by

$$S_\lambda(n) = \sum_{j=1}^n X_{\lambda(j)} \, ,$$

where $\lambda = [\lambda(1), \lambda(2), \ldots, \lambda(n)]$. Let us determine the mean and variance of $S_\lambda(n)$. We have

$$\langle S_\lambda(n) \rangle = \int \ldots \int \left(\sum_{j=1}^n x_j \right) P(x_1, x_2, \ldots, x_n) \, \mathrm{d}x_1 \mathrm{d}x_2 \ldots \mathrm{d}x_n \, ,$$

where $P(x_1, x_2, \ldots, x_n)$ is the joint probability density function of the random variables $X_{\lambda(1)}, X_{\lambda(2)}, \ldots, X_{\lambda(n)}$. Since

$$\int \cdots \int x_j P(x_1, x_2, \ldots, x_n) \, \mathrm{d}x_1 \mathrm{d}x_2 \ldots \mathrm{d}x_n = \langle X_{\lambda(j)} \rangle = m_j \ ,$$

we deduce that

$$\langle S_\lambda(n) \rangle = \sum_{j=1}^{n} \langle X_{\lambda(j)} \rangle = \sum_{j=1}^{n} m_j \ .$$

In order to analyze the behavior of $[S_\lambda(n)]^2$, we introduce the covariance $\Gamma_{ij} = \langle X_{\lambda(i)} X_{\lambda(j)} \rangle - m_i m_j$, which can also be written

$$\Gamma_{ij} = \int \cdots \int (x_i x_j - m_i m_j) \, P(x_1, x_2, \ldots, x_n) \, \mathrm{d}x_1 \mathrm{d}x_2 \ldots \mathrm{d}x_n \ .$$

The mean of the square of the sum is by definition

$$\langle [S_\lambda(n)]^2 \rangle = \int \cdots \int \left(\sum_{j=1}^{n} x_j \right)^2 P(x_1, x_2, \ldots, x_n) \, \mathrm{d}x_1 \mathrm{d}x_2 \ldots \mathrm{d}x_n \ ,$$

or

$$\langle [S_\lambda(n)]^2 \rangle = \sum_{i=1}^{n} \sum_{j=1}^{n} \int \cdots \int x_i x_j P(x_1, x_2, \ldots, x_n) \, \mathrm{d}x_1 \mathrm{d}x_2 \ldots \mathrm{d}x_n \ .$$

Given that $\left[\langle S_\lambda(n) \rangle \right]^2 = \left(\sum\limits_{j=1}^{n} m_j \right)^2 = \sum\limits_{i=1}^{n} \sum\limits_{j=1}^{n} m_i m_j$, it is easy to see that the variance of $S_\lambda(n)$ can be written

$$\langle [S_\lambda(n)]^2 \rangle - [\langle S_\lambda(n) \rangle]^2 = \sum_{i=1}^{n} \sum_{j=1}^{n} \Gamma_{ij} \ .$$

The second moments $\langle [X_{\lambda(i)}]^2 \rangle$ are not simply additive. However, if the random variables $X_{\lambda(1)}, X_{\lambda(2)}, \ldots, X_{\lambda(n)}$, are uncorrelated, then we have by definition that $\Gamma_{ij} = \sigma_i^2$ if $i = j$ and $\Gamma_{ij} = 0$ otherwise. We thus obtain

$$\langle [S_\lambda(n)]^2 \rangle - [\langle S_\lambda(n) \rangle]^2 = \sum_{j=1}^{n} \sigma_j^2 \ .$$

It is interesting to observe the result obtained when the random variables are uncorrelated and distributed according to the same probability density function with mean m and variance σ^2 :

$$\langle S_\lambda(n) \rangle = nm \quad \text{and} \quad \langle [S_\lambda(n)]^2 \rangle - [\langle S_\lambda(n) \rangle]^2 = n\sigma^2 \ .$$

The standard deviation of $S_\lambda(n)$ thus behaves as $\sigma \sqrt{n}$. This is an extremely important result arising in many problems. Let us analyze the simple case where we carry out n independent measurements $X_{\lambda(1)}, X_{\lambda(2)}, \ldots, X_{\lambda(n)}$, of a physical quantity. If the characteristics of the noise change little during the measurement time, we can describe $X_{\lambda(1)}, X_{\lambda(2)}, \ldots, X_{\lambda(n)}$ by random variables distributed according to the same probability density function with mean m and variance σ^2. We can define the empirical mean of the n measurements by $\mu_\lambda(n) = S_\lambda(n)/n$, usually referred to as the sample mean. The expectation value and standard deviation (denoted by σ_μ) of $\mu_\lambda(n)$ are thus simply $\langle \mu_\lambda(n) \rangle = m$ and $\sigma_\mu = \sigma/\sqrt{n}$.

In other words, the mathematical expectation of the sample mean of the n uncorrelated measurements is just the statistical mean of $X_{\lambda(j)}$, whilst the standard deviation is reduced, and so therefore is the spread about the mean, by a factor of $\sqrt{n}$. This is why it is useful to carry out several measurements and take the average. In practice, the situation is often as follows. From n independent measurements $X_{\lambda(1)}, X_{\lambda(2)}, \ldots, X_{\lambda(n)}$, we can estimate the mean by $\mu_\lambda(n) = S_\lambda(n)/n$ and the variance by

$$\eta_\lambda(n) = \frac{1}{n} \sum_{j=1}^{n} \left[X_{\lambda(j)} - \mu_\lambda(n) \right]^2 .$$

(We shall see in Chapter 7 that this estimator is biased, but that for large n the bias is low.) We should thus retain from this that $\eta_\lambda(n)$ is an estimate of the variance of $X_{\lambda(j)}$ and that the variance of $\mu_\lambda(n)$ is rather of the order of $\eta_\lambda(n)/n$. This is indeed a useful result for plotting error bars in experimental measurements.

We can determine the probability density function of a sum of independent random variables. Consider first the case of two random variables X_λ and Y_λ distributed according to probability density functions $P_X(x)$ and $P_Y(y)$, respectively. From two independent realizations $X_{\lambda(1)}$ and $Y_{\lambda(2)}$, we define a new random variable $Z_\mu = X_{\lambda(1)} + Y_{\lambda(2)}$, where we have set $\mu = (\lambda(1), \lambda(2))$. We shall now investigate the probability density function $P_Z(z)$ of Z_μ, using $F_Z(z)$ to denote its distribution function. To simplify the argument, we assume that $P_X(x)$ is a continuous function. For a fixed value x of $X_{\lambda(1)}$, the probability that Z_μ is less than z is equal to $F_Y(z - x)$. Now the probability that $X_{\lambda(1)}$ lies between $x - \mathrm{d}x/2$ and $x + \mathrm{d}x/2$ is equal to $P_X(x)\mathrm{d}x$. The probability that $X_{\lambda(1)}$ lies between $x - \mathrm{d}x/2$ and $x + \mathrm{d}x/2$ and that Z_μ is simultaneously less than z is then $F_Y(z - x)P_X(x)\mathrm{d}x$. The probability that Z_μ is less than z independently of the value of $X_{\lambda(1)}$ is thus

$$F_Z(z) = \int_{-\infty}^{\infty} F_Y(z - x)P_X(x)\mathrm{d}x .$$

Differentiating with respect to z, we obtain the probability density function

$$P_Z(z) = \int_{-\infty}^{\infty} P_Y(z - x)P_X(x)\mathrm{d}x .$$

We thus deduce that the probability density function of the sum variable is obtained by convoluting the probability density functions of each of the random variables in the sum. Note, however, that this result is no longer true if the summed variables are not independent.

4.2 Characteristic Function

It is well known that the Fourier transform of a convolution product of two functions is equal to the product of the Fourier transforms of these functions. This property is an important factor motivating the introduction of the characteristic function $\Psi_X(\nu)$ associated with the probability density function:

$$\Psi_X(\nu) = \int_{-\infty}^{\infty} P_X(x) \exp(i\nu x) \, dx \; .$$

When it exists, the inverse transformation is obtained by

$$P_X(x) = \frac{1}{2\pi} \int_{-\infty}^{\infty} \Psi_X(\nu) \exp(-i\nu x) \, d\nu \; .$$

It is easy to see that $\Psi_X(0) = 1$, since $\int_{-\infty}^{\infty} P_X(x) dx = 1$.

When $\langle (X_\lambda)^n \rangle$ is well defined, it can be shown that

$$\Psi_X(\nu) = \sum_{n=0}^{r} \frac{\langle (X_\lambda)^n \rangle}{n!} (i\nu)^n + o(\nu^r) \; ,$$

where $o(\nu^r)$ tends to 0 more quickly than ν^r when ν tends to 0. This is a consequence of the expansion

$$\exp(i\nu x) = \sum_{n=0}^{\infty} \frac{(i\nu x)^n}{n!} \; .$$

When the characteristic function is analytic at the origin, it can be expanded in a series

$$\Psi_X(\nu) = \sum_{n=0}^{\infty} \frac{\langle (X_\lambda)^n \rangle (i\nu)^n}{n!} \; .$$

Therefore, if $\psi_x(\nu)$ is analytic knowing the different moments $\langle (X_\lambda)^n \rangle$ (for integer values of n), we can determine the characteristic function and hence also the probability density function. Here is yet another motivation for finding the moments of integer order when we are dealing with random variables.

We have seen that the random variable Z_μ defined as the sum of two independent random variables, viz., $Z_\mu = X_{\lambda(1)} + Y_{\lambda(2)}$, has probability density function given by

$$P_Z(z) = \int\limits_{-\infty}^{\infty} P_Y(z - x) P_X(x) \mathrm{d}x \ .$$

Its characteristic function is then

$$\Psi_Z(\nu) = \Psi_Y(\nu)\Psi_X(\nu) \ ,$$

where $\Psi_X(\nu)$ and $\Psi_Y(\nu)$ are the characteristic functions of X_λ and Y_λ, respectively.

Table 4.1 shows several characteristic functions for the most commonly occurring probability density functions. From the table, note that the sum of n exponential random variables[1] is distributed according to a Gamma probability law with parameter $\alpha = n$. Indeed, if a^{-1} is the mean of the exponential variable, its characteristic function is $(1 - \mathrm{i}\nu/a)^{-1}$. The characteristic function of the sum of n exponential variables is then $(1 - \mathrm{i}\nu/a)^{-n}$. In the same way, it can be shown that the sum of two Gamma variables with coefficients (β, α_1) and (β, α_2) produces a Gamma variable with coefficients $(\beta, \alpha_1 + \alpha_2)$.

Table 4.1. Characteristic functions for a selection of probability laws

Name	Probability density function	Characteristic function
Bernoulli	$(1 - q)\delta(x) + q\delta(x - 1)$	$(1 - q) + q\mathrm{e}^{\mathrm{i}\nu}$
Poisson	$\sum\limits_{n=0}^{\infty} \mathrm{e}^{-\mu}\delta(x - n)\mu^n/n!$	$\exp\left[-\mu\left(1 - \mathrm{e}^{\mathrm{i}\nu}\right)\right]$
Uniform $[0, 1]$	$\begin{cases} 1 & \text{if } 0 \leqslant x \leqslant 1 \\ 0 & \text{otherwise} \end{cases}$	$\mathrm{e}^{-\mathrm{i}\nu/2}\dfrac{\sin(\nu/2)}{\nu/2}$
Gaussian	$\dfrac{1}{\sqrt{2\pi}\sigma}\exp\left[-\dfrac{(x - m)^2}{2\sigma^2}\right]$	$\mathrm{e}^{\mathrm{i}m\nu - \sigma^2\nu^2/2}$
Exponential	$\begin{cases} a\mathrm{e}^{-ax} & \text{if } x \geqslant 0 \\ 0 & \text{otherwise} \end{cases}$	$\left(1 - \dfrac{\mathrm{i}\nu}{a}\right)^{-1}$
Gamma	$\begin{cases} \dfrac{\beta^\alpha x^{\alpha-1}}{\Gamma(\alpha)}\mathrm{e}^{-\beta x} & \text{if } x \geqslant 0 \\ 0 & \text{otherwise} \end{cases}$	$\left(1 - \dfrac{\mathrm{i}\nu}{\beta}\right)^{-\alpha}$
Cauchy	$\dfrac{a}{\pi(a^2 + x^2)}$ with $a > 0$	$\mathrm{e}^{-a|\nu|}$

[1] When there is no risk of ambiguity, we will use this abbreviated manner of speaking to indicate that a random variable has an exponential probability density function.

4.3 Central Limit Theorem

We now consider a sequence of random variables $X_{\lambda(1)}, X_{\lambda(2)}, \ldots, X_{\lambda(n)}$ which we shall assume to be independent with finite mean and second moment. The mean and variance of $X_{\lambda(n)}$ are m_n and σ_n^2, respectively. We define the sum random variable

$$S_\lambda(n) = \sum_{j=1}^{n} X_{\lambda(j)} \,,$$

where $\lambda = [\lambda(1), \lambda(2), \ldots, \lambda(n)]$. We have seen that the mean of $S_\lambda(n)$ is

$$M_n = \sum_{j=1}^{n} m_j \,,$$

and that its variance is

$$V_n^2 = \sum_{j=1}^{n} \sigma_j^2 \,.$$

Moreover, suppose that for any j between 1 and n, σ_j^2/V_n^2 tends to 0 as n tends to infinity. The condition that each random variable has finite second moment tells us that it is not too widely scattered about its mean. The condition that, for any j between 1 and n, σ_j^2/V_n^2 tends to 0 as n tends to infinity ensures that the fluctuations of one random variable do not dominate the others.

The central limit theorem tells us that the random variable $Z_\lambda(n) = [S_\lambda(n) - M_n]/V_n$ converges in law toward a reduced Gaussian random variable (i.e., with zero mean and unit variance). Convergence in law toward a reduced Gaussian distribution means that, as n tends to infinity, the distribution function $F_{Z(n)}(z)$ of $Z_\lambda(n)$ tends pointwise to the distribution function $F_{RG}(z)$ of a reduced Gaussian law:

$$\lim_{n \to \infty} F_{Z(n)}(z) = F_{RG}(z) \,.$$

The result is shown using the characteristic functions of each random variable. The proof is simpler if we introduce the centered variables $Y_{\lambda(j)} = X_{\lambda(j)} - m_j$ and define $U_\lambda(n) = \sum_{j=1}^{n} Y_{\lambda(j)}$. We then note that $Z_\lambda(n) = U_\lambda(n)/V_n$. Let $\Psi_{Y,j}(\nu)$, $\Psi_{U,n}(\nu)$ and $\Psi_{Z,n}(\nu)$ be the characteristic functions of $Y_{\lambda(j)}$, $U_\lambda(n)$ and $Z_\lambda(n)$, respectively. Then we have

$$\Psi_{U,n}(\nu) = \Psi_{Y,1}(\nu)\Psi_{Y,2}(\nu)\ldots\Psi_{Y,n}(\nu) \,.$$

Moreover, if $P_{Z,n}(z)$ and $P_{U,n}(s)$ are the probability density functions of $Z_\lambda(n)$ and $U_\lambda(n)$, respectively, we have $P_{Z,n}(z) = V_n P_{U,n}(V_n z)$ and hence $\Psi_{Z,n}(\nu) = \Psi_{U,n}(\nu/V_n)$. Indeed, $\Psi_{Z,n}(\nu) = \int P_{Z,n}(z) \exp(iz\nu)\mathrm{d}z$, which can also be written $\Psi_{Z,n}(\nu) = V_n \int P_{U,n}(V_n z) \exp(iz\nu)\mathrm{d}z$. Making the change of variable $s = V_n z$, we can then write $\Psi_{Z,n}(\nu) = \int P_{U,n}(s) \exp(is\nu/V_n) \mathrm{d}s$, or finally, $\Psi_{Z,n}(\nu) = \Psi_{U,n}(\nu/V_n)$. Returning to the main argument, we thus have

$$\Psi_{Z,n}(\nu) = \Psi_{Y,1}\left(\frac{\nu}{V_n}\right)\Psi_{Y,2}\left(\frac{\nu}{V_n}\right)\dots\Psi_{Y,n}\left(\frac{\nu}{V_n}\right) .$$

For fixed ν and large enough n, we have

$$\Psi_{Y,j}\left(\frac{\nu}{V_n}\right) = 1 - \frac{\sigma_j^2\, \nu^2}{2\, V_n^2} + o\left(\frac{\sigma_j^2\, \nu^2}{2\, V_n^2}\right) .$$

Therefore, for fixed ν and large enough n, we may write

$$\ln\left[\Psi_{Z,n}(\nu)\right] = \sum_{j=1}^{n} \ln\left[1 - \frac{\sigma_j^2\, \nu^2}{2\, V_n^2} + o\left(\frac{\sigma_j^2\, \nu^2}{2\, V_n^2}\right)\right] ,$$

or

$$\ln\left[\Psi_{Z,n}(\nu)\right] = \sum_{j=1}^{n} \left[-\frac{\sigma_j^2\, \nu^2}{2\, V_n^2} + o'\left(\frac{\sigma_j^2\, \nu^2}{2\, V_n^2}\right)\right] ,$$

or again,

$$\ln\left[\Psi_{Z,n}(\nu)\right] = -\frac{\nu^2}{2} + o_n\left(\frac{\nu^2}{2}\right) ,$$

where

$$o_n\left(\frac{\nu^2}{2}\right) = \sum_{j=1}^{n} o'\left(\frac{\sigma_j^2\, \nu^2}{2\, V_n^2}\right) \quad \text{and} \quad \lim_{n\to\infty} o_n\left(\frac{\nu^2}{2}\right) = 0 .$$

We thus obtain

$$\lim_{n\to\infty} \Psi_{Z,n}(\nu) = \exp\left(-\frac{\nu^2}{2}\right) ,$$

and hence,

$$\lim_{n\to\infty} F_{Z,n}(z) = \int_{-\infty}^{z} \frac{1}{\sqrt{2\pi}} \exp\left(-\frac{\xi^2}{2}\right) \mathrm{d}\xi ,$$

where $F_{Z,n}(z)$ is the distribution function of $P_{Z,n}(z)$. It is sometimes claimed that

$$\lim_{n\to\infty} P_{Z,n}(z) = \frac{1}{\sqrt{2\pi}} \exp\left(-\frac{z^2}{2}\right) ,$$

which is not exactly the result we have proved. Indeed, only convergence in law, i.e., convergence of the distribution function, is obtained with

$$\lim_{n\to\infty} \Psi_{Z,n}(\nu) = \exp\left(-\frac{\nu^2}{2}\right) .$$

It is interesting to reformulate this basic result. We have $S_\lambda(n) = V_n Z_\lambda(n) + M_n$ so if we set $s_\lambda(n) = S_\lambda(n)/n = \sum_{j=1}^{n} X_{\lambda(j)}/n$, then $s_\lambda(n)$ tends to a

Gaussian random variable with mean $\lim_{n\to\infty} M_n/n$, i.e., with mean equal to $\lim_{n\to\infty} \sum_{j=1}^{n} m_j/n$, and standard deviation $\lim_{n\to\infty} V_n/n$, i.e., with variance equal to $\lim_{n\to\infty} \sum_{j=1}^{n} \sigma_j^2/n^2$.

Note also that, if the random variables are distributed according to the same probability density function with mean m and variance σ^2, then $\sum_{j=1}^{n} X_{\lambda(j)}/n$ tends toward a Gaussian random variable with mean m and variance σ^2/n.

We shall see that the central limit theorem is a very important result because it can help us to understand many physical phenomena. It proves the existence of a universal type of behavior which arises whenever we are dealing with a sum of independent random variables with comparable fluctuations (the latter being characterized by their variance). At this level, it is difficult to obtain a simple interpretation for this result, which raises at least two questions:

• Why is there a unique law?
• Why is this law Gaussian, or normal, as it is sometimes called?

We shall see in Chapter 5 that arguments based on information theory will help us to elucidate this problem.

The theorem can be extended in various ways to independent real stochastic vectors. Consider a sequence of real N-component stochastic vectors $\boldsymbol{X}_{\lambda(1)}, \boldsymbol{X}_{\lambda(2)}, \dots, \boldsymbol{X}_{\lambda(n)}$, with identical distribution. Let $\boldsymbol{m} = \langle \boldsymbol{X}_{\lambda(i)} \rangle$ denote the mean vector and define the covariance matrix $\overline{\overline{\varGamma}}$ by

$$\overline{\overline{\varGamma}} = \left\langle \left[\boldsymbol{X}_{\lambda(i)} - \boldsymbol{m} \right] \left[\boldsymbol{X}_{\lambda(j)} - \boldsymbol{m} \right]^{\dagger} \right\rangle .$$

We define the vector characteristic function by

$$\Psi_X(\boldsymbol{\nu}) = \int_{-\infty}^{+\infty} \int_{-\infty}^{+\infty} \cdots \int_{-\infty}^{+\infty} P_X(x_1, x_2, \dots, x_N)$$
$$\times \exp\left[i(x_1\nu_1 + x_2\nu_2 + \dots + x_N\nu_N) \right] dx_1 dx_2 \dots dx_N ,$$

which can be written

$$\Psi_{\boldsymbol{X}}(\boldsymbol{\nu}) = \int_{-\infty}^{+\infty} P_X(\boldsymbol{x}) \exp(i\boldsymbol{\nu}^{\dagger}\boldsymbol{x}) d\boldsymbol{x} .$$

We put $\boldsymbol{Y}_{\lambda(i)} = \boldsymbol{X}_{\lambda(i)} - \boldsymbol{m}$ and

$$\Psi_{\boldsymbol{Y}}(\boldsymbol{\nu}) = \int_{-\infty}^{+\infty} P_Y(\boldsymbol{y}) \exp(i\boldsymbol{\nu}^{\dagger}\boldsymbol{y}) d\boldsymbol{y} .$$

Then setting

$$\boldsymbol{Z}_\lambda(n) = \frac{1}{\sqrt{n}} \sum_{j=1}^{n} \boldsymbol{Y}_{\lambda(i)} ,$$

we obtain

$$\Psi_{\boldsymbol{Z},n}(\boldsymbol{\nu}) = \left[\Psi_{\boldsymbol{Y}} \left(\frac{\boldsymbol{\nu}}{\sqrt{n}} \right) \right]^n .$$

Now,

$$\Psi_{\boldsymbol{Y}} \left(\frac{\boldsymbol{\nu}}{\sqrt{n}} \right)$$

$$= \int_{-\infty}^{+\infty} P_Y(\boldsymbol{y}) \left\{ 1 + \frac{i}{\sqrt{n}} (\boldsymbol{\nu}^\dagger \boldsymbol{y}) - \frac{1}{2n} (\boldsymbol{\nu}^\dagger \boldsymbol{y})^2 + o \left[\left(\frac{1}{\sqrt{n}} \boldsymbol{\nu}^\dagger \boldsymbol{y} \right)^2 \right] \right\} \mathrm{d}\boldsymbol{y} .$$

We have $\int_{-\infty}^{+\infty} P_Y(\boldsymbol{y}) \mathrm{d}\boldsymbol{y} = 1$, and

$$\int_{-\infty}^{+\infty} P_Y(\boldsymbol{y}) \frac{i}{\sqrt{n}} (\boldsymbol{\nu}^\dagger \boldsymbol{y}) \mathrm{d}\boldsymbol{y} = \frac{i}{\sqrt{n}} \boldsymbol{\nu}^\dagger \int_{-\infty}^{+\infty} P_Y(\boldsymbol{y}) \boldsymbol{y} \mathrm{d}\boldsymbol{y} .$$

But $\int_{-\infty}^{+\infty} P_Y(\boldsymbol{y}) \boldsymbol{y} \mathrm{d}\boldsymbol{y} = 0$, so

$$\int_{-\infty}^{+\infty} P_Y(\boldsymbol{y}) \frac{i}{\sqrt{n}} (\boldsymbol{\nu}^\dagger \boldsymbol{y}) \mathrm{d}\boldsymbol{y} = 0 .$$

For the term in $(\boldsymbol{\nu}^\dagger \boldsymbol{y})^2$, we note that

$$(\boldsymbol{\nu}^\dagger \boldsymbol{y})^2 = \boldsymbol{\nu}^\dagger [\boldsymbol{y}\boldsymbol{y}^\dagger] \boldsymbol{\nu} ,$$

and hence,

$$\int_{-\infty}^{+\infty} P_Y(\boldsymbol{y}) \frac{1}{2n} (\boldsymbol{\nu}^\dagger \boldsymbol{y})^2 \mathrm{d}\boldsymbol{y} = \frac{1}{2n} \int_{-\infty}^{+\infty} P_Y(\boldsymbol{y}) \boldsymbol{\nu}^\dagger [\boldsymbol{y}\boldsymbol{y}^\dagger] \boldsymbol{\nu} \mathrm{d}\boldsymbol{y} .$$

We have $\int_{-\infty}^{+\infty} P_Y(\boldsymbol{y}) [\boldsymbol{y}\boldsymbol{y}^\dagger] \mathrm{d}\boldsymbol{y} = \overline{\overline{\Gamma}}$, or $\int_{-\infty}^{+\infty} P_Y(\boldsymbol{y}) (\boldsymbol{\nu}^\dagger \boldsymbol{y})^2 \mathrm{d}\boldsymbol{y} = \boldsymbol{\nu}^\dagger \overline{\overline{\Gamma}} \boldsymbol{\nu}$, and therefore

$$\Psi_{\boldsymbol{Y}} \left(\frac{\boldsymbol{\nu}}{\sqrt{n}} \right) = 1 - \frac{1}{2n} \boldsymbol{\nu}^\dagger \overline{\overline{\Gamma}} \boldsymbol{\nu} + \tilde{o} \left[\left(\frac{1}{\sqrt{n}} \boldsymbol{\nu}^\dagger \right)^2 \right] ,$$

where $\tilde{o} \left[(\boldsymbol{\nu}^\dagger / \sqrt{n})^2 \right]$ is a scalar tending to zero faster than $1/n$. Since $\Psi_{\boldsymbol{Z},n}(\boldsymbol{\nu}) = [\Psi_{\boldsymbol{Y}} (\boldsymbol{\nu}/\sqrt{n})]^n$, we obtain

$$\Psi_{\boldsymbol{Z},n}(\boldsymbol{\nu}) = \left\{ 1 - \frac{1}{2n} \boldsymbol{\nu}^\dagger \overline{\overline{\Gamma}} \boldsymbol{\nu} + \tilde{o} \left[\left(\frac{1}{\sqrt{n}} \boldsymbol{\nu}^\dagger \right)^2 \right] \right\}^n ,$$

and hence,

$$\lim_{n \to \infty} \Psi_{\boldsymbol{Z},n}(\boldsymbol{\nu}) = \lim_{n \to \infty} \left\{ 1 - \frac{1}{2n} \boldsymbol{\nu}^\dagger \overline{\overline{\Gamma}} \boldsymbol{\nu} + \tilde{o} \left[\left(\frac{1}{\sqrt{n}} \boldsymbol{\nu}^\dagger \right)^2 \right] \right\}^n ,$$

or

$$\lim_{n\to\infty} \Psi_{\boldsymbol{Z},n}(\nu) = \exp\left(-\frac{1}{2}\nu^{\dagger}\overline{\overline{\Gamma}}\nu\right) \ .$$

We recognize the characteristic function of a probability density of Gaussian stochastic vectors, with zero mean and covariance matrix $\overline{\overline{\Gamma}}$, viz.,

$$P_{\boldsymbol{Z}}(z) = \frac{1}{\left(\sqrt{2\pi}\right)^{N}\sqrt{|\overline{\overline{\Gamma}}|}} \exp\left(-\frac{1}{2}z^{\dagger}\overline{\overline{K}}z\right) \ ,$$

where $\overline{\overline{K}}$ is the matrix inverse to $\overline{\overline{\Gamma}}$.

4.4 Gaussian Noise and Stable Probability Laws

It is a common hypothesis in physics to assume that the noise accompanying a measured signal is Gaussian. If the noise results from the addition of a large number of independent random phenomena with finite second moment, the central limit theorem tends to support this hypothesis. However, a certain number of conditions have to be fulfilled.

We assumed that the random variables were independent. In practice, it suffices that they should be uncorrelated. On the other hand, it is quite clear that if there is a perfect correlation between the various realizations, e.g., if the modulus of the reduced covariance is unity, there is little hope of the sum converging toward a Gaussian random variable. We have already seen that, if the random variables we are summing are correlated, the variance of the sum will not simply be the sum of their variances. It should be emphasized that this does not imply that a sum of correlated random variables will not converge to a Gaussian random variable. If the variables are partially correlated and have finite second moment, satisfying the relation between variances specified in the last section, then the sum can converge toward a Gaussian random variable, but the convergence is slower than if they were uncorrelated.

If there is a non-linear element in the system output, i.e., after the random variables have been summed, then the probability density function will be modified, as described in the section dealing with change of variables, and will no longer be Gaussian.

The second moments of the summed random variables must be finite. This condition is absolutely essential. Indeed, suppose we had Cauchy probability density functions $P_X(x) = a/\left[\pi(a^2 + x^2)\right]$, where $a > 0$. These distributions have no moment of order greater than or equal to 1. In other words, $\langle (X_\lambda)^r \rangle$ is not defined for any value of r greater than or equal to 1. We have seen that the characteristic function of the Cauchy distribution is $\exp(-a|\nu|)$. If we consider a sequence of independent Cauchy random variables $X_{\lambda(1)}, X_{\lambda(2)}, \ldots, X_{\lambda(n)}$, the mean random variable defined by

$s_\lambda(n) = \sum_{j=1}^{n} X_{\lambda(j)}/n$, where $\lambda = [\lambda(1), \lambda(2), \ldots, \lambda(n)]$, will have characteristic function $\exp(-a|\nu|)$. Then $s_\lambda(n)$ will be a Cauchy variable with probability density function $P_{S(n)}(s) = a/\left[\pi(a^2 + s^2)\right]$, that is, with the same parameter as the summed variables. The mean $s_\lambda(n)$ therefore fluctuates as much as each of the summed random variables.

When we add together two identically distributed and independent Gaussian variables $X_{\lambda(1)}$ and $X_{\lambda(2)}$, the sum is still Gaussian. We thus say that the Gaussian law is stable. More precisely, a probability density function represents a stable probability law if, when we add two independent variables $X_{\lambda(1)}$ and $X_{\lambda(2)}$ identically distributed according to $P_X(x)$, there exist two numbers a and b such that $[X_{\lambda(1)} + X_{\lambda(2)}]/a + b$ is distributed according to $P_X(x)$. We have just seen that the Cauchy probability law is stable. The central limit theorem guarantees that if $\langle (X_\lambda)^r \rangle$ converges for any value of r greater than 2, the only stable probability laws are Gaussian. However, if $\langle (X_\lambda)^r \rangle$ does not converge for any value of r greater than α, where α is strictly less than 2, there may be other stable laws. This is the case for the Cauchy distribution, for which $\alpha = 1$.

The study of stable distributions is very interesting but goes somewhat beyond the scope of this book. However, if the noise is described by a random variable which is not too scattered, that is, which has finite second moment, then the Gaussian hypothesis may be acceptable, provided that a large number of independent phenomena add together to make up the noise. On the other hand, if the phenomena causing the fluctuations undergo large deviations which prevent us from defining a finite second moment, then we are compelled to reject the Gaussian noise hypothesis.

4.5 A Simple Model of Speckle

Signals or images obtained when an object is illuminated by a coherent wave involve a significant level of noise. This is the case for example when we shine a laser beam, a coherent electromagnetic wave, or an acoustic wave onto a surface. Radar images provide a perfect illustration of this phenomenon. Figure 4.1 shows an image acquired using a synthetic aperture radar (SAR), with its typical speckle noise.

This kind of noise gives the image its grainy appearance, as though sprinkled with pepper and salt. The size of the grains depends on the experimental setup. If the detector integrates the signal over a smaller region than the grain size, a simple model can be made which often leads to a good approximation, whereby the speckle noise is described by multiplicative noise with exponential probability density function. We can now construct such a model using the ideas described above.

A scalar monochromatic wave

$$A(t) = \frac{A_0}{|S|} \exp(i\omega t)$$

Fig. 4.1. Image of an agricultural area in the Caucasus acquired using the synthetic aperature radar (SAR) aboard the European Remote Sensing satellite ERS 1. The image shows characteristic speckle noise. (Image provided by the CNES, courtesy of the ESA)

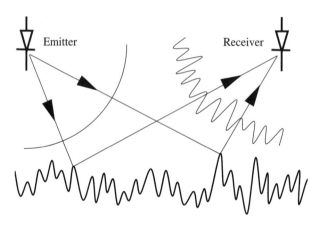

Fig. 4.2. Schematic illustration of the scattering of a wave by an irregular surface, which leads to the production of speckle

illuminates a rough surface S (see Fig. 4.2) with constant reflectivity. The surface S corresponds to the illuminated surface which forms the wave producing the field measured in a given pixel.

The factor $|S|$, which represents the measure of the surface S, is introduced for reasons of homogeneity with regard to physical units, as we shall see shortly. Other conventions could have been chosen, but this one is perhaps the simplest. At the detector, the amplitude of the field can be written

$$A_{\mathrm{R}}(t) = \iint\limits_S \frac{A_0}{|S|} \rho \exp\left[\mathrm{i}\omega(t - t_{x,y})\right] \mathrm{d}x\mathrm{d}y \ ,$$

where ρ is the square root of the reflection coefficient and $t_{x,y}$ describes the retardation of the ray leaving the emitter and arriving at the point with coordinates (x,y) on the surface before converging on the detector. $\omega t_{x,y}$ is thus a phase term $\Phi_{x,y}$, which can be chosen to lie between 0 and 2π. To be precise, we set $\Phi_{x,y} = \omega t_{x,y} - 2\pi n$, where n is a natural number chosen so that $\Phi_{x,y}$ lies between 0 and 2π. If the depth fluctuations on the surface are large, we may expect a significant spread of values for $\omega t_{x,y}$ relative to 2π, so that $\Phi_{x,y}$ is likely to be well described by a random variable uniformly distributed between 0 and 2π. We thus write

$$A_{\mathrm{R}}(t) = \iint\limits_S \frac{A_0}{|S|} \rho \exp(\mathrm{i}\omega t - \mathrm{i}\Phi_{x,y})\mathrm{d}x\mathrm{d}y \ ,$$

or $A_{\mathrm{R}}(t) = A_0 \rho \exp(\mathrm{i}\omega t) Z_\lambda$, where

$$Z_\lambda = \frac{1}{|S|} \iint\limits_S \exp(-\mathrm{i}\Phi_{x,y})\mathrm{d}x\mathrm{d}y \ .$$

Note that $A_{\mathrm{R}}(t)$ is a random variable but that its dependence on λ is not mentioned, to simplify the notation. Speckle thus amounts to multiplying the reflected amplitude $A_0\rho$ by Z_λ. It should be pointed out that the model is multiplicative because we assumed that the reflectivity ρ is constant, i.e., independent of x and y. If this were not so, the model would not necessarily be simply multiplicative. It is therefore important to specify the model precisely.

To proceed with this calculation, we now make a simplifying hypothesis, namely that we may cut the surface S up into N parts, each of which introduces an independent phase difference ϕ_j, for $j = 1, \ldots, N$. We can then write

$$Z_\lambda = \frac{1}{N} \sum_{j=1}^{N} \exp(-\mathrm{i}\phi_j) \ .$$

Decomposing Z_λ into real and imaginary parts $Z_\lambda = X_\lambda + \mathrm{i}Y_\lambda$, we find that X_λ and Y_λ are sums of independent random variables with finite second moment. In fact, we shall only show that X_λ and Y_λ are uncorrelated and that they have the same variance, and we shall then assume that they are independent. We have

$$X_\lambda Y_\lambda = \frac{1}{N^2} \sum_{i=1}^{N} \sin \phi_i \sum_{j=1}^{N} \cos \phi_j = \frac{1}{N^2} \sum_{i=1}^{N} \sum_{j=1}^{N} \sin \phi_i \cos \phi_j \ .$$

If we assume that the ϕ_ℓ are uniformly distributed between 0 and 2π, we obtain

$$\langle X_\lambda Y_\lambda \rangle = \frac{1}{N^2} \sum_{i=1}^{N} \sum_{j=1}^{N} \langle \sin \phi_i \cos \phi_j \rangle$$

and

$$\langle \sin \phi_i \cos \phi_j \rangle = 0 \ .$$

Hence, finally, $\langle X_\lambda Y_\lambda \rangle = 0$, which shows that the variables X_λ and Y_λ are uncorrelated. In the same way, it can be shown that

$$\langle (X_\lambda)^2 \rangle = \frac{1}{N^2} \sum_{\ell=1}^{N} \langle \cos^2 \phi_\ell \rangle$$

and

$$\langle (Y_\lambda)^2 \rangle = \frac{1}{N^2} \sum_{\ell=1}^{N} \langle \sin^2 \phi_\ell \rangle \ .$$

Now $\langle \cos^2 \phi_\ell \rangle = \langle \sin^2 \phi_\ell \rangle$ since the ϕ_ℓ are uniformly distributed between 0 and 2π, so that $\langle (X_\lambda)^2 \rangle = \langle (Y_\lambda)^2 \rangle$. This shows that the variables X_λ and Y_λ have the same variance.

When several elementary scatterers contribute, i.e., when N is large, X_λ and Y_λ are therefore Gaussian random variables with probability density functions

$$P_X(x) = \frac{1}{\sqrt{2\pi}\sigma} \exp\left(-\frac{x^2}{2\sigma^2}\right) \quad \text{and} \quad P_Y(y) = \frac{1}{\sqrt{2\pi}\sigma} \exp\left(-\frac{y^2}{2\sigma^2}\right) \ ,$$

respectively. The field intensity is given by $I_R = |A_R(t)|^2 = |A_0|^2 |\rho|^2 |Z_\lambda|^2$. This may also be written

$$I_R = I_0 |\rho|^2 B_\lambda \ ,$$

with $I_0 = |A_0|^2$ and $B_\lambda = |Z_\lambda|^2$, whose probability density function we shall now determine. In Sect. 2.7 of Chapter 2 we saw that, if we set $U_\lambda = |X_\lambda|^2$, the probability density function of U_λ is $P_U(u) = (\sqrt{2\pi}u\sigma)^{-1} \exp\left(-u/2\sigma^2\right)$. This is a Gamma distribution with parameters $2\sigma^2$ and $1/2$. The probability density function of $V_\lambda = |Y_\lambda|^2$ is the same as that of U_λ, and we have already seen that the sum of two Gamma variables with parameters $((2\sigma^2)^{-1}, 1/2)$ gives another Gamma variable, this time with parameters $((2\sigma^2)^{-1}, 1)$, i.e.,

$$P_B(b) = \frac{1}{2\sigma^2} \exp\left(-\frac{b}{2\sigma^2}\right) \ .$$

It is easy to see that $\langle B_\lambda \rangle = 2\sigma^2$, but $\langle I_R \rangle = I_0 |\rho|^2$ and hence, $\langle B_\lambda \rangle = 1$. (It may seem rather brutal to eliminate σ^2 like this. However, when we set $\langle I_R \rangle = I_0 |\rho|^2$, this means that we are defining the phenomenological factor $|\rho|$ in such a way that $|\rho|^2 = \langle I_R \rangle / I_0$ and hence $\langle B_\lambda \rangle = 1$.) We thus obtain $P_B(b) = \exp(-b)$ and hence,

$$P_{I_R}(I_R) = \frac{1}{I_0|\rho|^2} \exp\left(-\frac{I_R}{I_0|\rho|^2}\right) .$$

The general shape of the probability density function of I_R is depicted in Fig. 4.3.

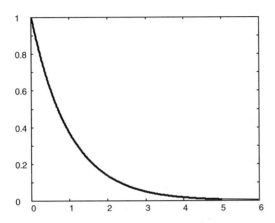

Fig. 4.3. Probability density function for the intensity of a scalar monochromatic wave with mean value 1, in the presence of fully developed speckle for a single look

The representation of intensity fluctuations in the form $I_R = I_0|\rho|^2 B_\lambda$, where B_λ is an exponential random variable, constitutes the simplest possible model of fully developed speckle, i.e., non-averaged speckle. This model follows very simply from the central limit theorem. Note also that an exponential variable involves large fluctuations since its standard deviation is equal to its mean.

Using the results of Sections 4.2 and 4.3, if the measured quantity results from the mean of n independent measurements, we then have

$$P_{I_R}(I_R) = \left(\frac{n}{I_0|\rho|^2}\right)^n \frac{I_R^{n-1}}{\Gamma(n)} \exp\left(-n\frac{I_R}{I_0|\rho|^2}\right) .$$

In other words, the multiplicative noise B_λ will be a Gamma variable of order n and mean 1.

For electromagnetic waves, and in particular, light, the wave is not scalar but vectorial, since it can be polarized (see Section 3.14). The above calculation is only valid for one of the two polarization components. In the general case, we have

$$\boldsymbol{A}_R(t) = A_R^X(t)\boldsymbol{e}_x + A_R^Y(t)\boldsymbol{e}_y .$$

We deduce that $I_R = |\boldsymbol{A}_R(t)|^2 = |A_R^X(t)|^2 + |A_R^Y(t)|^2$. Suppose that the surface is illuminated by a completely unpolarized wave and that the field $\boldsymbol{A}_R(t)$ is also unpolarized and is a complex Gaussian stochastic vector. The hypothesis that the wave is completely unpolarized can be formulated by

$$I_0^X = \langle |A_0^X(t)|^2 \rangle = \langle |A_0^Y(t)|^2 \rangle = I_0^Y \quad \text{and} \quad \langle [A_0^X(t)]^* A_0^Y(t) \rangle = 0 \, .$$

Hence, for a single measurement, $|\boldsymbol{A}_R(t)|^2$ is the sum of two random variables with probability density function

$$\frac{2}{I_0|\rho|^2} \exp\left(-\frac{2I}{I_0|\rho|^2}\right) \, .$$

Since the two intensities $|A_0^X(t)|^2$ and $|A_0^Y(t)|^2$ are independent, the measured quantity is the sum of two independent measurements and we have

$$P_{I_R}(I_R) = \left(\frac{2}{I_0|\rho|^2}\right)^2 I_R \exp\left(-2\frac{I_R}{I_0|\rho|^2}\right) \, .$$

To obtain this result, we use the fact that:

- $A_0^X(t)$ and $A_0^Y(t)$ are jointly Gaussian,
- $A_0^X(t)$ and $A_0^Y(t)$ are uncorrelated,

which amounts to saying that $A_0^X(t)$ and $A_0^Y(t)$ are independent. The general appearance of the probability density function of I_R is depicted in Fig. 4.4.

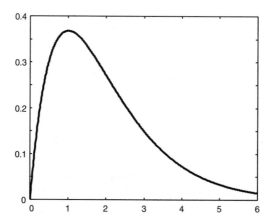

Fig. 4.4. Probability density function for the intensity of an unpolarized monochromatic wave with mean value equal to unity, in the presence of fully developed speckle for a single look

Let us return to the case of a scalar wave or, equivalently, consider one component of the electric field. The model presented above describes the probability density function for a given reflectivity. To be precise, the random event λ is associated with different surfaces of the same reflectivity but different roughness. In other words, experimentally, we would obtain an intensity histogram for the field at the detector that would be close to an exponential probability density function by calculating an ensemble average over several realizations of surfaces with the same statistical characteristics. If we are interested in an image such as the one in Fig. 4.1, the intensity histogram for the field at the detector due to a homogeneous region, i.e., a single cultivated field, will be close to this exponential probability density function if the reflectivity is constant over the whole region. Each pixel for a homogeneous region must therefore correspond to a region of the ground with the same reflectivity if the histogram of gray levels of pixels in this region is to correspond to an exponential probability density. We have represented this situation schematically in Fig. 4.5.

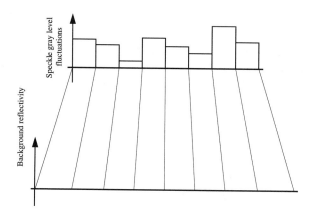

Fig. 4.5. Constant reflectivity model which leads to an exponential probability density function for the intensity in the presence of fully developed speckle for a single look

If the ground reflectivity is not constant, the gray level histogram for a homogeneous region will no longer correspond to an exponential distribution. We have represented this situation schematically in Fig. 4.6.

The random experiment we are referring to in order to define a probability density function now results from a combination of two phenomena. The first relates to speckle and results from interference as discussed above. Let X_λ be the random variable representing the value of the intensity for unit reflectivity. X_λ is thus distributed according to an exponential law $P_X(x)$ with mean I_0, which we shall assume to be 1. The second random phenomenon relates to the value of the reflectivity of the zone projected onto a pixel in the image. Let R_μ be the random variable which represents the value of this reflectivity, which

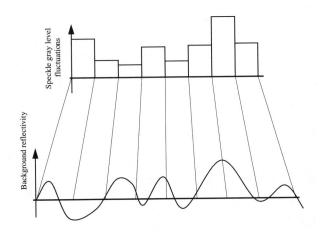

Fig. 4.6. Variable reflectivity model which no longer leads to an exponential probability density function for the intensity in the presence of speckle for a single look

we assume to be distributed with probability density function $P_R(R)$. λ and μ are treated as two independent random events since they arise from very different phenomena. The intensity $I_{(\lambda,\mu)}$ measured at the detector is thus a product of X_λ and the reflectivity R_μ, viz.,

$$I_{(\lambda,\mu)} = X_\lambda R_\mu \ .$$

For a given value of R, the probability density function for $I_{(\lambda,\mu)}$ is

$$P_{I|R}(I|R) = \frac{1}{R} P_X\left(\frac{I}{R}\right) ,$$

because $I_{(\lambda,\mu)} = X_\lambda R_\mu$ and $P_I(I)\mathrm{d}I = P_X(X)\mathrm{d}X$. Applying Bayes' relation, we then obtain $P_I(I)$

$$P_I(I) = \int P_{I|R}(I|R)P_R(R)\mathrm{d}R \ .$$

The probability density function $P_I(I)$ for $I_{(\lambda,\mu)}$ is thus

$$P_I(I) = \int \frac{1}{R} P_X\left(\frac{I}{R}\right) P_R(R)\mathrm{d}R \ .$$

Experimental observations in synthetic aperture radar imaging of textured zones have shown that a good model can be made by assuming that reflectivities have Gamma probability distributions:

$$P_R(R) = \left(\frac{\nu}{\beta}\right)^\nu \frac{R^{\nu-1}}{\Gamma(\nu)} \exp\left(-\nu\frac{R}{\beta}\right) \ .$$

The probability density function $P_I(I)$ for a single look is then

$$P_I(I) = \int \frac{1}{R} \exp\left(-\frac{I}{R}\right) \left(\frac{\nu}{\beta}\right)^\nu \frac{R^{\nu-1}}{\Gamma(\nu)} \exp\left(-\nu\frac{R}{\beta}\right) dR \ .$$

There is no simpler explicit form for this law which is known as the K distribution with parameter ν. A typical example is illustrated in Fig. 4.7 for a scalar monochromatic wave of mean 1 in the presence of speckle for a single look, described by a K distribution with parameter $\nu = 1$.

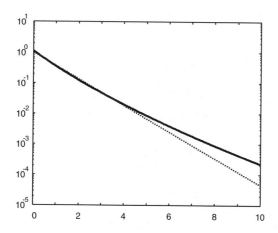

Fig. 4.7. Probability density function (*continuous curve*) for a K distribution with parameter $\nu = 1$. The scale on the *ordinate* is logarithmic. The *dotted curve* shows an exponential distribution with the same mean

4.6 Random Walks

The random walk model is interesting in several respects. It constitutes a remarkable application of the central limit theorem and we shall examine the main results. Suppose that a particle moves in a 3-dimensional space, making a random displacement at each tick of the clock. The unit of time is taken as τ. (This amounts to treating time as discrete.) More precisely, we consider that if the particle is at the point r at time t, where $r = (x, y, z)^{\mathrm{T}}$, the probability density for it to be at the point $r + dr$ at time $t + \tau$ will be denoted

$$P(r + dr, t + \tau \,|\, r, t) \ .$$

This random walk has no memory since the displacement at a given time t is independent of the positions occupied by the particle before it reached the point r at time t.

The random walk is said to be stationary if $P(r + dr, t + \tau \,|\, r, t)$ does not depend on t. It is said to be homogeneous if $P(r + dr, t + \tau \,|\, r, t)$ does not depend on r. In the following, we shall be concerned with homogeneous stationary random walks. In this case, we can write simply

$$P(r + dr, t + \tau \,|\, r, t] = W(dr, \tau) \,.$$

Moreover, we shall consider two further hypotheses:

- the displacements in each direction are independent so that we may write $W(d, \tau) = w_X(\delta x, \tau) w_Y(\delta y, \tau) w_Z(\delta z, \tau)$, with $d = (\delta x, \delta y, \delta z)$,
- the first two moments are finite and we shall write for the first component, for example,

$$m_X = \int x w_X(x, \tau) dx \quad \text{and} \quad \sigma_X^2 = \int (x - m_X)^2 w_X(x, \tau) dx \,.$$

If R_0 is the position of the particle at time $t = 0$ and $R_{n,\lambda}$ is its position at time $t = n\tau$, a realization of the random walk is described by

$$R_{n,\lambda} = R_0 + \sum_{j=1}^{n} \delta_{\lambda_j} \,,$$

where $\lambda = (\lambda_1, \lambda_2, \ldots, \lambda_n)$ and δ_{λ_j} represents the step of the walk made at time $t = j\tau$. δ_{λ_j} is therefore a realization of a stochastic vector distributed according to $W(d, \tau)$. $R_{n,\lambda}$, R_0 and δ_{λ_j} are vectors in the 3-dimensional space. We project this equation onto the Ox axis to give

$$X_{n,\lambda} = X_0 + \sum_{j=1}^{n} \delta_{\lambda_j}^X \,.$$

In order to simplify the following analysis, we assume that $X_0 = Y_0 = Z_0 = 0$ (see Fig. 4.8).

We thus see that $X_{n,\lambda}$ is a sum of n independent random variables. The conditions of the central limit theorem are all satisfied and we can therefore consider the following approximation:

$$P_{X,n}(x) = \frac{1}{\sigma_X \sqrt{2\pi n}} \exp\left[-\frac{(x - n m_X)^2}{2 n \sigma_X^2}\right] \,.$$

It is useful to introduce the notation $t = n\tau$, $v_X = m_X/\tau$ and $\chi_X^2 = \sigma_X^2/\tau$, and likewise for y and z. We may then write

$$P_{X,t}(x) = \frac{1}{\chi_X \sqrt{2\pi t}} \exp\left[-\frac{(x - v_X t)^2}{2 \chi_X^2 t}\right] \,.$$

Since we have assumed that the displacements are independent in each direction, the same can be said of the components of $R_{n,\lambda}$. We thus deduce

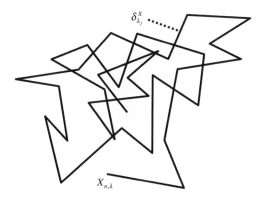

Fig. 4.8. Example of a random walk in a plane

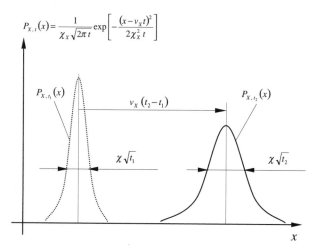

$$P_{X,t}(x) = \frac{1}{\chi_X \sqrt{2\pi t}} \exp\left[-\frac{(x - v_X t)^2}{2\chi_X^2 t}\right]$$

Fig. 4.9. Evolution of the probability density function for a 1-dimensional random walk

that the probability density function for $\boldsymbol{R}_{n,\lambda}$ is simply the product of these marginal distributions with respect to each component:

$$P_{R,t}(x, y, z) = \frac{1}{\chi_X \chi_Y \chi_Z} \left(\frac{1}{\sqrt{2\pi t}}\right)^3 \exp\left[-\frac{Q(x, y, z)}{2t}\right],$$

where

$$Q(x, y, z) = \frac{(x - v_X t)^2}{\chi_X^2} + \frac{(y - v_Y t)^2}{\chi_Y^2} + \frac{(z - v_Z t)^2}{\chi_Z^2}.$$

Figure 4.9 shows how the probability density evolves in the case of a 1-dimensional random walk.

In the case of isotropic diffusion, for which $\chi_X = \chi_Y = \chi_Z = \chi$ (only the diffusion part is assumed to be isotropic, for we may have $v_X \neq v_Y \neq v_Z$), we obtain simply

$$P_{R,t}(x, y, z) = \left(\frac{1}{\sqrt{2\pi\chi^2 t}} \right)^3 \exp \left[-\frac{A(x, y, z)}{2\chi^2 t} \right] ,$$

with

$$A(x, y, z) = (x - v_X t)^2 + (y - v_Y t)^2 + (z - v_Z t)^2 .$$

We can immediately generalize this result to the case of a random walk in D dimensions. Setting $\boldsymbol{R} = (r_1, r_2, \ldots, r_D)^{\mathrm{T}}$ and making the same hypotheses as above, i.e., stationarity, homogeneity, isotropic diffusion and finite second moments, we find that

$$P_{R,t}(\boldsymbol{R}) = \left(\frac{1}{\sqrt{2\pi\chi^2 t}} \right)^D \exp \left[-\frac{1}{2\chi^2 t} \left(\sum_{j=1}^{D} (r_j - v_j t)^2 \right) \right] .$$

Putting

$$\boldsymbol{m} = (m_1, m_2, \ldots, m_D)^{\mathrm{T}} , \quad \boldsymbol{v} = (v_1, v_2, \ldots, v_D)^{\mathrm{T}} , \quad \|\boldsymbol{R}\|^2 = \left| \sum_{j=1}^{D} r_j^2 \right| ,$$

we then observe the fundamental results for isotropic diffusion with finite second moment:

$$\langle \boldsymbol{R}_{n,\lambda} \rangle = n\boldsymbol{m} = \boldsymbol{v}t \quad \text{and} \quad \langle \|\boldsymbol{R}_{n,\lambda} - \boldsymbol{v}t\|^2 \rangle = D\chi^2 t ,$$

or

$$\sqrt{\langle \|\boldsymbol{R}_{n,\lambda} - \boldsymbol{v}t\|^2 \rangle} = \chi\sqrt{Dt} .$$

Before concluding this section, it is worth noting that speckle noise corresponds to the special case of an isotropic random walk without drift (i.e., with $\boldsymbol{v} = 0$) in two dimensions.

4.7 Application to Diffusion

The diffusion equation is a partial differential equation describing many macroscopic phenomena which correspond to the diffusion of a probability law. As an example, we may mention the diffusion of molecules in a solvent or the diffusion of heat in a solid. It is based upon a similar result to the central limit theorem and is thus very general. In this section, we shall be concerned with the example of particle diffusion, whilst bearing in mind that it has a considerably wider scope.

Diffusion is considered to result from random motions in a 3-dimensional space. Let $P(x, y, z, t)$ denote the probability density function for finding a particle at the point with coordinates (x, y, z) at time t and $W(\delta_x, \delta_y, \delta_z, \tau)$ the probability density function for observing a displacement of amplitude $(\delta_x, \delta_y, \delta_z)^{\mathrm{T}}$ during the time interval τ. We will assume that this displacement results from a great many small, independent, random displacements distributed according to stationary laws with finite second moments and independent along each of the three axes. $W(\delta_x, \delta_y, \delta_z, \tau)$ is a transition rate density and we may thus write

$$W(\delta_x, \delta_y, \delta_z, \tau) = P\left(x + \delta_x, y + \delta_y, z + \delta_z, t + \tau \,|\, x, y, z, t\right) .$$

Applying Bayes' relation (Chapter 2), we can write

$$P\left(x', y', z', t + \tau\right) = \iiint P\left(x', y', z', t + \tau \,|\, x, y, z, t\right) P(x, y, z, t) \mathrm{d}x \mathrm{d}y \mathrm{d}z .$$

It is important to note that t and $t + \tau$ are parameters which do not play the same role as x, y, z and x', y', z'. The latter are associated with the random variables constituted by the particle positions. We may rewrite the last equation in the form

$$P(x, y, z, t + \tau) = \iiint P\left(x, y, z, t + \tau \,|\, x - \delta_x, y - \delta_y, z - \delta_z, t\right)$$
$$\times P\left(x - \delta_x, y - \delta_y, z - \delta_z, t\right) \mathrm{d}\delta_x \mathrm{d}\delta_y \mathrm{d}\delta_z ,$$

or alternatively,

$$P(x, y, z, t + \tau)$$
$$= \iiint W\left(\delta_x, \delta_y, \delta_z, \tau\right) P\left(x - \delta_x, y - \delta_y, z - \delta_z, t\right) \mathrm{d}\delta_x \mathrm{d}\delta_y \mathrm{d}\delta_z .$$

This is known as the Chapman–Kolmogorov equation, interpreted schematically in Fig. 4.10. Expanding $P\left(x - \delta_x, y - \delta_y, z - \delta_z, t\right)$ to second order in δ_x, δ_y and δ_z, we obtain

$$P\left(x - \delta_x, y - \delta_y, z - \delta_z, t\right) \approx P(x, y, z, t) - \frac{\partial}{\partial x} P(x, y, z, t)\delta_x$$
$$- \frac{\partial}{\partial y} P(x, y, z, t)\delta_y - \frac{\partial}{\partial z} P(x, y, z, t)\delta_z$$
$$+ \frac{1}{2} \frac{\partial^2}{\partial x^2} P(x, y, z, t)\delta_x^2 + \frac{1}{2} \frac{\partial^2}{\partial y^2} P(x, y, z, t)\delta_y^2$$
$$+ \frac{1}{2} \frac{\partial^2}{\partial z^2} P(x, y, z, t)\delta_z^2 + \frac{\partial^2}{\partial x \partial y} P(x, y, z, t)\delta_x \delta_y$$
$$+ \frac{\partial^2}{\partial y \partial z} P(x, y, z, t)\delta_y \delta_z + \frac{\partial^2}{\partial x \partial z} P(x, y, z, t)\delta_x \delta_z .$$

We have assumed that the transition rate density $W\left(\delta_x, \delta_y, \delta_z, \tau\right)$ results from a great many small, independent, random displacements distributed according

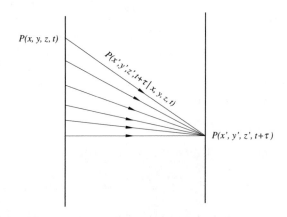

Fig. 4.10. Schematic representation of the Chapman–Kolmogorov equation

to stationary laws with finite second moments and independent along each of the three axes. We therefore choose to denote the various moments as follows:

$$\iiint \delta_x W\left(\delta_x, \delta_y, \delta_z, \tau\right) \mathrm{d}\delta_x \mathrm{d}\delta_y \mathrm{d}\delta_z = v_X \tau \ ,$$

and

$$\iiint \left(\delta_x - v_X \tau\right)^2 W\left(\delta_x, \delta_y, \delta_z, \tau\right) \mathrm{d}\delta_x \mathrm{d}\delta_y \mathrm{d}\delta_z = \chi_X^2 \tau \ ,$$

and likewise for the other coordinates. Integrating both sides of the truncated expansion with respect to

$$\iiint \ldots W\left(\delta_x, \delta_y, \delta_z, \tau\right) \mathrm{d}\delta_x \mathrm{d}\delta_y \mathrm{d}\delta_z \ ,$$

we deduce that

$$
\begin{aligned}
\Delta(x,y,z,t+\tau,t) \approx &-\frac{\partial}{\partial x} P(x,y,z,t) v_x \tau - \frac{\partial}{\partial y} P(x,y,z,t) v_y \tau \\
&-\frac{\partial}{\partial z} P(x,y,z,t) v_z \tau + \frac{1}{2} \frac{\partial^2}{\partial x^2} P(x,y,z,t) \chi_x^2 \tau \\
&+\frac{1}{2} \frac{\partial^2}{\partial y^2} P(x,y,z,t) \chi_y^2 \tau + \frac{1}{2} \frac{\partial^2}{\partial z^2} P(x,y,z,t) \chi_z^2 \tau \\
&+ \text{ terms in } \tau^2 \ ,
\end{aligned}
$$

where $\Delta(x,y,z,t+\tau,t) = P(x,y,z,t+\tau) - P(x,y,z,t)$. Moreover,

$$P(x,y,z,t+\tau) - P(x,y,z,t) \approx \frac{\partial}{\partial \tau} P(x,y,z,t)\tau \ .$$

Since τ is taken to be small, we can neglect terms in τ^2. We thereby obtain the diffusion equation:

$$\frac{\partial}{\partial t}P(x,y,z,t) = -v_x\frac{\partial}{\partial x}P(x,y,z,t) - v_y\frac{\partial}{\partial y}P(x,y,z,t)$$

$$-v_z\frac{\partial}{\partial z}P(x,y,z,t) + \frac{\chi_x^2}{2}\frac{\partial^2}{\partial x^2}P(x,y,z,t)$$

$$+\frac{\chi_y^2}{2}\frac{\partial^2}{\partial y^2}P(x,y,z,t) + \frac{\chi_z^2}{2}\frac{\partial^2}{\partial z^2}P(x,y,z,t) \ .$$

The two terms in this equation correspond to two different physical phenomena. The first,

$$-v_x\frac{\partial P(x,y,z,t)}{\partial x} - v_y\frac{\partial P(x,y,z,t)}{\partial y} - v_z\frac{\partial P(x,y,z,t)}{\partial z} \ ,$$

represents the drift due to the nonzero value of the mean motion, whilst the second,

$$\frac{\chi_x^2}{2}\frac{\partial^2 P(x,y,z,t)}{\partial x^2} + \frac{\chi_y^2}{2}\frac{\partial^2 P(x,y,z,t)}{\partial y^2} + \frac{\chi_z^2}{2}\frac{\partial^2 P(x,y,z,t)}{\partial z^2} \ ,$$

represents true diffusion.

To simplify the following discussion, we assume that the diffusion is isotropic, i.e., $\chi_X = \chi_Y = \chi_Z = \chi$, and that there is no drift, i.e., $v_X = v_Y = v_Z = 0$. We then have

$$\frac{\partial P(x,y,z,t)}{\partial t} = \frac{\chi^2}{2}\Delta P(x,y,z,t) \ ,$$

where

$$\Delta P(x,y,z,t) = \frac{\partial^2}{\partial x^2}P(x,y,z,t) + \frac{\partial^2}{\partial y^2}P(x,y,z,t) + \frac{\partial^2}{\partial z^2}P(x,y,z,t) \ .$$

If we multiply $P(x,y,z,t)$ by the number N of particles, we obtain the diffusion equation for the local concentration of particles. The heat propagation equation is completely equivalent. Indeed, under analogous hypotheses, the temperature $T(x,y,z,t)$ at the coordinate point (x,y,z,t) at time t evolves according to

$$\frac{\partial T(x,y,z,t)}{\partial t} = \frac{\chi^2}{2}\Delta T(x,y,z,t) \ .$$

It is of great practical importance to be able to write the dynamical equation for the concentration as a partial differential equation. Indeed, in a formal sense, it becomes possible to solve any problem once the boundary conditions have been specified. We may say that this equation provides a model for the macroscopic evolution, whereas the microscopic equations are only involved through the appearance of (v_x, v_y, v_z) and (χ_x, χ_y, χ_z). It is important to grasp the phenomenological nature of this equation. Indeed, if the diffusion equation is applied to periods of time too short or distances that are too large certain paradoxes arise. For suppose that, at time $t = 0$, the molecule concentration is represented by a Dirac distribution $\delta(x, y, z)$ centered at the origin. The solution of the diffusion equation shows us that, for any positive time ε, no matter how small, there is a nonzero probability of finding molecules at a distance L as large as we wish. This equation is thus compatible with molecular displacements at speeds faster than the speed of light. The diffusion equation must therefore be applied only within its field of validity. It should thus be remembered that, in order to justify taking only first and second order terms in our expansion, we assumed that the time step was not too small and the displacements not too large.

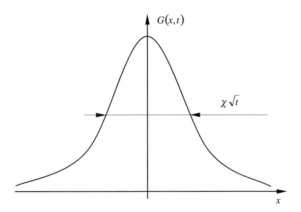

Fig. 4.11. Green function for the diffusion equation in the 1-dimensional case

We saw in Section 3.12 that, given the Green function for a partial differential equation, it is easy to determine the solution for arbitrary initial conditions in an infinite medium. For the case of an infinite, homogeneous and isotropic medium without drift, the Green function of the diffusion equation (see Fig. 4.11) can be written

$$G(x, y, z, t, x', y', z', t') = \frac{1}{\left[\chi\sqrt{2\pi(t - t')}\right]^3} \exp\left[-\frac{\|\boldsymbol{r} - \boldsymbol{r}'\|^2}{2\chi^2(t - t')}\right],$$

where we assume that $t > t'$ and $\|\boldsymbol{r} - \boldsymbol{r}'\|^2 = (x - x')^2 + (y - y')^2 + (z - z')^2$.

4.8 Random Walks and Space Dimensions

Just as for the random walk, we can immediately generalize the last result to the case of diffusion in a space of dimension D. We consider the homogeneous and isotropic case without drift. Putting $\boldsymbol{R} = (r_1, r_2, \ldots, r_D)^{\mathrm{T}}$, we obtain

$$\frac{\partial P(\boldsymbol{R}, t)}{\partial t} = \frac{\chi^2}{2} \Delta P(\boldsymbol{R}, t) \ ,$$

with

$$\Delta P(\boldsymbol{R}, t) = \Delta P(r_1, r_2, \ldots, r_D, t) = \sum_{j=1}^{D} \frac{\partial^2 P(r_1, r_2, \ldots, r_D, t)}{\partial r_j^2} \ .$$

The probability density function $P(\boldsymbol{R}, t)$ for finding a particle at the point with coordinates $\boldsymbol{R} = (r_1, r_2, \ldots, r_D)^{\mathrm{T}}$ at time t given the initial conditions

$$P(r_1, r_2, \ldots, r_D, 0) = \delta(r_1)\delta(r_2)\ldots\delta(r_D) \ ,$$

where $\delta(x)$ represents the Dirac distribution, is

$$P(\boldsymbol{R}, t) = \left(\frac{1}{\sqrt{2\pi\chi^2 t}}\right)^D \exp\left(-\frac{1}{2\chi^2 t}\|\boldsymbol{R}\|^2\right) \ ,$$

where

$$\|\boldsymbol{R}\|^2 = \sum_{j=1}^{D}(r_j)^2 \ .$$

This is the approximation obtained at the end of the section on random walks. It leads to the same result as the Green function of the diffusion equation. There is nothing surprising about this, since the underlying physical model chosen to describe diffusion was indeed the random walk model. We will reconsider this approximation below.

Let us analyze a random walk experiment starting out from the origin and in which we take a snapshot of the particle in its current position every dt seconds, where the time interval dt is chosen arbitrarily (see Fig. 4.12). Imagine now that we repeat this experiment N times, where N is some large number. We therefore expect the probability P_n of finding the particle in the neighborhood dV (assumed small) of the origin on one of the N photos at a time ndt to be approximately

$$P_n = \left(\frac{1}{\sqrt{2\pi\chi^2 ndt}}\right)^D dV \ .$$

We assume that dV is small enough to ensure that

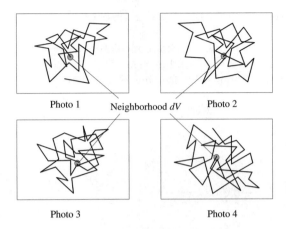

Fig. 4.12. Random walk experiment starting at the origin and taking photos of the particle position at time intervals dt

$$\exp\left[-\left(\sum_{j=1}^{D}(r_j)^2\right)/2\chi^2 t\right] \approx 1$$

for any point $\boldsymbol{R}$ in dV. NP_n represents the number of experiments for which the particle turns up in the neighborhood dV of the origin at time ndt. The number of times $N(t_0)$ the particle returns to the origin for times t greater than or equal to $t_0 = n$dt is thus $\sum_{j=n}^{\infty} NP_j$. Assuming now that dt is small enough to ensure that

$$N(t_0) = \sum_{j=n}^{\infty} NP_j \approx \frac{N\mathrm{d}V}{\mathrm{d}t}\int_{t_0}^{\infty}\left(\frac{1}{\sqrt{2\pi\chi^2 t}}\right)^{D}\mathrm{d}t\,,$$

we then obtain

$$N(t_0) \approx \frac{N\mathrm{d}V}{\mathrm{d}t\left(\sqrt{2\pi\chi^2}\right)^{D}}\int_{t_0}^{\infty} t^{-D/2}\mathrm{d}t\,.$$

Setting

$$I_D(t_0) = \int_{t_0}^{\infty} t^{-D/2}\mathrm{d}t\,,$$

we see finally that there are two cases:

- if $D \leqslant 2$, then $I_D(t_0)$ diverges, i.e., $I_D(t_0)$ tends to infinity,
- if $D > 2$, then $I_D(t_0)$ converges and we have $\lim_{t_0\to\infty} N(t_0) = 0$.

We therefore conclude that the number of times $N(t_0)$ that the particle returns to the origin for times t greater than or equal to t_0 diverges if $D \leqslant 2$ and converges if $D > 2$. So this feature depends on the dimension of the space in which the random walk takes place. Consequently, we expect qualitatively and quantitatively different results if $D \leqslant 2$ or $D > 2$. It can indeed be shown more rigorously that, for a random walk on a discrete lattice in a space of dimension D, the probability that the particle returns to the origin is equal to 1 if $D \leqslant 2$ and is strictly less than 1 if $D > 2$.

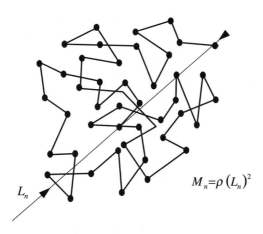

Fig. 4.13. A mass m is placed at each step of a random walk and the total mass M_n then varies as the square of the characteristic size L_n of the cluster

The fact that the dimension $D = 2$ plays a special role can also be understood in a slightly different way. Consider a homogeneous and isotropic random walk in discrete time steps with no drift. Suppose that a mass m is placed at each site occupied during the random walk. Suppose also that successively occupied sites are linked by a line segment. After n steps, the mass M_n of the chain formed in this way is therefore nm and its characteristic length is

$$L_n \approx \sqrt{\langle \|R_{n,\lambda}\|^2 \rangle} = \sigma \sqrt{n} \,,$$

where σ is a constant (see Fig. 4.13). We thus obtain

$$M_n = \frac{m}{\sigma^2} (L_n)^2 \,.$$

In a space of dimension D, the mass of a homogeneous cube of density ρ and side L is ρL^D. For the random walk, setting $\rho = m/\sigma^2$, we have $M_n = \rho(L_n)^2$. From this standpoint, we can thus say that the intrinsic dimension of the random walk is 2, regardless of the dimension of the space in which it takes place (if $D > 2$).

We may thus put forward the following picture of the results we have just described. If the dimension of the space is strictly greater than the intrinsic dimension of the random walk, then this random walk intersects itself much less often than in the other case.

4.9 Rare Events and Particle Noise

The Poisson distribution provides a simple model describing the number of particles reaching a detector in a given time interval τ. This result turns up in many different areas of physics. In optics, for example, we can describe the number of photons detected in a given time interval, or in radioactivity, the number of particles reaching the detector per unit time. To construct this distribution, we assume that the particle flux is stationary, i.e., independent of time. More precisely, if $N(t, t + \mathrm{d}t)$ represents the mean number of detected particles (in the sense of expectation values) over the time interval $[t, t + \mathrm{d}t]$, the flux at time t is expressed in the form

$$\Phi(t) = \lim_{\mathrm{d}t \to 0} \frac{\mathrm{d}N(t, t + \mathrm{d}t)}{\mathrm{d}t} .$$

The flux is therefore stationary if $\Phi(t)$ is independent of t and it will then be written simply as Φ. Over a small time interval $\mathrm{d}t$, the mean number of detected particles is $\Phi\mathrm{d}t$. Let p_n be the probability of detecting n particles during this time interval. When $\mathrm{d}t$ is very small, we expect to have $p_0 \gg p_1 \gg p_2$, etc. In the limit as $\mathrm{d}t \to 0$, we may consider that only the two events corresponding to detection of no particles or detection of a single particle are not negligible (see Fig. 4.14). In this case, the number of detected particles in the time interval $\mathrm{d}t$ obeys a Bernoulli law with parameter $\Phi\mathrm{d}t$. (Indeed, we have already seen that the parameter of a Bernoulli distribution is equal to its mean.) The characteristic function can thus be written

$$\Psi_{\mathrm{d}t}(\nu) = (1 - \Phi\mathrm{d}t) + \Phi\mathrm{d}t\, \mathrm{e}^{\mathrm{i}\nu} .$$

We may now determine the probability $P_\tau(n)$ that n particles are detected in a time interval of length τ. Indeed, we divide this interval into N segments of very short length $\mathrm{d}t$, so that $\tau = N\mathrm{d}t$. The number of particles detected in the interval of duration τ is the sum of the Bernoulli variables which describe the number of particles detected in each time interval of length $\mathrm{d}t$. In the limit as $\mathrm{d}t$ tends to zero, the characteristic function of $P_\tau(n)$ is therefore

$$\Psi_\tau(\nu) = \lim_{N \to \infty} \left[\Psi_{\tau/N}(\nu) \right]^N ,$$

or

$$\Psi_\tau(\nu) = \lim_{N \to \infty} \left[\left(1 - \Phi\frac{\tau}{N} \right) + \Phi\frac{\tau}{N}\mathrm{e}^{\mathrm{i}\nu} \right]^N .$$

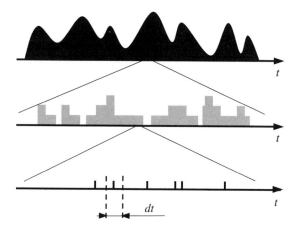

Fig. 4.14. The time interval is chosen small enough to ensure that there is a negligible probability of the number of detection events being greater than 1

Now $\lim_{N\to\infty}\left(1-a/N\right)^N = e^{-a}$ so, using the fact that

$$\left[\left(1-\frac{\Phi\tau}{N}\right)+\frac{\Phi\tau e^{i\nu}}{N}\right]^N = \left[1+\frac{\Phi\tau\left(e^{i\nu}-1\right)}{N}\right]^N ,$$

this implies that

$$\Psi_\tau(\nu) = \exp\left[-\Phi\tau\left(1-e^{i\nu}\right)\right] .$$

This corresponds to the characteristic function of a Poisson distribution with parameter $\Phi\tau$. We thus have

$$P_\tau(n) = e^{-\Phi\tau}\frac{\left(\Phi\tau\right)^n}{n!} ,$$

where $n! = n(n-1)(n-2)\ldots 3 \times 2 \times 1$.

The mean and variance of the Poisson distribution defined by $P(n) = e^{-\mu}\mu^n/n!$ are both equal to μ. We have seen that the characteristic function associated with this law is $\Psi(\nu) = \exp\left[-\mu\left(1-e^{i\nu}\right)\right]$. We thus see that, if we add ℓ random variables that are identically distributed according to $p(n)$, we obtain a Poisson variable with mean and variance equal to $\ell\mu$. This result is easy to understand in the context of the physical interpretation we have just discussed. It should not be thought, however, that the Poisson distribution is stable. Indeed, we know that a probability density function $P_X(x)$ corresponds to a stable probability law if, when we add together two independent variables $X_{\lambda(1)}$ and $X_{\lambda(2)}$, identically distributed according to $P_X(x)$, there are two numbers a and b such that $\left(X_{\lambda(1)}+X_{\lambda(2)}\right)/a + b$ is distributed according to $P_X(x)$. To obtain the same mean when we add together two identically distributed Poisson variables, we would have to divide the result of the sum by 2. In this case, we would no longer have a Poisson variable since the result would not necessary be integer-valued.

However, when μ is large, a good approximation to the characteristic function $\Psi(\nu) = \exp\left[-\mu\left(1 - e^{i\nu}\right)\right]$ is $\exp\left(i\mu\nu - \mu\nu^2/2\right)$ which corresponds to the characteristic function of a Gaussian variable with mean and variance equal to μ. Indeed, if μ is large, $\exp\left[-\mu\left(1 - e^{i\nu}\right)\right]$ is only non-negligible if ν is close to 0. Now in this case, $e^{i\nu} \approx 1 + i\nu - \nu^2/2$ and hence, $\Psi(\nu) = \exp\left[-\mu\left(1 - e^{i\nu}\right)\right] \approx \exp\left[-\mu\left(-i\nu + \nu^2/2\right)\right]$. A sum of Poisson variables thus converges to a Gaussian variable. In other words, in high fluxes, Poisson noise is equivalent to Gaussian noise. Note, however, that the variance of this Gaussian distribution will be equal to its mean, as always happens with a Poisson distribution. It is interesting to compare this result with the random walk where the variance and mean are also proportional.

4.10 Low Flux Speckle

In this section, we consider the example of fully developed speckle measured in the presence of a low photon flux. This will clearly illustrate how we should apply the ideas introduced above.

In Section 4.5, we saw that speckle noise is obtained when we illuminate an object with a coherent wave, such as a laser beam. This noise is manifested through a very grainy appearance in which the grain size is determined by the experimental setup. Consider the case where the detector integrates the signal over a region that is smaller than the grain size. We have seen that a good approximation is obtained by describing this speckle noise, assumed homogeneous, by a multiplicative noise factor with exponential probability density function. The probability density function of the intensity is then

$$P_I(I) = \frac{1}{I_0} \exp\left(-\frac{I}{I_0}\right) .$$

Suppose further that the photon flux is stationary and very low. In Section 4.9, we saw that the number of photons measured over a time interval τ is a random variable N_μ whose probability distribution can be accurately described by a Poisson law, viz.,

$$P_\tau(n) = e^{-\Phi\tau}\frac{(\Phi\tau)^n}{n!} .$$

We now write this law with the notation $\Phi\tau = \gamma I$, where I is the mean intensity received at the detector and γ is a coefficient depending on the surface properties, the efficiency of the detector and the time interval τ. We now observe that, if we consider that the speckle pattern is projected onto a detector whose position is random, then the intensity I is itself a random variable that we shall write in the form I_λ. The random variables N_μ and I_λ arise from very different phenomena. The Poisson noise reflects fluctuations in the number of photons detected as a function of time, whilst speckle noise reflects fluctuations as a function of the spatial coordinates. Moreover, the Poisson noise is assumed to be stationary, whereas the speckle is assumed to be homogeneous. We thus write

$$P_\tau(n|I) = e^{-\gamma I}\frac{(\gamma I)^n}{n!} .$$

The probability law resulting from the combination of these two random phenomena is therefore

$$P_\tau(n) = \int_0^{+\infty} P_\tau(n|I)P_I(I)\mathrm{d}I ,$$

where the argument here is analogous to the one in Section 4.5. We can then write

$$P_\tau(n) = \int_0^{+\infty} e^{-\gamma I}\frac{(\gamma I)^n}{n!}\frac{1}{I_0}\exp\left(-\frac{I}{I_0}\right)\mathrm{d}I .$$

It is now easy to obtain

$$P_\tau(n) = \frac{1}{1+\gamma I_0}\left(\frac{\gamma I_0}{1+\gamma I_0}\right)^n .$$

We do this as follows. Setting $x = I/I_0$ and $\alpha = \gamma I_0$, we have

$$P_\tau(n) = \frac{\alpha^n}{n!}\int_0^{+\infty}\exp(-\alpha x)x^n\exp(-x)\mathrm{d}x .$$

If we put $J_n = \int_0^{+\infty}\exp\left[-(1+\alpha)x\right]x^n\mathrm{d}x$, we find that

$$P_\tau(n) = \frac{\alpha^n}{n!}J_n .$$

Moreover, it is easy to check that $J_n = n!/(1+\alpha)^{n+1}$ and hence that

$$P_\tau(n) = \frac{1}{1+\alpha}\left(\frac{\alpha}{1+\alpha}\right)^n ,$$

as required.

A simple calculation[2] is enough to show that

$$\langle N\rangle = \gamma I_0 ,$$

and that

$$\sigma_N^2 = \langle N^2\rangle - \langle N\rangle^2 = \gamma I_0(1+\gamma I_0) .$$

At very low fluxes, we have $\gamma I_0 \ll 1$ and hence $\sigma_N^2 \approx \gamma I_0$, which corresponds to the variance of the Poisson noise. The main source of fluctuations is thus the Poisson noise. In high fluxes, we have $\gamma I_0 \gg 1$ and hence $\sigma_N^2 \approx (\gamma I_0)^2$, which corresponds to the variance of the speckle noise. The Poisson noise is then negligible in comparison with the fluctuations due to speckle.

---

[2] A more complete notation would be $N_{\mu,\lambda}$. In order to simplify, we have not indicated the random events μ and λ.

Exercises

To simplify the notation, we drop explicit mention of the dependence of random variables on random events λ in these exercises.

Exercise 4.1. Sum of Gaussian Variables

Consider two independent realizations X_1 and X_2 of two random variables that are identically distributed according to the same Gaussian probability density function. Determine the probability of $X = (X_1 + X_2)/2$.

Exercise 4.2. Noise and Filtering

Let B_i be noise samples, where $i \in \mathbb{Z}$ and $\mathbb{Z}$ is the set of positive and negative integers. Assume that B_i is a sequence of independent random numbers with zero mean and values uniformly distributed over the interval $[-\alpha, \alpha]$. Hence,

$$\langle B_i \rangle = 0 , \quad \langle B_i B_j \rangle = \sigma_B^2 \delta_{(i-j)} \quad \text{where} \quad \delta_{(i-j)} = \begin{cases} 1 & \text{if } i = j , \\ 0 & \text{otherwise} , \end{cases}$$

and where $\langle \ \rangle$ represents the expectation value operator with respect to the various realizations B_i of the noise.

(1) Calculate σ_B^2 as a function of α.

Suppose that another sequence S_i is obtained by averaging the B_i according to the rule

$$S_i = \frac{1}{\sqrt{N}} \sum_{j=1}^{N} B_{i+j} .$$

(2) What is the covariance function of S_i ?
(3) What is the probability density function of S_i when $N = 2$? Express the result as a function of σ_B^2.
(4) What is the probability density function of S_i as $N \to +\infty$?

Consider now

$$S_i = \frac{1}{\sqrt{N}} \sum_{j=1}^{N} a_j B_{i+j} ,$$

where $\sum_{j=1}^{N} |a_j|^2 = N$.

(5) What is the probability density function of S_i when $N \to +\infty$ if $\forall j$, $|a_j|^2 / \left(\sum_{i=1}^{N} |a_i|^2 \right) \to 0$ when $N \to +\infty$?
(6) S_i results from a linear filtering of B_i. $Y_i = |S_i|^2$ represents the power of the signal after filtering. What is the probability density function of Y_i when $N \to +\infty$?

Exercise 4.3. Particle Noise

Consider a highly simplified system consisting of a vacuum tube containing an anode and a cathode. An ammeter is connected in series and we wish to characterize the fluctuations we will measure as a function of the mean value of the current passing through the tube, denoted $\langle I \rangle$. We also assume that the ammeter carries out the measurement over a time interval τ. To devise a simple model, we divide this time interval τ into N sub-intervals of duration $\delta\tau$, so that $\tau = N\delta\tau$. The duration $\delta\tau$ is assumed to be small enough to ensure that $p + q = 1$, where p is the probability that one electron goes through and q the probability that no electrons go through.

(1) Explain why

$$p(m) = \frac{N!}{(N-m)!m!}p^m q^{N-m}$$

is the probability that, during the time interval τ, exactly m electrons will go through the tube, where $m! = m \times (m-1)\ldots 3 \times 2$.
(2) Calculate the first two moments $\langle m \rangle$ and $\langle m^2 \rangle$.
(3) What happens to $\langle m \rangle$ and $\langle m^2 \rangle$ when we take the limit $N \to \infty$ whilst holding Np constant (which thus implies that $p \to 0$ as $\langle m \rangle/N$).
(4) Given that $I = me/\tau$, where e is the charge of the electron, calculate $\langle I^2 \rangle - \langle I \rangle^2$ in terms of e, $\langle I \rangle$ and τ.

Exercise 4.4. Polarization and Speckle

Consider a point optical detector which can measure the intensity in the vertical or horizontal polarization states of a light signal. We assume that these two intensities are described by independent random variables I_{hor} and I_{ver} with the same exponential probability density function, viz.,

$$P_{I_{\mathrm{hor}}}(I) = P_{I_{\mathrm{ver}}}(I) = \begin{cases} \dfrac{1}{a}\exp\left(-\dfrac{I}{a}\right) & \text{if } I > 0 \,, \\ 0 & \text{otherwise} \,. \end{cases}$$

(1) What does a represent?
(2) Defining the total intensity X and the polarization intensity Y by

$$X = I_{\mathrm{hor}} + I_{\mathrm{ver}} \,, \quad Y = I_{\mathrm{hor}} - I_{\mathrm{ver}} \,,$$

determine the probability density functions of X and Y.

Exercise 4.5. Random Walk

Consider a stationary random walk on a lattice in 1 dimension. For each step, let p be the probability of taking a step of size 1 unit, s the probability of staying put (i.e., taking a step of amplitude 0), and q the probability of making a step of amplitude -1. Then $p + q + s = 1$.

(1) Determine the characteristic function of the random variable R_n representing the position at step n.
(2) Suggest two ways of determining the variance of R_n.

Exercise 4.6. Random Walk

Consider a continuous random walk in discrete time steps. Let $P(r)$ be the probability of taking a step of amplitude r and R_n the position at step n. Discuss the difference in the asymptotic behavior for large n when

(1) $P(r) = \dfrac{1}{2} \exp(-|r|)$, where $|r|$ represents the absolute value of r,

(2) $P(r) = \dfrac{1}{\pi}(1 + r^2)^{-1}$.

Exercise 4.7. Diffusion

Consider a particle diffusion problem in 1 dimension. Let $P(x,t)$ be the probability density function for finding a particle at point x at time t. The Green function for the problem is

$$G_X(x,t) = \frac{1}{\sqrt{2\pi t}\sigma} \exp\left(-\frac{x^2}{2\sigma^2 t}\right) .$$

Initial conditions are defined by the sum of Dirac distributions

$$P(x,0) = \sum_{n=-\infty}^{+\infty} \delta(x - na) .$$

(1) Determine the probability density as a function of time, but without calculating the sums.
(2) What would be the probability density as a function of time if the particles diffused over a circle of radius R and the initial conditions were

$$P(x,0) = \delta(x) .$$

Exercise 4.8. Random Walk with Jumps

The aim here is to model random walks in which large jumps may occur sporadically. For example, we might think of a flea which walks for a while, then takes a jump, walks a bit more, then takes another jump, and so on. To simplify, we consider this random walk in 1 dimension. We write simply

$$Z_i = X_i + \sum_{\ell=1}^{L} Y_{i,\ell} ,$$

where the X_i are random variables distributed according to the Cauchy probability density function

$$P_X(x) = \frac{a}{\pi} \frac{1}{x^2 + a^2} \ ,$$

and where the $Y_{i,\ell}$ are random variables distributed according to the Gaussian probability density function

$$P_Y(y) = \frac{1}{\sqrt{2\pi}\sigma} \exp\left(-\frac{y^2}{2\sigma^2}\right) \ .$$

(1) Express the characteristic function of Z_i in terms of the characteristic functions of X_i and $Y_{i,\ell}$.
(2) Setting

$$R_n = \frac{1}{n} \sum_{i=1}^{n} Z_i \ ?$$

express the characteristic function of R_n in terms of the characteristic functions of X_i and $Y_{i,\ell}$.
(3) What happens when $n \to +\infty$?

Exercise 4.9. Product of Random Variables

In this example, we multiply together strictly positive random variables. We write simply

$$Y_n = \prod_{i=1}^{n} X_i \ ,$$

where the X_i are random variables distributed according to the probability density function $P_X(x)$.

(1) Determine the asymptotic probability distribution of Y_n and give a condition for the validity of this expression.
(2) Generalize this result to the case of nonzero random variables for which the probability of the sign is independent of the probability density of the modulus.

5

Information and Fluctuations

It is hardly necessary to point out that information has become a pillar of modern society. The concept of information made its entry into the exact sciences only relatively recently, since it was formalized in the years 1945–1948 by Shannon in order to tackle the technical problems of communication. In actual fact, it was already implicitly present in the idea of entropy introduced by Boltzmann at the end of the nineteenth century. As we shall see, like the idea of stationarity, entropy does not characterize a particular realization, but rather the whole set of possible realizations. In contrast, Kolmogorov complexity is defined for each realization and we may give an intuitive meaning to the idea of fluctuation or randomness that we would sometimes like to attribute to a series of observations.

5.1 Shannon Information

Shannon sought to define an objective measure of information that would prove useful in the exact sciences. The basic idea consists in quantifying the information carried by a realization of an event in such a way that the measure depends only on the probability of that event actually occurring. In everyday life, it is easy to see that a piece of information seems to gain in importance as its probability becomes smaller. In reality, our interest in a piece of information is intimately tied up with social and psychological factors, or simply the profit we may obtain from it, but these features are not taken into account in Shannon's theory. Consider the following two random events:

- the temperature at the base of the Eiffel tower was 5°C at 12h on 1 January,
- the temperature at the base of the Eiffel tower was 20°C at 12h on 1 January.

We shall simply say that the realization of the second event contains more information than the realization of the second, because it is less probable.

In order to construct a rigorous theory, we must begin by analyzing the simple case in which the set Ω of all possible events is finite. As in the first chapter, these events will be denoted λ_i, where $i = 1, 2, \ldots, N$, if there are N possible events, and hence $\Omega = \{\lambda_1, \lambda_2, \ldots, \lambda_N\}$. We assume that we may assign a probability p_i to each random event λ_i, where $\sum_{i=1}^{N} p_i = 1$. Shannon suggested defining the quantity of information contained in the realization of the random event λ_i by

$$Q_i = - \ln p_i \ .$$

We shall simply call this the information content of λ_i. If ln is the natural logarithm, the units of the measure are nats; if it is the logarithm to base 2, the units are bits.

This definition does indeed satisfy the condition that Q_i should increase as p_i decreases. It has a second advantage if we consider the simultaneous realization of independent events. To this end, consider two sets of random events:

$$\Omega^{(1)} = \left\{\lambda_1^{(1)}, \lambda_2^{(1)}, \ldots, \lambda_N^{(1)}\right\} \quad \text{and} \quad \Omega^{(2)} = \left\{\lambda_1^{(2)}, \lambda_2^{(2)}, \ldots, \lambda_N^{(2)}\right\} \ ,$$

each equipped with a probability law $p_j^{(1)}$ for the events $\lambda_j^{(1)}$ and $p_j^{(2)}$ for the events $\lambda_j^{(2)}$. As the events are assumed independent, the probability of observing $\lambda_j^{(1)}$ and $\lambda_\ell^{(2)}$ simultaneously is simply $p_j^{(1)} p_\ell^{(2)}$. The information content of the joint event $\left(\lambda_j^{(1)}, \lambda_\ell^{(2)}\right)$ is thus

$$Q_{j,\ell} = - \ln \left[p_j^{(1)} p_\ell^{(2)} \right] \ .$$

Since the events are independent, we expect the total information content to be the sum of the information contents of the two components, i.e., $Q_{j,\ell} = Q_j + Q_\ell$, and this is indeed the case with Shannon's definition.

It is worth noting that the information content carried by an event which is certain in the probabilistic sense, so that $p_i = 1$, is actually zero, i.e., $Q_i = 0$. In contrast, the information content carried by a very unlikely event can be arbitrarily large, i.e.,

$$\lim_{p_i \to 0} Q_i = +\infty \ .$$

At this stage, the definition of Shannon information may look somewhat arbitrary. Here again, if we gauge the interest of a definition by the relevance of the results it generates, there can be no doubt that this definition is extremely productive. It underlies the mathematical theory of information transmission and coding systems. It can be used to set up highly efficient techniques for optimizing such systems. However, since our objective is to characterize the fluctuations in physical systems and estimate physical quantities in the presence of fluctuations, we shall not emphasize this interesting and important aspect of information theory. We shall instead focus on applications in classical physics.

5.2 Entropy

The entropy of a set of random events Ω is defined as the mean quantity of information it can provide, viz.,

$$S(\Omega) = \sum_{j=1}^{N} p_j Q_j = - \sum_{j=1}^{N} p_j \ln p_j \ .$$

It is easy to see that the entropy is a positive quantity. It is zero if only one event has a nonzero probability. Indeed, we have $\lim_{p_j \to 0}(p_j \ln p_j) = 0$ and $1 \ln 1 = 0$. Note also that, if the events are equiprobable, we have $p_j = 1/N$ for each value of j and hence

$$S(\Omega) = \ln N \ .$$

The entropy is an extensive quantity. Indeed, the entropy of a pair of independent random variables is equal to the sum of their respective entropies. Consider the random event Λ comprising the pair of independent random events (λ, μ). Let Ω_1 and Ω_2 be the sets of random events λ and μ, respectively. The set Ω_T of random events $\Lambda = (\lambda, \mu)$ is the Cartesian product of Ω_1 and Ω_2, i.e., $\Omega_T = \Omega_1 \times \Omega_2$. By the independence of λ and μ, $P(\Lambda) = P(\lambda)P(\mu)$. The entropy of Ω_T is

$$S(\Omega_T) = - \sum_{\Lambda \in \Omega_T} P(\Lambda) \ln P(\Lambda) \ .$$

We thus deduce that

$$S(\Omega_T) = - \sum_{\lambda \in \Omega_1} \sum_{\mu \in \Omega_2} P(\lambda, \mu) \ln P(\lambda, \mu) \ ,$$

or

$$S(\Omega_T) = - \sum_{\lambda \in \Omega_1} \sum_{\mu \in \Omega_2} P(\lambda, \mu) \left[\ln P(\lambda) + \ln P(\mu) \right] \ .$$

Using

$$P(\lambda) = \sum_{\mu \in \Omega_2} P(\lambda, \mu) \quad \text{and} \quad P(\mu) = \sum_{\lambda \in \Omega_1} P(\lambda, \mu) \ ,$$

we then have

$$S(\Omega_T) = S(\Omega_1) + S(\Omega_2) \ ,$$

where $S(\Omega_1) = - \sum_{\lambda \in \Omega_1} P(\lambda) \ln P(\lambda)$ and $S(\Omega_2) = - \sum_{\mu \in \Omega_2} P(\mu) \ln P(\mu)$.

We can now give a simple interpretation of the Shannon entropy by considering experiments in which we observe the realization of L independent random events arising from Ω. We thus form a new random event denoted Λ_L, which takes its values in the set $\Theta = \Omega \times \Omega \times \ldots \times \Omega$, where $\Omega \times \Omega$ is the Cartesian product of Ω with itself and where Ω appears L times in the expression for Θ. In other words, and to put it more simply, the new random events are the sequences $\Lambda_L = \{\lambda(1), \lambda(2), \ldots, \lambda(L)\}$, where $\lambda(n)$ is the nth independent realization of an event in Ω. [It is important to distinguish the notation $\lambda(j)$ and λ_j. Indeed, λ_j is the jth element of Ω, where $\Omega = \{\lambda_1, \lambda_2, \ldots, \lambda_N\}$, whereas $\lambda(j)$ is the jth realization in the sequence $\Lambda_L = \{\lambda(1), \lambda(2), \ldots, \lambda(L)\}$.] The number N_L of different sequences Λ_L for which each event λ_j, representing the jth event of Ω, appears ℓ_j times is

$$N_L = \frac{L!}{\ell_1! \ell_2! \ldots \ell_N!} ,$$

where $\ell! = 1 \times 2 \times 3 \times \ldots \times (\ell - 1) \times \ell$. Indeed, consider first the event λ_1. There are L positions in which to place the first event λ_1. There then remain $L - 1$ for the second event λ_1, and so on. We thus see that there are

$$L(L-1)\ldots(L - \ell_1 + 1) = \frac{L!}{(L - \ell_1)!}$$

in which to place the ℓ_1 events λ_1 in the sequence of L realizations of independent events. However, with the previous argument, two sequences which differ only by a permutation of the events λ_1 are considered to be different. Since there exist $\ell_1!$ permutations of the events λ_1, the number of sequences that are truly different, in which the event λ_1 appears ℓ_1 times, is

$$\frac{L!}{(L - \ell_1)! \ell_1!} .$$

There are now $L - \ell_1$ places for the events λ_2. For the given positions of the events λ_1, there are therefore

$$\frac{(L - \ell_1)!}{(L - \ell_1 - \ell_2!)! \ell_2!}$$

possibilities for placing the events λ_2. There are thus

$$\frac{L!}{(L - \ell_1)! \ell_1!} \frac{(L - \ell_1)!}{(L - \ell_1 - \ell_2!)! \ell_2!} = \frac{L!}{(L - \ell_1 - \ell_2)! \ell_1! \ell_2!}$$

possibilities for placing the events λ_1 and λ_2. Repeating this argument, it is easy to convince oneself that the number of different sequences is

$$\frac{L!}{\ell_1! \ell_2! \ldots \ell_N!} .$$

We now analyze the case where L is very large. For this we shall need the simplified Stirling approximation

$$\ln(\ell!) \approx \left(\ell + \frac{1}{2}\right) \ln \ell - \ell ,$$

whence

$$\ln N_L \approx \left(L + \frac{1}{2}\right)\ln L - L - \sum_{j=1}^{N}\left[\left(\ell_j + \frac{1}{2}\right)\ln\ell_j - \ell_j\right] .$$

Now $L = \sum_{j=1}^{N}\ell_j$ and hence,

$$\ln N_L \approx \left(L\ln L - \sum_{j=1}^{N}\ell_j \ln\ell_j\right) + \left(\frac{1}{2}\ln L - \frac{1}{2}\sum_{j=1}^{N}\ln\ell_j\right) .$$

When L is large, the second term is negligible compared with the first and hence

$$\ln N_L \approx \sum_{j=1}^{N}\ell_j \ln\left(\frac{L}{\ell_j}\right) .$$

We may thus write

$$\lim_{L\to\infty}\frac{\ln N_L}{L} = \lim_{L\to\infty}\left[-\sum_{j=1}^{N}\frac{\ell_j}{L}\ln\left(\frac{\ell_j}{L}\right)\right] .$$

In this case, the law of large numbers allows us to assert that the event λ_j will occur approximately $p_j L$ times, where p_j is the probability of the event λ_j occurring. We will thus have

$$\lim_{L\to\infty}\frac{\ln N_L}{L} = -\sum_{j=1}^{N}p_j \ln p_j ,$$

or

$$\lim_{L\to\infty}\frac{\ln N_L}{L} = S(\Omega) ,$$

or

$$N_L \approx \exp\left[S(\Omega)L\right] .$$

We now consider two extreme cases. Suppose to begin with that only the event λ_1 has nonzero probability of occurring. We have seen that the entropy is then minimal and zero. In this case, the number N_L of different sequences of random events is obviously equal to 1, since only the sequence $\lambda_1, \lambda_1, \lambda_1, \ldots, \lambda_1$ then has nonzero probability of occurring. Note that this result is consistent with what was said before, since $\ln N_L = 0$ and $S(\Omega) = 0$.

In contrast, if the set of random events is made up of n equiprobable events, we have $p_i = 1/n$ and the number N_L of different sequences of random events is then $N_L = n^L$. The entropy is easy to determine, and we obtain $S(\Omega) = -\sum_{j=1}^{N}(1/n)\ln(1/n) = \ln n$, which is in good agreement with $N_L \approx \exp\left[S(\Omega)L\right]$, since $N_L = n^L$ or $n^L = \exp(L\ln n)$.

To sum up, we observe that the entropy defined as Shannon's mean information content is indeed a characteriztic measure of the potential disorder of Ω, for it directly determines the number of different potential realizations that Ω can generate. We thus see that Shannon's choice for the information content leads to a useful definition of entropy. Indeed, the entropy is defined as the mean quantity of information that the source Ω of random events can generate. Moreover, we have just shown that it is directly related to the number of different sequences of independent realizations that we can observe.

5.3 Kolmogorov Complexity

As already mentioned, we sought to define a measure of information that is objective and depends only on the probabilities of the possible random events. The Shannon information content achieves this aim, but nevertheless suffers from a rather unfortunate limitation when we are concerned with some particular realization. In the 1960s, Kolmogorov and Chaitin found a way of elucidating this point by defining the concept of complexity for a sequence of characters. This idea is, however, of a rather theoretical nature. Indeed, it is not generally possible to determine, hence to measure, the Kolmogorov complexity for a given sequence of characters and we shall be concerned only with the basic ideas here.

Consider the four figures making up the secret code of a bank card. Suppose that these figures are drawn at random with the same probability, so that each figure has a probability of $1/10$ of being drawn. The two code numbers 9035 and 0000 thus contain the same quantity of information, since they have the same probability of occurring. Common sense would nevertheless lead us to consider that the second code was more likely to be discovered than the first. In reality, these codes will have the same probability of being discovered during a random search in which the figures are chosen with probability $1/10$. We shall see, however, that they do not have the same complexity in the sense of Kolmogorov. In other words, it is not because the first name of one of your friends and relations has the same probability as any other sequence of characters (where each character is chosen at random with the same probability) that it is a good idea to use it as the password on your computer.

In this section, we will be concerned only with sequence of binary numbers. Suppose that each binary sequence corresponds to a series of realizations of Bernoulli variables with parameter q, so that the probability of observing 0 is $1 - q$ and the probability of observing 1 is q. The two sequences

$$\Lambda_1 = 101010101010101010101010101010101010101010101010$$

and

$$\Lambda_2 = 110010001110101010011100010101000111010011010011$$

have the same probability $q^{24}(1 - q)^{24}$. However, it would be much easier to memorize the first sequence (Λ_1) than the second (Λ_2). This is simply because there is a very simple algorithm for generating the first sequence. This algorithm might be, for example,

> `Write 48 figures alternating between 0 and 1 and starting`
> `with 1.`

It is on the basis of this observation that we can define the notion of Kolmogorov complexity for a sequence of characters. The Kolmogorov complexity $K(\Lambda)$ of a sequence Λ is the length of the shortest program able to generate it. We may thus reasonably expect to obtain $K(\Lambda_1) < K(\Lambda_2)$, whereas if $Q(\Lambda_1)$ and $Q(\Lambda_2)$ denote the information contents of the sequences Λ_1 and $Q(\Lambda_2)$, we have $Q(\Lambda_1) = Q(\Lambda_2)$.

The Kolmogorov complexity of a sequence is maximal when there is no algorithm simpler than a full description of it term by term. In this case, if the length of the sequence Λ is n, we have $K(\Lambda) = n + \text{Const.}$, where the constant is independent of n. The complexity of a password corresponding to the first name of one of your friends and relations will certainly rate lower than Dr45k;D. It is because a hacker assumes that you will have chosen a password of low complexity that he or she will begin by trying the first names of those dear to you. The algorithm applied by the potential intruder could consist in trying all the first names of your friends and relations, together with their family names and dates of birth. If the hacker is well organized, he or she will run through all the nouns in a dictionary.

We also observe that, if we find a program of length p which can generate a sequence of length n with $p < n$, this means that we have managed to compress the sequence by a factor of n/p. For a given length n, the compression will be all the more efficient as p is small.

A good example of a situation in which we must seek a simple algorithm are the so-called logic tests which consist in completing a series of numbers. To begin with, the algorithm must reproduce the series of numbers, whereupon we may use it to predict the following numbers. Imagine now a more sophisticated test in which, as well as predicting the rest of the sequence, we first ask whether this will even be possible. The fact that we have not found a solution does not mean that there does not exist a solution. Hence, for these series, we cannot obviously assert at first sight that there does not exist a solution. It can be shown that finding the program of minimal length is a problem with no simple solution in general. Unfortunately, this significantly reduces the practical relevance of Kolmogorov complexity. However, it is worth noting that this definition still retains several points of interest. In the next section, we will be able to use it to define a degree of randomness, which is interesting even if we cannot always calculate it.

The other significant feature is the "philosophical" implication contained in the definition of Kolmogorov complexity. Indeed, we may consider that the aim of a physical theory is to sum up in the most concise manner possible everything that we observe. Let us take the case of electrostatics and imagine that we are interested in measuring the electric potential $V(x, y, z)$ in a region $\mathcal{D}$ of space located in a vacuum and bounded by a surface $\mathcal{S}$. If we do not know the law $\Delta V(x, y, z) = 0$, where $\Delta V(x, y, z) = \partial^2 V(x, y, z)/\partial^2 x + \partial^2 V(x, y, z)/\partial^2 y + \partial^2 V(x, y, z)/\partial^2 z$, then we must provide the whole set of values for the potential $V(x, y, z)$ in $\mathcal{D}$, in order to describe our experimental observations. On the other hand, if we know the theory contained so concisely in the equation $\Delta V(x, y, z) = 0$, we need only provide the values of the potential and its derivatives on the surface $\mathcal{S}$, together with a numerical program capable of solving the equation. We are thus faced with two options:

- to transmit the whole set of data comprising the potentials $V(x, y, z)$ in $\mathcal{D}$,
- to transmit the compressed form comprising the values of the potential $V(x, y, z)$ and its derivatives on $\mathcal{S}$ together with the program for solving the equation $\Delta V(x, y, z) = 0$ in $\mathcal{D}$.

Moreover, suppose that in these two cases we encode the data using binary numbers. We see then that the theory associated with $\Delta V(x, y, z) = 0$ reduces the length of the binary sequence to be transmitted.

We may thus say that the binary sequence representing the whole set of values of the potential $V(x, y, z)$ in $\mathcal{D}$ has a lower Kolmogorov complexity than the one we would have obtained if the observed data did not possess some kind of internal structure, in other words, if they could not be summed up concisely using a mathematical law. The theory $\Delta V(x, y, z) = 0$ thus shows that the complexity (in the sense of Kolmogorov) of the experimental observation is less than the complexity of data without internal structure.

The notion of physical theory is often linked to the idea of predictability. To be precise, it is generally considered that a good theory must not only allow us to describe results already obtained, but that it must also be able to predict new ones. In fact, the reasoning developed above applies once again. This is because, in physics, in order to be able to predict, we must develop a mathematical model which allows us to reduce the description of our observations, thus allowing us to reconstruct them from a smaller number of values. In the Kolmogorov approach, the complexity will be lower if the description is reduced. This agrees with the widely accepted principle according to which a good theory is all the better if it can describe the experimental results (and also predict new ones) and if it is simple. We might thus say that it leads us to attribute a low Kolmogorov complexity to the observations.

How many sequences of length n can be compressed with a program of length p? To answer this, we note that the number of different programs of length k is equal to 2^k. The number of different programs of length less than p is thus equal to $1 + 2 + 4 + \cdots + 2^{p-1}$, which is just $2^p - 1$. The number of different sequences which can be generated by programs of length less than p is at most equal to the number of different program of length less than p, which is less than 2^p. The number of different sequences of length n is 2^n. Hence, amongst all those sequences of length n, fewer than 2^p can be generated by a program of length less than p. We thus find that the fraction of all sequences of length n which have Kolmogorov complexity smaller than p is less than $2^p/2^n$, or 2^{p-n}. This result indicates that most random sequences have Kolmogorov complexity close to their own length. For example, the fraction of all sequences of length 1000 which have Kolmogorov complexity less than 700 is less than 0.5×10^{-90}. Sequences with low Kolmogorov complexity are therefore exceptional.

5.4 Information and Stochastic Processes

By simply looking at the functions graphed in Fig. 5.1, it is not possible to decide whether they are random or deterministic. Our intuition might nevertheless lead us to consider that $X_\lambda(t)$ looks more like the realization of a stochastic process, whilst $Y_\lambda(t)$ looks more like a deterministic function.

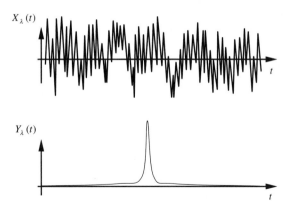

Fig. 5.1. Either of the two functions $X_\lambda(t)$ and $Y_\lambda(t)$ could represent a deterministic function or a stochastic process

According to the definition we have adopted, a stochastic process is a function whose value is determined by a random experiment. To be precise, we consider a set Ω of random events λ and associate a function $X_\lambda(t)$ with each event λ. A realization of a stochastic process is a deterministic function and there is therefore no way of distinguishing between a realization of a stochastic process and a deterministic function within the framework of the present approach.

If we observe the same function $X_\lambda(t)$ in each experiment, it is certainly more appropriate to model it with a deterministic function $x(t)$. However, if

$Y_\lambda(t)$ is centered on a time τ_λ which varies in an unpredictable way from one observation to another, it is more useful to model this family of functions by a stochastic process. (In reality, it is enough not to be trying to predict it. Nothing requires there to be anything unpredictable about it.) Consequently, we note once again that the idea of stochastic process corresponds to a certain standpoint we have adopted, rather than some intrinsic property of the signals we observe. Since the property of stationarity is defined in terms of the mean, i.e., the expectation value, of the various possible realizations, it is of course quite impossible to decide whether or not a stochastic process is stationary on the basis of a single realization.

It is nevertheless clear that $X_\lambda(t)$ seems much more irregular than $Y_\lambda(t)$, and we might be tempted to declare that $X_\lambda(t)$ is "more random" than $Y_\lambda(t)$. The power of $X_\lambda(t)$ is more widely spread out across the observation period than the power of $Y_\lambda(t)$, and this might suggest that $X_\lambda(t)$ is "more stationary" than $Y_\lambda(t)$. In order to analyze this kind of intuition in more detail, we shall examine the case of binary-valued sampled functions. The sampling theorem asserts that, provided we are dealing with signals having a bounded spectrum, there is no loss of generality in considering only sampled signals. Binary-valued processes nevertheless constitute a less general class. Later, we shall analyze the problems raised by continuously varying random variables.

In the present case, the stochastic processes are simply random binary sequences, which we shall denote by $X_\lambda(n)$ and $Y_\lambda(n)$, where $n \in [1, N]$. Let us examine two random sequences analogous to the functions represented in Fig. 5.1.

We thus define $X_\lambda(n)$ as a sequence of 0s and 1s drawn randomly and independently from each other with probability $1/2$. The sequence $Y_\lambda(n)$ will be identically zero except for one sample j, where it will equal 1. In other words, $Y_\lambda(n) = 0$ if $n \neq j$ and $Y_\lambda(n) = 1$ if $n = j$. It is fairly clear that a realization of $X_\lambda(n)$ will generally have a much greater Kolmogorov complexity than a realization of $Y_\lambda(n)$. In the latter case, the algorithm to construct $Y_\lambda(n)$ is rather simple:

Write N 0s and replace the jth term by 1.

We may then say that the sequence $Y_\lambda(n)$ is not algorithmically random. A more precise definition of this idea consists in considering a sequence as algorithmically random if its Kolmogorov complexity is equal to its length. This definition is particularly attractive but unfortunately turns out to be rather impractical owing to the difficulty we have already mentioned, namely the difficulty in determining the program of minimal length able to describe the sequence.

The Kolmogorov complexity characterizes a given realization and this is indeed what interests us here. We may also analyze the complexity of each of the random sequences $X_\lambda(n)$ and $Y_\lambda(n)$, i.e., the set of all possible realizations. The approach adopted here then consists in calculating their entropy. With regard to $X_\lambda(n)$, there are 2^N different possible sequences, all of which are equally probable. The entropy is thus $S(X) = N \ln 2$. There are only N different possible sequences for $Y_\lambda(n)$, once again assumed equiprobable. Its entropy is therefore $S(Y) = \ln N$. We thus have $S(Y) \ll S(X)$, which means that, from the entropy standpoint, $Y_\lambda(n)$ is also simpler than the sequence $X_\lambda(n)$.

We now analyze the stationarity properties of these two sequences. We can make no assertions on the basis of a single realization. However, it is very easy to show that, if we neglect edge effects,[1] the two sequences both possess first and second moments that are invariant under time translations. Indeed, we have $\langle X_\lambda(n) \rangle = 1/2$ and $\langle Y_\lambda(n) \rangle = 1/N$ and, in addition, $\langle X_\lambda(n)X_\lambda(m) \rangle = (1 + \delta_{n-m})/4$ and $\langle Y_\lambda(n)Y_\lambda(m) \rangle = \delta_{n-m}/N$, where δ_{n-m} is the Kronecker symbol. These two sequences are therefore stationary.

5.5 Maximum Entropy Principle

Let N_L be the number of different sequences of length L in which the frequency of occurrence of each random event is equal to its probability. The larger the entropy of a probability law, the greater will be the number N_L of different sequences it can generate during independent realizations. Let us suppose that we measure the complexity of a probability law by the number N_L of different sequences it can generate. In this case, choosing from amongst a set of possible laws the one which maximizes the entropy amounts to choosing the most "complex" law from a stochastic point of view. In other words, the law with the biggest entropy, which therefore has the greatest mean information content, is the one containing the maximal potential disorder.

The choice of probability laws to represent various physical phenomena may rest upon a range of different arguments, as illustrated by the simple examples in Chapter 4. It is not always possible to proceed in this way and the information available is often incomplete, as happens when we only know the expectation values of certain quantities. It may be, for example, that we only know certain statistical moments. One strategy then would be to choose the probability law which maximizes the entropy, whilst maintaining compatibility with the knowledge we have of these expectation values. The associated mathematical problem will then be one of optimization in the presence of constraints, since we shall be seeking the probability law which maximizes the entropy whilst imposing the values of certain statistical means.

To determine the probability law which maximizes the entropy $S(\Omega) = -\sum_{j=1}^{N} p_j \ln p_j$ with no other constraint than $\sum_{i=1}^{N} p_i = 1$, we can use the Lagrange multiplier method. This is explained in Section 5.10. The variables are then $P = (p_1, p_2, \ldots, p_N)^{\mathrm{T}}$, the criterion is $S(\Omega) = S(P) = -\sum_{j=1}^{N} p_j \ln p_j$, and the constraint is $g(P) = \sum_{i=1}^{N} p_i - 1 = 0$. The Lagrange function is thus

---

[1] Neglecting edge effects amounts to assuming that we only consider $n, m \in [1, N]$.

$$\Psi_\mu(P) = -\sum_{j=1}^{N} p_j \ln p_j - \mu \left(\sum_{j=1}^{N} p_j - 1 \right) .$$

We obtain the optimal solution by writing

$$\frac{\partial}{\partial p_j} \Psi_\mu(P) = 0 ,$$

which leads to $- \ln p_j - 1 - \mu = 0$, or $p_j = \exp(-1 - \mu)$. Then the constraint $\sum_{i=1}^{N} p_i = 1$ clearly leads to $p_j = 1/N$. We thus see that the probability law maximizing the entropy is the uniform distribution. To be precise, the condition $\partial \Psi_\mu(P)/\partial p_j = 0$ does not guarantee that we obtain the maximum value of $\Psi_\mu(P)$ but only an extremum or a saddle point. To check that the solution obtained is indeed a maximum, we must check that the Hessian matrix H with elements

$$H_{i,j} = \frac{\partial^2}{\partial p_j \partial p_i} \Psi_\mu(P)$$

is negative definite. We have $\partial \Psi_\mu(P)/\partial p_j = - \ln p_j - 1 - \mu$, and hence $\partial^2 \Psi_\mu(P)/\partial p_j^2 = -1/p_j$ and $\partial^2 \Psi_\mu(P)/\partial p_j \partial p_i = 0$ if $i \neq j$. The solution we have found does therefore correspond to a maximum. In what follows, we shall leave this check to the reader.

We now seek the probability law of the discrete random variables X_λ (i.e., taking values in some countable set) which maximizes the entropy under the two constraints $\langle X_\lambda \rangle = m$ and $\langle X_\lambda^2 \rangle = \mathcal{M}_2$. Let p_j be the probability that the value of the random variable X_λ is x_j. We must therefore take into account the constraints:

$$g_0(P) = \sum_{j=1}^{N} p_j - 1 , \quad g_1(P) = \sum_{j=1}^{N} x_j p_j - m , \quad g_2(P) = \sum_{j=1}^{N} x_j^2 p_j - \mathcal{M}_2 ,$$

and the Lagrange function is thus

$$\Psi_{\mu_0,\mu_1,\mu_2}(P) = -\sum_{j=1}^{N} p_j \ln p_j - \mu_0 \left(\sum_{j=1}^{N} p_j - 1 \right) - \mu_1 \left(\sum_{j=1}^{N} x_j p_j - m \right)$$

$$- \mu_2 \left(\sum_{j=1}^{N} x_j^2 p_j - \mathcal{M}_2 \right) .$$

The optimum situation is achieved when

$$\frac{\partial \Psi_{\mu_0,\mu_1,\mu_2}(P)}{\partial p_j} = 0 ,$$

which implies that $- \ln p_j - 1 - \mu_0 - \mu_1 x_j - \mu_2 x_j^2 = 0$, or

$$p_j = \frac{\exp\left(-\mu_1 x_j - \mu_2 x_j^2\right)}{Z(\mu_1, \mu_2)} \ ,$$

where $Z(\mu_1, \mu_2) = \sum_{j=1}^{N} \exp\left(-\mu_1 x_j - \mu_2 x_j^2\right)$. It is a more delicate matter to identify the parameters μ_1 and μ_2 than to identify μ_0. Note, however, that the mathematical form obtained is analogous to a Gaussian distribution (although not the same, because we have been discussing a discrete probability law).

It is interesting to relate this result to the central limit theorem. Suppose that N is very large, that the x_j are regularly spaced ($x_j = jd$) and that the index j runs over the set of integers (see Fig. 5.2). We will thus have $p_j = \exp\left(-\mu_1 jd - \mu_2 j^2 d^2\right) / Z(\mu_1, \mu_2)$ and in the limit as d becomes very small, we will obtain a good approximation to the Gaussian distribution $p_j = P_G(jd)d$ with

$$P_G(x) = \frac{1}{\sqrt{2\pi}\sigma} \exp\left[-\frac{(x-m)^2}{2\sigma^2}\right] \ ,$$

where m and σ are the mean and standard deviation of the distribution, respectively.

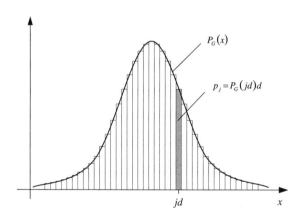

Fig. 5.2. Approximating a discrete probability distribution by a continuous probability density function

The central limit theorem was discussed in Chapter 4. It tells us that, if there is no random variable whose variance dominates over the others, the sum of independent random variables with finite second moment converges toward a Gaussian random variable. More precisely, when we sum P independent random variables, the mean and variance of the sum are directly determined by the means and variances of each of the summed random variables. The central limit theorem allows us to say that, amongst all probability distributions with fixed mean and variance, it is to the Gaussian distribution that the probability density function of the sum variable must finally converge. We obtain here a new interpretation of this result. For given mean and variance, the Gaussian distribution is the one which generates the largest number of different sequences in which the frequency of occurrence of each random event is equal to its probability during independent realizations. Indeed, it has maximum entropy relative to all other distributions with the same mean and variance. Note that we must be given a resolution d for distinguishing two values before we can speak of the probability of each random event. For a given variance, it thus contains the maximal potential disorder. We might say that it is the most complex law from the stochastic point of view. In other words, the universal character of the Gaussian distribution in the context of the central limit theorem corresponds to convergence toward the probability distribution containing the maximal potential disorder, where the measure of disorder is the number of different sequences a law can generate during independent realizations. We may say schematically that, when we sum random variables with finite second moment, the result has maximal complexity or maximal disorder. However, it is important to understand the exact meaning attributed to the notions of complexity and disorder when we make such a claim.

Finally, for a given power, the assumption that the fluctuations in a physical quantity are Gaussian can be understood as a hypothesis of maximal disorder or *a priori* minimal knowledge. The entropy is the mean information content that the source of random events can supply during independent realizations. In the case of a source with high entropy, each realization will tend to bring a lot of information, and this is compatible with the interpretation whereby the information available to us *a priori*, i.e., before the trials, is itself minimal.

5.6 Entropy of Continuous Distributions

For continuous random variables, we speak rather of probability density than just probability. We must therefore ask whether it is possible to define the entropy of a continuous probability distribution. The answer is affirmative, although we must be careful not to attribute the deep meaning to it that we were able to in the case of discrete probability laws.

We consider a continuous variable X_λ, which simply means that X_λ can take a continuous set of values, with a probability density function $P_X(x)$ that is itself continuous. Let us quantify the range of variations of this random variable with a step δ, as shown in Fig. 5.3. This amounts to applying the transformation

$$X_\lambda \mapsto Y_\lambda = J_\lambda \delta ,$$

where J_λ is a random variable with positive or negative integer values defined from X_λ by $X_\lambda \in [J_\lambda \delta - \delta/2, J_\lambda \delta + \delta/2]$. Y_λ is thus a discrete random variable isomorphic to J_λ and the probability distribution of J_λ is

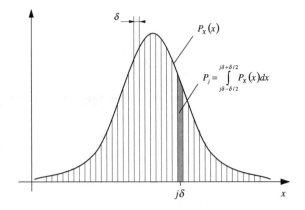

Fig. 5.3. Approximating a continuous probability density function by a discrete probability law

$$P_j = \int\limits_{j\delta-\delta/2}^{j\delta+\delta/2} P_X(x)\mathrm{d}x \; .$$

The entropy of J_λ and hence of Y_λ is $S(Y) = S(J) = -\sum_j P_j \ln P_j$. If δ is small enough, we may write $P_j \approx P_X(j\delta)\delta$ so that

$$S(Y) \approx -\sum_j P_X(j\delta)\delta \ln\big[P_X(j\delta)\delta\big] \; .$$

Using the Riemann approximation to the integral, we obtain

$$S(Y) \approx -\int P_X(x)\ln\big[P_X(x)\delta\big]\mathrm{d}x \; ,$$

or

$$S(Y) \approx -\int P_X(x)\ln\big[P_X(x)\big]\mathrm{d}x - \ln\delta \; .$$

We thus note that, when δ tends to 0, the entropy $S(Y)$ diverges. This result is easily understood if we remember that the entropy is a measure of the number N_L of different sequences of length L that can be generated from independent realizations in such a way that the frequency of occurrence of each random event is equal to its probability. As the value of δ decreases, the number of different sequences increases, until it diverges as δ tends to 0.

In the limit as δ tends to 0, X_λ and Y_λ become identical. We may thus say that the entropy of any continuous random variable is formally infinite. This is hardly a practical result! We therefore define the entropy of continuous random variables in terms of the probability density function by

$$S(X) = -\int P_X(x) \ln \left[P_X(x) \right] \mathrm{d}x \ .$$

It is important to remember that, with this definition, we lose certain features of the entropy as it applies to discrete random variables. In particular, there is no longer any guarantee that the entropy of a continuous random variable will be positive.

In order to illustrate this notion, let us determine the entropy of a Gaussian distribution. We have

$$P_X(x) = \frac{1}{\sqrt{2\pi}\sigma} \exp \left[-\frac{(x-m)^2}{2\sigma^2} \right] \ ,$$

whereupon

$$S(X) = -\int \frac{1}{\sqrt{2\pi}\sigma} \exp \left[-\frac{(x-m)^2}{2\sigma^2} \right] \left[-\frac{(x-m)^2}{2\sigma^2} - \ln \left(\sqrt{2\pi}\sigma \right) \right] \mathrm{d}x \ .$$

A simple calculation leads to $S(X) = 1/2 + \ln \left(\sqrt{2\pi}\sigma \right)$, which can also be written

$$S(X) = \ln \left(\sqrt{2e\pi}\sigma \right) \ .$$

Note that the entropy can be positive or negative, depending on the value of σ.

5.7 Entropy, Propagation and Diffusion

In this section we shall study the dynamical evolution of the entropy in two very similar, but nevertheless different cases, namely, propagation and diffusion. To keep the discussion simple, we restrict to the case of 1-dimensional signals (see Fig. 5.4).

We begin with the evolution of the entropy during propagation of optical signals. In the 1-dimensional case, the equation obeyed by the field $A(x,t)$ is simply

$$\frac{\partial^2 A(x,t)}{\partial x^2} - \frac{1}{c^2} \frac{\partial^2 A(x,t)}{\partial t^2} = 0 \ .$$

We denote the intensity of the field $A(x,t)$ by $I(x,t)$ at the point with coordinate x and at time t, so that $I(x,t) = |A(x,t)|^2$. A simple model consists in considering that, at the point with coordinate x and at time t, the number of photons that can be detected is proportional to $I(x,t)$. Suppose that the light pulse we are interested in has finite energy E, where $E = \int_{-\infty}^{\infty} I(x,t)\mathrm{d}x$. Moreover, if we assume that there is no absorption during propagation, the energy will be constant in time. We may thus consider that for a pulse with one photon, the density probability of detecting a photon at the point with coordinate x and at time t is given by

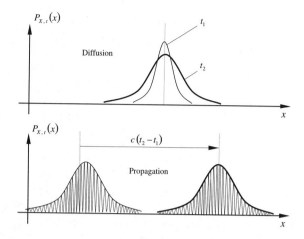

Fig. 5.4. Comparing the dynamical evolution of probability densities for systems undergoing diffusion and propagation

$$P_{X,t}(x) = \frac{I(x,t)}{E} \ .$$

Here we have argued by analogy with diffusion phenomena, where we study the probability of finding a particle at a point with coordinate x and at time t.

Let us now show that, if the initial conditions are $A(x,0) = A_0(x)$, then a solution is $A(x,t) = A_0(x - ct)$. Indeed, we have

$$\frac{\partial^2}{\partial x^2} A(x,t) = \frac{\partial^2}{\partial x^2} A_0(x - ct) \quad \text{and} \quad \frac{\partial^2}{\partial t^2} A(x,t) = \frac{\partial^2}{\partial t^2} A_0(x - ct) \ .$$

Putting $u = x - ct$, it follows that

$$\frac{\partial^2}{\partial x^2} A(x,t) = \frac{\partial^2}{\partial u^2} A_0(u) \frac{\partial^2 u}{\partial x^2} \quad \text{and} \quad \frac{\partial^2}{\partial t^2} A(x,t) = \frac{\partial^2}{\partial u^2} A_0(u) \frac{\partial^2 u}{\partial t^2} \ .$$

Now $\partial^2 u/\partial t^2 = c^2 \partial^2 u/\partial x^2$, so we can deduce that, whatever the function $A_0(x)$,

$$\frac{\partial^2 A_0(x - ct)}{\partial x^2} - \frac{1}{c^2} \frac{\partial^2 A_0(x - ct)}{\partial t^2} = 0 \ .$$

In the same way, it can be shown that $A(x,t) = A_0(x + ct)$ is also a solution of the propagation equation. In this case, the wave moves toward negative x values, whereas with $A(x,t) = A_0(x - ct)$, the wave moves toward positive x values. We consider only the last solution. We thus have $I(x,t) = I_0(x - ct)$, where $I_0(x) = |A_0(x)|^2$, and $P_{X,t}(x) = P_{X,0}(x - ct) = I_0(x - ct)/E$.

Let us now investigate how the entropy changes in time. At a given time t, the definition of the entropy for a continuous probability distribution implies that

$$S_t = - \int P_{X,t}(x) \ln\left[P_{X,t}(x)\right] \mathrm{d}x \ .$$

This may be rewritten as

$$S_t = - \int P_{X,0}(x - ct) \ln\left[P_{X,0}(x - ct)\right] \mathrm{d}x \ ,$$

or, using the definition of u,

$$S_t = - \int P_{X,0}(u) \ln\left[P_{X,0}(u)\right] \mathrm{d}u \ .$$

This shows that $S_t = S_0$.

For 1-dimensional signals, propagation therefore occurs at constant entropy. Consequently, there is no irreversibility during propagation, for this would be reflected by an increase in entropy. This result is consistent with the existence of two solutions $A(x,t) = A_0(x - ct)$ and $A(x,t) = A_0(x + ct)$, related by simply changing the sign of the time argument. Note that the analysis of propagation in higher dimensional spaces would not necessarily lead to the same conclusion.

We now investigate the way the entropy evolves during particle diffusion. In the 1-dimensional case, the diffusion equation is

$$\frac{\partial P_{X,t}(x)}{\partial t} - \frac{\chi^2}{2}\frac{\partial^2 P_{X,t}(x)}{\partial x^2} = 0 \ .$$

To bring out the analogy with the propagation equation, we can also write

$$\frac{\partial^2 P_{X,t}(x)}{\partial x^2} - \frac{2}{\chi^2}\frac{\partial P_{X,t}(x)}{\partial t} = 0 \ .$$

Once again, the expression for the entropy is $S_t = - \int P_{X,t}(x) \ln\left[P_{X,t}(x)\right] \mathrm{d}x$ and its derivative with respect to time is therefore

$$\frac{\partial}{\partial t}S_t = - \int \frac{\partial}{\partial t}P_{X,t}(x) \ln\left[P_{X,t}(x)\right] \mathrm{d}x - \int P_{X,t}(x)\frac{\partial P_{X,t}(x)/\partial t}{P_{X,t}(x)}\mathrm{d}x \ ,$$

or

$$\frac{\partial}{\partial t}S_t = - \int \frac{\partial}{\partial t}P_{X,t}(x)\left\{1 + \ln\left[P_{X,t}(x)\right]\right\} \mathrm{d}x \ .$$

Using the diffusion equation

$$\frac{\partial P_{X,t}(x)}{\partial t} = \frac{\chi^2}{2}\frac{\partial^2 P_{X,t}(x)}{\partial x^2} \ ,$$

we can write

$$\frac{\partial}{\partial t}S_t = -\frac{\chi^2}{2} \int \frac{\partial^2}{\partial x^2}P_{X,t}(x)\left\{1 + \ln\left[P_{X,t}(x)\right]\right\} \mathrm{d}x \ .$$

Integrating by parts, we obtain

$$\frac{\partial}{\partial t} S_t = -\frac{\chi^2}{2} \left[F_1(x,t) + F_2(x,t) \right] ,$$

where

$$F_1(x,t) = \left. \frac{\partial}{\partial x} P_{X,t}(x) \left\{ 1 + \ln \left[P_{X,t}(x) \right] \right\} \right|_{-\infty}^{\infty} ,$$

and

$$F_2(x,t) = - \int \frac{1}{P_{X,t}(x)} \left[\frac{\partial P_{X,t}(x)}{\partial x} \right]^2 dx .$$

If we assume that $P_{X,t}(x)$ decreases monotonically as $x \to \infty$ and also as $x \to -\infty$, the first term is then zero. Indeed, let us examine the limit of this term when $x \to \infty$. If we set $f(x) = P_{X,t}(x)$ and $f'(x) = df(x)/dx$, we have

$$f'(x) \left[1 + \ln f(x) \right] = \frac{d}{dx} \left[f(x) \ln f(x) \right] .$$

Now $\lim_{x \to \infty} f(x) \ln f(x) = 0$, since we have $\lim_{x \to \infty} f(x) = 0$. Furthermore, $f'(x) \left[1 + \ln f(x) \right] \geqslant 0$ since $f'(x) < 0$ and $\lim_{x \to \infty} \ln f(x) = -\infty$. This is a reasonable hypothesis if the initial conditions are only nonzero within a bounded region. It was shown in Section 3.12 that the general solution can be written $\int P_0(x')G(x - x', t)dx'$, where $P_0(x)$ is the probability at time $t = 0$. In addition, we have $dG(x - x', t)/dx < 0$ if $x > x'$. Hence, if $P_0(x) = 0$ when $x > x_0$, we have

$$\frac{d}{dx} \int P_0(x')G(x - x', t)dx' < 0 .$$

The case in which the range of integration is not infinite is also interesting. The first term would then only be zero if there were no concentration gradient at the edges. Returning to the problem at hand, we now have

$$\lim_{x \to \infty} \frac{d}{dx} \left[f(x) \ln f(x) \right] = 0 .$$

This implies that

$$\lim_{x \to \infty} \left\{ f'(x) \left[1 + \ln f(x) \right] \right\} = 0.$$

We obtain the same result for $x \to -\infty$. We thus find that

$$\frac{\partial}{\partial t} S_t = \frac{\chi^2}{2} \int \frac{1}{P_{X,t}(x)} \left[\frac{\partial P_{X,t}(x)}{\partial x} \right]^2 dx > 0 .$$

The entropy thus increases with time during a diffusion process. This result would also be true in dimensions greater than 1. Diffusion is therefore an irreversible process and, unlike the propagation equation, the diffusion equation is not invariant under time reversal.

5.8 Multidimensional Gaussian Case

In this section, we shall be concerned with real-valued, zero-mean stochastic processes sampled at a finite number of times. Such processes are simply random sequences, denoted $X_\lambda(n)$, where $n \in [1, \ldots, N]$. We define the covariance matrix $\overline{\overline{\Gamma}}$ by $\Gamma_{ij} = \langle X_\lambda(i) X_\lambda(j) \rangle$ and its inverse is denoted $\overline{\overline{K}}$. If the stochastic process is Gaussian, the joint probability density function $P_X(x_1, x_2, \ldots, x_N)$ of $X_\lambda(1), X_\lambda(2), \ldots, X_\lambda(N)$ is then (see Chapter 2)

$$P_X(x_1, x_2, \ldots, x_N) = \frac{1}{\left(\sqrt{2\pi}\right)^N |\overline{\overline{\Gamma}}|^{1/2}} \exp\left(-\frac{1}{2} \sum_{i=1}^{N} \sum_{j=1}^{N} x_i K_{ij} x_j \right) ,$$

where $|\overline{\overline{\Gamma}}|$ is the determinant of $\overline{\overline{\Gamma}}$. In the multidimensional case, the entropy of a probability density function is simply

$$S = -\int \ldots \int P_X(x_1, x_2, \ldots, x_N) \ln\left[P_X(x_1, x_2, \ldots, x_N) \right] \mathrm{d}x_1 \mathrm{d}x_2 \ldots \mathrm{d}x_N .$$

For our Gaussian process, this becomes

$$S = \frac{1}{2} \int \ldots \int Q_X(x_1, x_2, \ldots, x_N) P_X(x_1, x_2, \ldots, x_N) \mathrm{d}x_2 \ldots \mathrm{d}x_N ,$$

where

$$Q_X(x_1, x_2, \ldots, x_N) = \sum_{i=1}^{N} \sum_{j=1}^{N} x_i K_{ij} x_j + \ln\left[(2\pi)^N |\overline{\overline{\Gamma}}| \right] .$$

Of course, $\Gamma_{ij} = \iint x_i x_j P_X(x_1, x_2, \ldots, x_N) \mathrm{d}x_1 \mathrm{d}x_2 \ldots \mathrm{d}x_N$, whereupon we may deduce that

$$S = \frac{1}{2} \sum_{i=1}^{N} \sum_{j=1}^{N} \Gamma_{ij} K_{ij} + \ln\left[\left(\sqrt{2\pi}\right)^N |\overline{\overline{\Gamma}}|^{1/2} \right] .$$

Now

$$\sum_{i=1}^{N} \sum_{j=1}^{N} \Gamma_{ij} K_{ij} = N ,$$

and hence,

$$S = \frac{1}{2} \ln\left[(2\pi \mathrm{e})^N |\overline{\overline{\Gamma}}| \right] .$$

We shall now express this entropy in terms of the spectral density of the sequence $X_\lambda(n)$. To this end, we must consider stationary random sequences. However, it is difficult to define stationarity for a finite-dimensional random sequence. Our task is made easier by constructing an infinitely long periodic sequence from $X_\lambda(n)$, viz.,

$$X^{\mathrm{P}}_\lambda(n) = X_\lambda\big[\mathrm{mod}_N(n)\big] \;,$$

where the function $\mathrm{mod}_N(n)$ is defined by $\mathrm{mod}_N(n) = n - pN$ and p is a whole number chosen so that $n - pN \in [1, N]$. Recall that the sequence $X_\lambda(n)$ is said to be weakly cyclostationary if $X^{\mathrm{P}}_\lambda(n)$ is weakly stationary (up to second order moments). This means that $\langle X^{\mathrm{P}}_\lambda(n)\rangle$ and $\langle X^{\mathrm{P}}_\lambda(n)X^{\mathrm{P}}_\lambda(n+m)\rangle$ must be independent of n. In this case, the covariance matrix $\overline{\overline{\Gamma}}$ has a special mathematical structure. In fact,

$$\begin{aligned}\Gamma_{nm} &= \langle X^{\mathrm{P}}_\lambda(n)X^{\mathrm{P}}_\lambda(m)\rangle - \langle X^{\mathrm{P}}_\lambda(n)\rangle\langle X^{\mathrm{P}}_\lambda(m)\rangle\\ &= \Gamma\big[\mathrm{mod}_N(m-n)\big] = \Gamma\big[\mathrm{mod}_N(n-m)\big]\;.\end{aligned}$$

We saw in Section 3.5 that the power spectral density $\hat\Gamma(\nu)$ of $X_\lambda(n)$ satisfies the relations

$$\hat\Gamma(\nu) = \frac{1}{N}\sum_{m=0}^{N-1}\Gamma(m)\exp\left(\frac{\mathrm{i}2\pi\nu m}{N}\right)\;,$$

and

$$\langle \hat X_\lambda(\nu_1)\hat X^*_\lambda(\nu_2)\rangle = N^2\hat\Gamma(\nu_1)\delta_{\nu_1-\nu_2}\;.$$

We introduce the matrix $\overline{\overline{F}}$ whose ν, n entry is

$$F_{\nu n} = \frac{1}{\sqrt{N}}\exp\left(-\frac{\mathrm{i}2\pi\nu n}{N}\right)\;.$$

This matrix is unitary, i.e., $\overline{\overline{F}}^\dagger\overline{\overline{F}} = \overline{\overline{F}}\,\overline{\overline{F}}^\dagger = \mathrm{Id}_N$, where $\overline{\overline{F}}^\dagger$ is the transposed complex conjugate of $\overline{\overline{F}}$ and Id_N the identity matrix in N dimensions. We thus find that

$$\left(\overline{\overline{F}}\,\overline{\overline{\Gamma}}\,\overline{\overline{F}}^\dagger\right)_{\nu_1\nu_2} = N\hat\Gamma(\nu_1)\delta_{\nu_1-\nu_2}\;.$$

noting that $\hat\Gamma(\nu)$ are the eigenvalues of $\overline{\overline{\Gamma}}$, hence real and positive, as explained in Section 2.8. Now

$$\left|\overline{\overline{F}}\,\overline{\overline{\Gamma}}\,\overline{\overline{F}}^\dagger\right| = |\overline{\overline{F}}|\,|\overline{\overline{\Gamma}}|\,|\overline{\overline{F}}^\dagger| = |\overline{\overline{\Gamma}}|\;,$$

and we thus deduce that

$$|\overline{\overline{\Gamma}}| = N^N\prod_{\nu=0}^{N-1}\hat\Gamma(\nu)\;.$$

We have

$$S = \frac{1}{2}\ln\left[(2\pi\mathrm{e}N)^N\prod_{\nu=0}^{N-1}\hat\Gamma(\nu)\right]\;,$$

so that

$$S = \frac{1}{2} \sum_{\nu=0}^{N-1} \ln \left[2\pi e N \hat{\Gamma}(\nu) \right] .$$

Consider the trivial case of a white sequence, i.e., an uncorrelated sequence, with power σ^2. We thus have $\hat{\Gamma}(\nu) = \sigma^2/N$, which implies that

$$S = \frac{N}{2} \ln \left(\frac{2\pi e N \sigma^2}{N} \right) = \frac{N}{2} \ln(2\pi e \sigma^2) .$$

This result is consistent with the entropy value $(1/2)\ln(2\pi e\sigma^2)$ of a scalar Gaussian variable. (The entropy of N independent scalar Gaussian variables is simply the sum of the entropies of each random variable.)

We now analyze the evolution of the entropy when noise is transformed by a convolution filter. To keep this simple, we assume once again that the noise is sampled and cyclostationary. The noise before filtering is denoted by $X_\lambda(n)$ and after filtering by $Y_\lambda(n)$. We have seen that we must have

$$\hat{\Gamma}_{YY}(\nu) = |\hat{\chi}(\nu)|^2 \hat{\Gamma}_{XX}(\nu) ,$$

where $\hat{\Gamma}_{XX}(\nu)$ and $\hat{\Gamma}_{YY}(\nu)$ are the spectral densities of $X_\lambda(n)$ and $Y_\lambda(n)$, respectively. $\hat{\chi}(\nu)$ is the transfer function characterizing the convolution filter, i.e., the discrete Fourier transform of the impulse response (or kernel) of the convolution filter. Let $S(X)$ and $S(Y)$ be the entropies of $X_\lambda(n)$ and $Y_\lambda(n)$, respectively. We have

$$S(Y) = \frac{1}{2} \sum_{\nu=0}^{N-1} \ln \left[2\pi e N \hat{\Gamma}_{YY}(\nu) \right] = \frac{1}{2} \sum_{\nu=0}^{N-1} \ln \left[2\pi e N |\hat{\chi}(\nu)|^2 \hat{\Gamma}_{XX}(\nu) \right] ,$$

and hence,

$$S(Y) = S(X) + \sum_{\nu=0}^{N-1} \ln \left[|\hat{\chi}(\nu)| \right] .$$

If there is no amplification of the signals, we have $|\hat{\chi}(\nu)| \leqslant 1$ and the filter therefore produces a reduction in entropy. We may say that filtering creates order, in the sense that entropy measures disorder. Note, however, that during the operation the power has decreased since

$$\sum_{\nu=0}^{N-1} \hat{\Gamma}_{YY}(\nu) = \sum_{\nu=0}^{N-1} |\hat{\chi}(\nu)|^2 \hat{\Gamma}_{XX}(\nu) \leqslant \sum_{\nu=0}^{N-1} \hat{\Gamma}_{XX}(\nu) .$$

5.9 Kullback–Leibler Measure

In many applications it is important to be able to compare two probability laws defined on the same set of random events. Various empirical approaches are available, such as the quadratic distance or other measures of distance. Let p_j and q_j be the two laws we seek to compare. The quadratic distance is then $d_2 = \sum_{j=1}^{N} |p_j - q_j|^2$, where we have summed over the N possible random events. However, there is no guarantee that such an approach will prove useful and others have been put forward by statisticians. In this section, we shall discuss the Kullback–Leibler measure, which has a very interesting probabilistic interpretation.

We obtained a simple interpretation of the Shannon entropy by considering experiments in which we observed the realizations of L independent random events arising from Ω. We then studied the random events Λ_L defined as the sequences $\Lambda_L = \{\lambda(1), \lambda(2), \ldots, \lambda(L)\}$, where $\lambda(n)$ is the nth realization of an event in Ω. Let us consider these random events Λ_L and let q_j denote the probability of the jth event λ_j of Ω. The probability W_L of observing a sequence Λ_L for which each event λ_j appears ℓ_j times is thus

$$W_L = \frac{L!}{\ell_1! \ell_2! \ldots \ell_N!} q_1^{\ell_1} q_2^{\ell_2} \cdots q_N^{\ell_N} .$$

Using the simplified Stirling approximation when L is very large, we can carry out the same analysis for the entropy. We then obtain

$$\lim_{L \to \infty} \frac{\ln W_L}{L} = \lim_{L \to \infty} \left[-\sum_{j=1}^{N} \frac{\ell_j}{L} \ln \left(\frac{\ell_j}{q_j L} \right) \right] .$$

The probability of finding a sequence such that the event λ_j occurs approximately $p_j L$ times is then

$$W_L \approx \exp\left[-L K_{\mathrm{u}}\left(P \,\|\, Q\right)\right] ,$$

where we define the Kullback–Leibler measure by

$$K_{\mathrm{u}}\left(P \,\|\, Q\right) = \sum_{j=1}^{N} p_j \ln \frac{p_j}{q_j} .$$

The main point is this: the larger the value of the Kullback–Leibler measure, the smaller the probability of observing a sequence with frequencies of occurrence p_i if the probability law happens to be q_i. In addition, it is clear that the approximation $W_L \approx \exp\left[-L K_{\mathrm{u}}\left(P \,\|\, Q\right)\right]$ is valid when L is large and thus that W_L tends exponentially to 0 unless $K_{\mathrm{u}}\left(P \,\|\, Q\right) = 0$. The Kullback–Leibler measure then characterizes the rate of decrease of the exponential.

The total number of different sequences is N^L. The number N_L of different sequences Λ_L for which each event λ_j occurs $\ell_j = p_j L$ times is approximately $W_L N^L$. When L is large, we have

$$W_L N^L \approx \exp\left\{ L \left[\ln N - K_{\mathrm{u}}\left(P \,\|\, Q\right)\right] \right\} .$$

These sequences are clearly equiprobable and their entropy is thus

$$S \approx L\big[\ln N - K_{\mathrm{u}}\left(P\,\|\,Q\right)\big]\ .$$

It is easy to show that $K_{\mathrm{u}}\left(P\,\|\,Q\right)$ is positive or zero and that $K_{\mathrm{u}}\left(P\,\|\,Q\right)=0$ if and only if $p_j = q_j$, $\forall j = 1,\ldots,N$. Indeed, for a given law $q_{j\in[1,N]}$, let us seek the law $p_{j\in[1,N]}$ which minimizes $K_{\mathrm{u}}\left(P\,\|\,Q\right)$. The Lagrange function for this problem is

$$\Psi(P) = K_{\mathrm{u}}\left(P\,\|\,Q\right) - \mu \sum_{j=1}^{N} p_j\ ,$$

or

$$\Psi(P) = \sum_{j=1}^{N} p_j \ln \frac{p_j}{q_j} - \mu \sum_{j=1}^{N} p_j\ .$$

Now $\partial\Psi(P)/\partial p_j = 0$ implies that $1 + \ln p_j - \ln q_j - \mu = 0$. The constraint $\sum_{j=1}^{N} p_j = 1$ leads to $\mu = 1$ and hence $p_j = q_j$, $\forall j = 1,\ldots,N$. In this case, we find immediately that $K_{\mathrm{u}}\left(P\,\|\,Q\right) = 0$. To check that this is indeed a minimum, we note that $\partial^2\Psi(P)/\partial p_j^2 = 1/p_j \geqslant 0$ and that $\partial^2\Psi(P)/\partial p_j \partial p_i = 0$ if $i \neq j$.

This shows that $K_{\mathrm{u}}\left(P\,\|\,Q\right)$ characterizes the separation between the laws $p_{j\in[1,N]}$ and $q_{j\in[1,N]}$. For this reason it is often referred to as the Kullback–Leibler distance. However, it should be noted that, from a mathematical point of view, this quantity does not satisfy the axioms required of a true definition of distance, i.e., it is not symmetric and does not satisfy a triangle inequality.

It is interesting to determine the Kullback–Leibler measure of a law $p_{j\in[1,N]}$ with respect to a uniform distribution $q_{j\in[1,N]} = 1/N$, recalling that it is the uniform probability law that maximizes the entropy. We have

$$K_{\mathrm{u}}\left(P\,\|\,Q_{\mathrm{unif}}\right) = \sum_{j=1}^{N} p_j \ln(Np_j)\ ,$$

or

$$K_{\mathrm{u}}\left(P\,\|\,Q_{\mathrm{unif}}\right) = \ln N - S(P)\ , \quad \text{where} \quad S(P) = -\sum_{j=1}^{N} p_j \ln p_j\ .$$

This is a special case of the result stated above:

$$S \approx L\big[\ln N - K_{\mathrm{u}}\left(P\,\|\,Q\right)\big]\ ,$$

recalling that, if S is the entropy of a given probability law, the entropy of the law associated with the sequences made up of L independent observations is LS.

5.10 Appendix: Lagrange Multipliers

We often seek the probability law which maximizes the entropy under certain constraints. For example, we may be looking for the probability law which maximizes the entropy $S(\Omega) = -\sum_{j=1}^{N} p_j \ln p_j$ under the constraint that $\sum_{i=1}^{N} p_i = 1$. To achieve this, we can use the Lagrange multiplier technique. Many mathematical works specialized in optimization rigorously establish the situations in which this technique can be applied and where it guarantees the existence and relevance of the solutions produced. We shall now show how to use this technique, whilst proposing a non-rigorous interpretation which nevertheless allows us to obtain a simple physical intuition.

We consider a function $F(X)$ of the vector variable $X = (x_1, x_2, \ldots, x_N)^{\mathrm{T}}$. Let us suppose that we seek the value for which this function reaches its maximum when the variable X satisfies the constraint $g(X) = 0$. In order to apply the Lagrange multiplier technique, we define the Lagrange function $\Psi_\mu(X) = F(X) - \mu g(X)$, where μ is a real parameter, also known as the Lagrange multiplier. We then seek the value of X which maximizes $\Psi_\mu(X)$, denoted symbolically by

$$X_\mu^{\mathrm{opt}} = \underset{X}{\mathrm{argmax}} \left[\Psi_\mu(X) \right] .$$

We next determine the value μ_0 of μ for which X_μ^{opt} satisfies the constraint, i.e., for which $g(X_{\mu_0}^{\mathrm{opt}}) = 0$. It is easy to see that, if we find a value $X_{\mu_0}^{\mathrm{opt}}$ such that $g(X_{\mu_0}^{\mathrm{opt}}) = 0$, it must correspond to the solution which maximizes $F(X)$ under the constraint $g(X) = 0$. Indeed, suppose that there were a value $\tilde{X}$ such that $F(\tilde{X}) > F(X_{\mu_0}^{\mathrm{opt}})$ and $g(\tilde{X}) = 0$. Then it is clear that we would have $\Psi_\mu(\tilde{X}) > \Psi_\mu(X_{\mu_0}^{\mathrm{opt}})$, which would contradict the hypothesis that X_μ^{opt} maximizes $\Psi_\mu(X)$.

This technique is easily generalized to the case where there are several constraints. For example, let these constraints be $g_1(X) = 0$ and $g_2(X) = 0$. We then define the Lagrange function

$$\Psi_{\mu_1,\mu_2}(X) = F(X) - \mu_1 g_1(X) - \mu_2 g_2(X) ,$$

where μ_1 and μ_2 are the two Lagrange multipliers. We then seek the value of X which maximizes $\Psi_{\mu_1,\mu_2}(X)$, denoted symbolically by

$$X_{\mu_1,\mu_2}^{\mathrm{opt}} = \underset{X}{\mathrm{argmax}} \left[\Psi_{\mu_1,\mu_2}(X) \right] .$$

We then determine the values of μ_1 and μ_2 which satisfy the constraints $g_1(X_{\mu_1,\mu_2}^{\mathrm{opt}}) = 0$ and $g_2(X_{\mu_1,\mu_2}^{\mathrm{opt}}) = 0$. It is clear that this approach can be generalized to an arbitrary number of constraints.

There is a simple physical interpretation of this technique. Let X_F^{opt} be the value of X which maximizes $F(X)$ and X_g^{opt} the one which minimizes $g(X)$. Suppose that $g(X_F^{\text{opt}}) > 0$ and that $g(X_g^{\text{opt}}) < 0$. [In the opposite case, that is, $g(X_F^{\text{opt}}) < 0$ and $g(X_g^{\text{opt}}) > 0$, the argument would be equivalent.] Note, however, that nothing guarantees the existence of X_F^{opt} and X_g^{opt}. Let $\tilde{X}_\alpha^{\text{opt}}$ be the value of X which maximizes $A_\alpha(X) = (1 - \alpha)F(X) - \alpha g(X)$. Maximizing $A_\alpha(X)$ is equivalent to maximizing $\Psi_\mu(X)$ if we set $\mu = \alpha/(1 - \alpha)$. Clearly, $\tilde{X}_0^{\text{opt}} = X_F^{\text{opt}}$ and $\tilde{X}_1^{\text{opt}} = X_g^{\text{opt}}$, and by varying α continuously between 0 and 1, we give more and more weight to the minimization of $g(X)$, because we favor solutions for which $-g(X)$ is large. Maximization of $A_\alpha(X)$ therefore achieves a compromize between maximization of $F(X)$ and minimization of $g(X)$. If there is a value α_0 of α for which $g(X_{\alpha_0}^{\text{opt}}) = 0$, this value α_0 then corresponds to the weighting which maximizes $F(X)$ whilst at the same time achieving $g(X) = 0$.

Exercises

Exercise 5.1

Consider a system with N states $e_0, e_1, \ldots, e_{N-1}$. The probability of finding the system in the state e_0 is

$$P_0 = \frac{1}{N} + (N - 1)\alpha ,$$

whilst for the other states $e_1, e_2, \ldots, e_{N-1}$, this probability is

$$P = \frac{1}{N} - \alpha .$$

(1) Specify the domain of definition of α.
(2) Calculate the entropy of the system.
(3) For what value of α is the entropy maximal?

Exercise 5.2. Entropy of Light Polarization

Consider a system with a complex, 2-dimensional Gaussian electric field vector $\boldsymbol{E}$. Let $\overline{\overline{\Gamma}}$ by its covariance matrix and assume its mean to be zero.

(1) Write down the probability density function of $\boldsymbol{E}$.
(2) Calculate the entropy of the system.
(3) Express the entropy in terms of the degree of polarization.
(4) Generalize to the case where the electric field vector $\boldsymbol{E}$ is 3-dimensional.

Exercise 5.3. Kullback–Leibler Measure for Probability Densities

Suggest a generalization of the Kullback–Leibler measure to the case of continuous probability distributions.

Exercise 5.4. Kullback–Leibler Distance

Using the results of the last exercise, determine the Kullback–Leibler distances between the following continuous probability distributions:

(1) scalar Gaussian distributions with different means and variances,
(2) Gamma distributions with different means and orders,
(3) Poisson distributions,
(4) geometric distributions.

Exercise 5.5. Chernov Measure

Consider two probability laws $P_a(n)$ and $P_b(n)$, where n is a positive integer. The aim here will be to determine the probability law $P_{s^*}(n)$ which lies at equal Kullback–Leibler distance from both $P_a(n)$ and $P_b(n)$ and which are the closest to them, where

$$K_u(P_s \| P_a) = \sum_{n=1}^{+\infty} P_s(n) \ln \left[\frac{P_s(n)}{P_a(n)} \right]$$

and

$$K_u(P_s \| P_b) = \sum_{n=1}^{+\infty} P_s(n) \ln \left[\frac{P_s(n)}{P_b(n)} \right] .$$

(1) Among all those probability laws that possess a Kullback–Leibler measure $K_u(P_s \| P_a)$ with respect to $P_a(n)$, show that the one closest to $P_b(n)$ can be written in the form

$$P_s(n) = \frac{1}{C(s)} P_b^s(n) P_a^{1-s}(n) ,$$

where

$$C(s) = \sum_{n=1}^{+\infty} P_b^s(n) P_a^{1-s}(n) ,$$

but without seeking to determine the parameter s.
(2) Show that, among the above probability laws, the one which has the same Kullback–Leibler measure with respect to both $P_a(n)$ and $P_b(n)$, i.e., such that $K_u(P_{s^*} \| P_a) = K_u(P_{s^*} \| P_b)$, corresponds to the value of s that minimizes $C(s)$, i.e.,

$$s^* = \operatorname*{argmin}_{s} C(s) .$$

(3) Show that

$$K_{\mathrm{u}}(P_{s^*} \| P_a) = K_{\mathrm{u}}(P_{s^*} \| P_b) = -\ln[C(s^*)] \ .$$

(4) Generalize this result to the case of continuous probability distributions.

Exercise 5.6. Chernov–Battacharyya measure

Define

$$C(s) = \sum_{n=1}^{+\infty} P_b^s(n) P_a^{1-s}(n) \ .$$

(1) Express

$$\frac{\mathrm{d}}{\mathrm{d}s} \ln[C(s)]_{s=0} \quad \text{and} \quad \frac{\mathrm{d}}{\mathrm{d}s} \ln[C(s)]_{s=1}$$

in terms of the Kullback–Leibler measures $K_{\mathrm{u}}(P_a \| P_b)$ and $K_{\mathrm{u}}(P_b \| P_a)$.

(2) Consider a second order approximation to $\ln[C(s)]$ as a function of s, and impose the constraints $C(0) = C(1) = 1$. Deduce that, to this approximation, $s^* \simeq 1/2$. [Recall that s^* is the value of s that minimizes $C(s)$.] From this approximation one can deduce the Battacharyya measure, which can be written

$$\mathcal{B}(P_a \| P_b) = -\ln\left[\sum_{n=1}^{+\infty} \sqrt{P_b(n) P_a(n)}\right] \ .$$

6

Thermodynamic Fluctuations

For a macroscopic system, any physical quantity fluctuates in space and time. These fluctuations are due to thermal agitation and we shall see that it is possible to analyze some of their properties without having to determine the exact configuration of all the particles in the system. For this purpose, when the physical system is in thermodynamic equilibrium, we must first determine the probability law for finding it in a given state. We shall then focus more closely on the fluctuations of macroscopic quantities associated with thermodynamic systems, although we shall restrict the discussion to cases described by classical physics.

6.1 Gibbs Statistics

Many macroscopic properties of a physical system are determined by the whole set of its microscopic characteristics. Now the number of particles in a macroscopic object is quite enormous, of the order of 6×10^{23}. There is thus no hope of determining the macroscopic properties of the system by determining the exact state of each particle included in it. However, we shall see that it is possible to determine certain thermodynamic properties of the system in a simpler way. It is precisely the aim of statistical physics to obtain such information. Indeed, statistical physics provides a way of calculating thermodynamic quantities from the Hamiltonian, and the Hamiltonian describes microscopic properties of the system. This approach is useful in many ways. In particular, and as we have just said, we do not need to know the the dynamical evolution of the exact state of each particle as a function of time. We do not therefore need to integrate the dynamical equations for each particle, and as a consequence we do not need to know those equations.

We shall only analyze the case of Gibbs canonical statistics in which we consider a system Γ in contact with a thermostat or heat bath. (We are only concerned here with systems described by classical physics. Quantum statistical systems such as Fermi–Dirac or Bose–Einstein statistics will not be discussed.) We also assume that the system can exchange energy with the thermostat and that its instantaneous energy content can therefore fluctuate. The number of particles N in the system remains constant, however. At thermodynamic equilibrium, we assume that the thermostat and the system have been in contact for a sufficiently long time to ensure that the average macroscopic properties are no longer time dependent. In other words, the macroscopic properties are stationary quantities. In particular, although the energy in Γ can fluctuate, its mean value, understood in the sense of the expectation value, will be assumed constant and fixed by the thermal energy of the thermostat.

A state of the system, denoted by X, is defined by giving the whole set of coordinates of all the particles included in the system. The set Ω of states of Γ constitutes the phase space of the system. For example, if we are interested in the magnetic properties of a solid, we can set up a simple model by assuming that the solid is made up of N magnetic moments. There exist materials in which it is reasonable to assume that the projection of this magnetic moment along a given axis can only take on a finite number of values, which we shall denote by $m_j\mu$ for the jth magnetic moment, where m_j is a whole number and μ is a constant determined by microscopic properties. A state of the system is then defined by giving the whole set of values of the numbers m_j, i.e., $X = \{m_1, m_2, \ldots, m_N\}$. We can thus see that Ω is generally embedded in a space of very high dimension indeed (see Fig. 6.1).

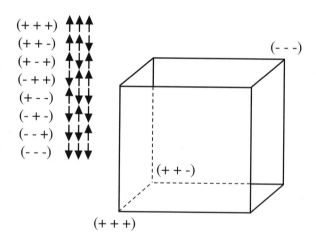

Fig. 6.1. Schematic illustration of the phase space when there are three particles and the state of each particle is represented by a magnetic moment which can only take on two values

We shall only be concerned with the probability at thermodynamic equilibrium of finding the system in a given state X. In order to determine this probability distribution, generally referred to as the Gibbs distribution, we shall formulate two hypotheses:

- the mean energy of the system is fixed by the thermostat,
- the system has evolved toward the probability distribution which maximizes its entropy.

We thus consider that, at thermodynamic equilibrium, the system has evolved toward "maximal disorder" in the entropic sense which we described in the last chapter. More precisely, we assume that at equilibrium, and for a fixed mean energy value, the probability law for the various states of the system is the one which leads to the maximal number of different sequences that can be generated during independent realizations. It is worth noting that we can also arrive at Gibbs statistics by applying arguments of limit theorem type. In the special case of a sum of random variables, we have already seen that the probability distribution obtained is the one which maximizes the entropy. There should therefore be no surprise in finding that these two methods (of limit theorem type and the maximum entropy principle) lead to the same result.

Let $P(X)$ be the probability of observing the system Γ in the state X and $H(X)$ the energy (or Hamiltonian) of this state. In order for $P(X)$ to be a probability distribution, it must satisfy $\sum_{X \in \Omega} P(X) = 1$. The mean energy $\langle H(X) \rangle$ is denoted by U, assumed fixed by the thermostat. Consequently, we must also take into account the constraint

$$\sum_{X \in \Omega} H(X) P(X) = U .$$

The problem now is simply to maximize the Shannon entropy:

$$S_{\text{Shannon}} = - \sum_{X \in \Omega} P(X) \ln P(X) ,$$

whilst maintaining the two constraints mentioned above. The Lagrange function is thus

$$\Psi_{\alpha,\beta}(P) = - \sum_{X \in \Omega} P(X) \ln P(X) - \alpha \sum_{X \in \Omega} P(X) - \beta \sum_{X \in \Omega} H(X) P(X) ,$$

where α and β are the Lagrange multipliers. The law P which maximizes $\Psi_{\alpha,\beta}(P)$ therefore satisfies

$$\frac{\partial}{\partial P(X)} \Psi_{\alpha,\beta}(P) = 0 ,$$

which reads $-1 - \ln P(X) - \alpha - \beta H(X) = 0$, or alternatively,

$$P(X) = \exp\left[-1 - \alpha - \beta H(X) \right] .$$

Identification of the parameter β is slightly delicate. It is common practice to express the Gibbs distribution in terms of this parameter, which is therefore used to index the probability law, but without substituting in its expression as a function of the mean energy. However, we can express α in terms of β, so that Gibbs statistics[1] is given by

$$P_\beta(X) = \frac{\exp\left[-\beta H(X)\right]}{Z_\beta} ,$$

where the partition function Z_β is defined by

$$Z_\beta = \sum_{X \in \Omega} \exp\left[-\beta H(X)\right] .$$

In thermodynamics, the entropy is defined by

$$S = -k_\mathrm{B} \sum_{X \in \Omega} P(X) \ln P(X) ,$$

where k_B is the Boltzmann constant, equal to approximately 1.38×10^{-23} J/K. We adopt this convention in what follows. The temperature in kelvins is defined [1, 15] by $1/T = \partial S / \partial U$ and it is easy to show that $\beta = 1/(k_\mathrm{B} T)$. Indeed, at equilibrium, we have

$$S = -k_\mathrm{B} \sum_{X \in \Omega} \left[-\beta H(X) - \ln Z_\beta\right] \frac{\exp\left[-\beta H(X)\right]}{Z_\beta} ,$$

so that

$$\frac{1}{k_\mathrm{B}} S = \beta \langle H(X) \rangle + \ln Z_\beta .$$

We thus find that

$$\frac{1}{k_\mathrm{B}} \frac{\partial S}{\partial \beta} = \langle H(X) \rangle + \beta \frac{\partial \langle H(X) \rangle}{\partial \beta} + \frac{\partial}{\partial \beta} \ln Z_\beta .$$

Clearly, $\partial \ln Z_\beta / \partial \beta = -\langle H(X) \rangle = -U$, and hence

$$\frac{1}{k_\mathrm{B}} \frac{\partial S}{\partial \beta} = \beta \frac{\partial U}{\partial \beta} .$$

β is a function of U and $\partial S / \partial U = 1/T$. We thus deduce that

$$\frac{\partial S}{\partial \beta} = \frac{\partial S}{\partial U} \frac{\partial U}{\partial \beta} = \frac{1}{T} \frac{\partial U}{\partial \beta} ,$$

and hence,

---

[1] It is common practice to speak of Gibbs statistics, although it is not a statistic, but rather a family of probability laws.

$$\frac{\partial S}{\partial \beta} = \frac{1}{T}\frac{\partial U}{\partial \beta} = k_{\mathrm B}\beta\frac{\partial U}{\partial \beta} \ ,$$

which implies finally that $k_{\mathrm B}\beta = 1/T$.

The absolute temperature $T_{\mathrm a}$ is equal to $k_{\mathrm B}$ times the temperature in degrees kelvin, i.e., $T_{\mathrm a} = k_{\mathrm B}T$. At thermodynamic equilibrium and at the temperature $1/\beta = T_{\mathrm a} = k_{\mathrm B}T$, we denote the thermodynamic entropy by S_β and the mean energy by U_β. The latter is also called the internal energy.

6.2 Free Energy

The free energy plays an important role in thermodynamics. In this section, we shall show that the differences of free energy between two macroscopic states are very simply related to the Kullback–Leibler measure between the probability laws associated with them.

We define the free energy of a system at thermodynamic equilibrium by

$$F_\beta = U_\beta - TS_\beta \ ,$$

where T is the temperature, S_β the entropy, and U_β the internal energy. At thermodynamic equilibrium, we have $S = k_{\mathrm B}\beta\langle H(X)\rangle + k_{\mathrm B}\ln Z_\beta$, which clearly implies

$$F_\beta = -k_{\mathrm B}T\ln Z_\beta = -\frac{1}{\beta}\ln Z_\beta \ .$$

Consider now a system Γ out of equilibrium and let $P(X)$ denote the probability law of its states X. The definition of the entropy is still applicable, so that $S = -k_{\mathrm B}\sum_{X\in\Omega} P(X)\ln P(X)$ and the mean energy is $U = \sum_{X\in\Omega} H(X)P(X)$. The free energy of the system out of equilibrium is then simply $F = U - TS$. (In this case, T is chosen equal to the temperature of the thermostat.)

The Kullback–Leibler measure between the law $P(X)$ and the equilibrium law $P_\beta(X) = \exp\left[-\beta H(X)\right]/Z_\beta$ is

$$K(P\,\|P_\beta) = \sum_{X\in\Omega} P(X)\ln\left[P(X)\frac{Z_\beta}{\exp\left[-\beta H(X)\right]}\right] \ ,$$

which can be written

$$K(P\,\|P_\beta) = \sum_{X\in\Omega} P(X)\ln P(X) + \beta\sum_{X\in\Omega} P(X)H(X) + \ln Z_\beta \ .$$

This means that $K(P\,\|P_\beta) = -S/k_{\mathrm B} + \beta U - \beta F_\beta$. We thus observe that the difference between the free energies can be simply expressed in terms of the Kullback–Leibler measure between the laws $P(X)$ and $P_\beta(X)$:

$$F - F_\beta = k_{\mathrm B}TK(P\,\|P_\beta) \ .$$

We know that the Kullback–Leibler measure is positive and that it is zero if $P(X) = P_\beta(X)$. We thus see that the free energy is minimal at thermodynamic equilibrium. In other words, if we consider a system out of equilibrium and place it in contact with a thermostat at temperature T, it will evolve toward the equilibrium state which corresponds to the minimum free energy. We can nevertheless imagine that the entropy might decrease if, for example, the system has a much higher internal energy than the one that would be imposed on it by contact with the thermostat.

6.3 Connection with Thermodynamics

The partition function can be used to deduce thermodynamic quantities at equilibrium such as the entropy, the internal energy and the free energy. We have already seen that the free energy is given by

$$F_\beta = -\frac{1}{\beta} \ln Z_\beta = -k_{\mathrm{B}} T \ln Z_\beta .$$

It is easy to see that the internal energy can be written

$$U_\beta = -\frac{\partial \ln Z_\beta}{\partial \beta} ,$$

and that the entropy is given by

$$S_\beta = -k_{\mathrm{B}} \beta \frac{\partial}{\partial \beta} \ln Z_\beta + k_{\mathrm{B}} \ln Z_\beta .$$

The internal energy and entropy can also be expressed in terms of the free energy:

$$U_\beta = \frac{\partial(\beta F_\beta)}{\partial \beta} \quad \text{and} \quad S_\beta = -\frac{\partial F_\beta}{\partial T} .$$

In Section 3.10, we gave several examples of conjugate intensive and extensive quantities. Two quantities are said to be conjugate if their product has units of energy and if they appear in the various thermodynamic energy functions. To be more precise, let $H_0(X)$ be the energy of the configuration X of the system Γ when there is no applied external field. Consider an applied field h which corresponds to the intensive quantity conjugate to the extensive quantity $M(X)$. To fix ideas, we consider the example in which $M(X)$ is the total magnetization for the configuration X and h is an applied magnetic field. We assume that, in the presence of the field, the energy $H_h(X)$ of the configuration X can be written

$$H_h(X) = H_0(X) - h M(X) .$$

The mean value M_β of $M(X)$ at equilibrium and the static susceptibility at equilibrium defined by $\chi_\beta = \partial M_\beta / \partial h$ are then easily determined as a function of the free energy. Indeed, we have

$$Z_\beta = \sum_{X \in \Omega} \exp\left[-\beta H_0(X) + \beta h M(X) \right] ,$$

and hence,

$$\frac{\partial Z_\beta}{\partial h} = \sum_{X \in \Omega} \beta M(X) \exp\left[-\beta H_0(X) + \beta h M(X) \right] .$$

(To simplify the notation, the dependence on the field h will be omitted when there is no risk of ambiguity.) Since,

$$\frac{\partial \ln Z_\beta}{\partial h} = \frac{1}{Z_\beta} \frac{\partial Z_\beta}{\partial h} ,$$

we obtain

$$\frac{\partial}{\partial h}\left(\frac{1}{\beta} \ln Z_\beta \right) = \sum_{X \in \Omega} M(X) \frac{\exp\left[-\beta H_0(X) + \beta h M(X) \right]}{Z_\beta} ,$$

or

$$M_\beta = \frac{\partial}{\partial h}\left(\frac{1}{\beta} \ln Z_\beta \right) \quad \text{and} \quad M_\beta = -\frac{\partial F_\beta}{\partial h} .$$

It follows immediately that

$$\chi_\beta = \frac{\partial^2}{\partial h^2}\left(\frac{1}{\beta} \ln Z_\beta \right) \quad \text{and} \quad \chi_\beta = -\frac{\partial^2 F_\beta}{\partial h^2} .$$

In Section 6.4, we will show that the susceptibility χ_β is positive. We then have

$$\frac{\partial^2 F_\beta}{\partial h^2} \leqslant 0 ,$$

which means that the free energy is a concave function of its argument h. Table 6.1 sums up the main properties we have just shown.

6.4 Covariance of Fluctuations

At nonzero temperatures, any extensive macroscopic thermodynamic quantity fluctuates due to thermal agitation. Indeed, each element making up the system will change its state as time goes by, so that the global state of the system will also vary with time.

Without loss of generality, we will treat magnetization as our example of an extensive macroscopic quantity. The conjugate intensive field will then be the applied magnetic field h. From a microscopic point of view, we define the local magnetization $m(r, t)$ at the point r at time t. As mentioned above, there is no hope of integrating the equations governing the dynamical evolution of the local magnetization. On the other hand, we can try to

Table 6.1. The main relations between the partition function and thermodynamic quantities

Partition function	$Z_\beta = \sum_{X \in \Omega} \exp\left[-\beta H(X) \right]$
Free energy	$F_\beta = -\dfrac{1}{\beta} \ln Z_\beta = -k_B T \ln Z_\beta$
Internal energy	$U_\beta = -\dfrac{\partial}{\partial \beta} \ln Z_\beta = \dfrac{\partial}{\partial \beta}(\beta F_\beta)$
Entropy	$S_\beta = -k_B \beta \dfrac{\partial}{\partial \beta} \ln Z_\beta + k_B \ln Z_\beta = -\dfrac{\partial}{\partial T} F_\beta$
Magnetization	$M_\beta = \dfrac{\partial}{\partial h}\left(\dfrac{1}{\beta} \ln Z_\beta \right) = -\dfrac{\partial}{\partial h} F_\beta$
Susceptibility	$\chi_\beta = \dfrac{\partial^2}{\partial h^2}\left(\dfrac{1}{\beta} \ln Z_\beta \right) = -\dfrac{\partial^2}{\partial h^2} F_\beta$

obtain information about the total magnetization $M(t)$. Let L be the lattice on which the magnetic atoms are located. The total magnetization is then $M(t) = \sum_{r \in L} m(r, t)$. We thus see that $M(t)$ is a sum of random variables and so, according to our investigations in Chapter 4, it is no surprise that we can obtain interesting information about this quantity.

In order to simplify our analysis, we assume that the system is homogeneous and ergodic. Moreover, we assume that it is in thermodynamic equilibrium. The quantities characterising it are therefore stationary. We define the total covariance function by

$$\Gamma_{mm}(r_1, t_1, r_2, t_2) = \langle m(r_1, t_1) m(r_2, t_2) \rangle - \langle m(r_1, t_1) \rangle \langle m(r_2, t_2) \rangle \ .$$

Since the system is stationary and homogeneous, we may write

$$\Gamma_{mm}(r, t, r + d, t + \tau) = \Gamma_{mm}(d, \tau) \ .$$

The spatial covariance function is obtained by putting $\tau = 0$ in the last expression:

$$\Gamma_{mm}(d) = \Gamma_{mm}(d, 0) = \Gamma_{mm}(r, t, r + d, t) \ .$$

Once again, there is an abuse of notation here, using the same symbol for functions of one, two or four variables. This simplifies things, provided that no ambiguity is thereby introduced.

The homogeneity hypothesis implies that $\langle m(r) \rangle = \langle m(r + d) \rangle$ and hence that

$$\Gamma_{mm}(d) = \langle m(r) m(r + d) \rangle - \langle m(r) \rangle^2 \ .$$

Because this spatial covariance does not depend on the time, it can be simply expressed in terms of Gibbs statistics. For this purpose, let $m(r, X)$ be the magnetization of the atom located at site r when the system is in state X. Then,

$$\Gamma_{mm}(\boldsymbol{d}) = \sum_{X \in \Omega} \left[m(\boldsymbol{r}, X) m(\boldsymbol{r} + \boldsymbol{d}, X) - \langle m(\boldsymbol{r}, X) \rangle^2 \right] P_\beta(X) \, ,$$

where $P_\beta(X) = \exp \left[-\beta H(X) \right] / Z_\beta$. As the system is homogeneous, we have $\langle m(\boldsymbol{r}, X) \rangle = \langle M(X) \rangle / N$, where N is the total number of magnetic atoms in the system. The spatial covariance can also be written

$$\Gamma_{mm}(\boldsymbol{d}) = \sum_{X \in \Omega} \left[m(\boldsymbol{r}, X) m(\boldsymbol{r} + \boldsymbol{d}, X) P_\beta(X) \right] - \frac{\langle M(X) \rangle^2}{N^2} \, .$$

The spatial covariance function can often be described by an exponential function:

$$\Gamma_{mm}(\boldsymbol{d}) = \Gamma_0 \exp \left(-\frac{\|\boldsymbol{d}\|}{\xi} \right) \, ,$$

where ξ then defines the correlation length of the fluctuations.

The total power of the fluctuations is $\langle \left[M(X) - M_\beta \right]^2 \rangle$, where we have used the notation $M_\beta = \langle M(X) \rangle$ to emphasize the dependence on β. We have seen that

$$M_\beta = \frac{\partial}{\partial h} \left(\frac{1}{\beta} \ln Z_\beta \right) \, ,$$

where

$$Z_\beta = \sum_{X \in \Omega} \exp \left[-\beta H_0(X) + \beta h M(X) \right] \, .$$

This expression was obtained by noting that

$$\frac{1}{\beta} \left(\frac{\partial \ln Z_\beta}{\partial h} \right) = \sum_{X \in \Omega} M(X) \frac{\exp \left[-\beta H_0(X) + \beta h M(X) \right]}{Z_\beta} \, .$$

Differentiating this expression a second time with respect to h, we obtain

$$\frac{1}{\beta^2} \left(\frac{\partial^2 \ln Z_\beta}{\partial h^2} \right) = \sum_{X \in \Omega} [M(X)]^2 P_\beta(X) - \left[\sum_{X \in \Omega} M(X) P_\beta(X) \right]^2 \, .$$

The details of the calculation are perfectly analogous to those given in Section 6.6. We thus observe that

$$\frac{1}{\beta^2} \left(\frac{\partial^2}{\partial h^2} \ln Z_\beta \right) = \langle [M(X)]^2 \rangle - [\langle M(X) \rangle]^2 \, .$$

We have seen that

$$\chi_\beta = \frac{\partial^2}{\partial h^2} \left(\frac{1}{\beta} \ln Z_\beta \right) \, ,$$

and we thus deduce that the total power of the fluctuations has the following simple expression in terms of the susceptibility:

$$\langle [M(X)]^2 \rangle - [\langle M(X) \rangle]^2 = \frac{\chi_\beta}{\beta} \ .$$

This result, which we shall call the total fluctuation theorem, shows above all that the susceptibility is a positive quantity.

Let us write this relation as a function of the Boltzmann constant and the absolute temperature in kelvins. We have $\beta = 1/(k_B T)$ so that

$$\langle [M(X)]^2 \rangle - [\langle M(X) \rangle]^2 = k_B T \chi_\beta \ .$$

The power of the fluctuations in the extensive quantity $M(X)$ is thus proportional to T and χ_β. The dependence on T is easily understood. The higher the value of T, the greater the thermal agitation, and this is what favors large fluctuations. A large value of χ_β implies that the system is "flexible" in the sense that it can react vigorously to the application of the intensive field h conjugate to $M(X)$. This flexibility in the system also leads to a high reactivity with regard to thermal agitation and so to big fluctuations. Note, however, that nothing proves that χ_β will be an increasing function of temperature. It can be shown that the susceptibility is in fact a decreasing function of temperature in the case of perfect paramagnetizm, which corresponds to a system of magnetic moments without interactions. We shall see later that the susceptibility is also a decreasing function of temperature when it is slightly above the critical temperature during a second order phase transition. It may happen that the product $T\chi_\beta$ is a decreasing function of T, in which case the same is true of the fluctuation power. Note finally that an analogous calculation leads to

$$\langle [H(X)]^2 \rangle - [\langle H(X) \rangle]^2 = -\frac{\partial U_\beta}{\partial \beta} \ .$$

We may also examine temporal correlations:

$$\Gamma_{MM}(\tau) = \langle M(t) M(t+\tau) \rangle - \langle M(t) \rangle \langle M(t+\tau) \rangle \ .$$

Let $M(t)$ be the magnetization in the state X^t at time t, so that $M(t) = M(X^t)$, and hence, $M(t) = \sum_{r \in R} m(r, t)$. If $P_{\beta, t, t+\tau}(X, Y)$ is the joint probability that the system is in the state X at time t and in the state Y at time $t + \tau$, we can write

$$\Gamma_{MM}(\tau) = \sum_{X \in \Omega} \sum_{Y \in \Omega} [M(X) M(Y) - \langle M(X) \rangle \langle M(Y) \rangle] \, P_{\beta, t, t+\tau}(X, Y) \ .$$

Although the situation here is analogous to the one discussed just previously, it is in fact slightly more involved, because we do not yet have a simple expression for $P_{\beta, t, t+\tau}(X, Y)$. This point will be analyzed further in Section 6.6.

6.5 A Simple Example

As an illustration of the above ideas, we consider a system comprising N particles, each of which has two energy levels E_1 and E_2. It is easy to deduce the thermodynamic quantities at equilibrium from the partition function

Fig. 6.2. Schematic representation of a system of particles with two energy levels

$$Z_\beta = \sum_{X \in \Omega} \exp\left[-\beta H(X)\right] .$$

Here $H(X)$ is simply $\sum_{i=1}^{N}\left[(1 - n_i)E_1 + n_i E_2\right]$, where n_i is equal to 0 if the ith particle is the state with energy E_1 and 1 if it is in the state with energy E_2 (see Fig. 6.2).

We can thus write

$$Z_\beta = \sum_{X \in \Omega} \exp\left\{-\beta \sum_{i=1}^{N}\left[(1 - n_i)E_1 + n_i E_2\right]\right\} .$$

A configuration is defined by the set of numbers n_i which characterizes the state of each particle. We thus have $X = (n_1, n_2, \ldots, n_N)$ and

$$Z_\beta = \sum_{(n_1, n_2, \ldots, n_N)} \exp\left\{-\beta \sum_{i=1}^{N}\left[(1 - n_i)E_1 + n_i E_2\right]\right\} ,$$

or alternatively,

$$Z_\beta = \sum_{n_1=0}^{1} \sum_{n_2=0}^{1} \cdots \sum_{n_N=0}^{1} \exp\left\{-\beta \sum_{i=1}^{N}\left[(1 - n_i)E_1 + n_i E_2\right]\right\} .$$

Putting

$$z_\beta = \sum_{n=0}^{1} \exp\left\{-\beta\left[(1 - n)E_1 + n E_2\right]\right\} ,$$

we find that $Z_\beta = (z_\beta)^N$. Now $z_\beta = \exp(-\beta E_1) + \exp(-\beta E_2)$, and we can deduce that

$$Z_\beta = \left[\exp(-\beta E_1) + \exp(-\beta E_2)\right]^N ,$$

or

$$\ln Z_\beta = N \ln\left[\exp(-\beta E_1) + \exp(-\beta E_2)\right] .$$

Let $P_n(0)$ and $P_n(1)$ be the probabilities of finding a particle in the states with energy E_1 and E_2, respectively. Writing

$$P_n(n_1) = \sum_{n_2=0}^{1} \sum_{n_3=0}^{1} \cdots \sum_{n_N=0}^{1} \frac{e^{-\beta H(X)}}{Z_\beta} \,,$$

we then have

$$P_n(0) = \frac{\exp(-\beta E_1)}{\exp(-\beta E_1) + \exp(-\beta E_2)} \,,$$

and

$$P_n(1) = \frac{\exp(-\beta E_2)}{\exp(-\beta E_1) + \exp(-\beta E_2)} \,.$$

The mean numbers of particles $\langle N_1 \rangle$ and $\langle N_2 \rangle$ in states with energies E_1 and E_2 are therefore

$$\langle N_1 \rangle = \frac{N \exp(-\beta E_1)}{\exp(-\beta E_1) + \exp(-\beta E_2)} \,,$$

and

$$\langle N_2 \rangle = \frac{N \exp(-\beta E_2)}{\exp(-\beta E_1) + \exp(-\beta E_2)} \,.$$

Since $F_\beta = -\beta^{-1} \ln Z_\beta$, the free energy is

$$F_\beta = -N\beta^{-1} \ln \left[\exp(-\beta E_1) + \exp(-\beta E_2) \right] \,.$$

We can rewrite this expression in terms of $\Delta E = E_2 - E_1$ to give

$$F_\beta = NE_1 - N\beta^{-1} \ln \left[1 + \exp(-\beta \Delta E) \right] \,.$$

To determine the internal energy, we can use the relation

$$U_\beta = -\frac{\partial}{\partial \beta} \ln Z_\beta \,,$$

which gives

$$U_\beta = N \frac{E_1 \exp(-\beta E_1) + E_2 \exp(-\beta E_2)}{\exp(-\beta E_1) + \exp(-\beta E_2)} \,.$$

Note that we can also write

$$U_\beta = \langle N_1 \rangle E_1 + \langle N_2 \rangle E_2 \,.$$

We leave the calculation of the entropy as an exercise for the reader.

Let us now calculate the fluctuations in the mean energy. We have

$$\langle [H(X)]^2 \rangle - [\langle H(X) \rangle]^2 = -\frac{\partial U_\beta}{\partial \beta} \,.$$

After a short calculation, this implies that

$$\langle [H(X)]^2 \rangle - [\langle H(X) \rangle]^2 = N \frac{[\Delta E]^2 \exp(-\beta \Delta E)}{\left[1 + \exp(-\beta \Delta E) \right]^2} \,.$$

This result shows that the relative energy fluctuation

$$\frac{\langle [H(X)]^2 \rangle - [\langle H(X) \rangle]^2}{(\Delta E)^2}$$

is a function of

$$\frac{\Delta E}{k_{\mathrm{B}} T} .$$

In other words, energy fluctuations become significant if $k_{\mathrm{B}} T > \Delta E$.

6.6 Fluctuation–Dissipation Theorem

In Section 6.4, we found the total power of the fluctuations of an extensive thermodynamic quantity. In the present section, we shall establish a more precise result which shows that the relaxation function and the temporal covariance function of fluctuations at equilibrium are related by a simple equation, known as the fluctuation–dissipation theorem.

To begin with, consider a totally deterministic experiment in which we investigate the return to the equilibrium value M_0 of an extensive macroscopic quantity $M(t)$ such as the magnetization. This return to equilibrium is usually measured when, having applied the conjugate intensive field for long enough to reach equilibrium, that field is suddenly switched off. We thus obtain the response $R_{h_0}(t)$ to an input of the form $h(t) = h_0[1 - \theta(t)]$, where $\theta(t)$ is the Heaviside step function (see Section 3.10 for more details). We may thus determine the response or relaxation function

$$\sigma(t) = \lim_{h_0 \to 0} \frac{R_{h_0}(t) - M_0}{h_0} ,$$

or

$$\sigma(t) = \left. \frac{\partial R_{h_0}(t)}{\partial h_0} \right|_{h_0 = 0} .$$

The linear limit is obtained when h_0 is small enough to ensure that $R_{h_0}(t) - M_0 \approx h_0 \sigma(t)$. This situation, represented schematically in Fig. 6.3, corresponds to the experimental configuration for measuring $\sigma(t)$.

The spontaneous fluctuations in the extensive quantity $M(t)$ are characterized by their covariance function $\Gamma_{MM}(\tau)$. This is shown schematically in Fig. 6.4. Formally, we should write $M_\lambda(t)$, for it is indeed a stochastic process. (The random event λ is related to the state X of the system at a given time, such as $t = 0$.) However, to simplify the notation, we shall drop explicit mention of this dependence on the random event λ. In the stationary case, the covariance function can be written

$$\Gamma_{MM}(\tau) = \langle M(t) M(t + \tau) \rangle - \langle M(t) \rangle \langle M(t + \tau) \rangle .$$

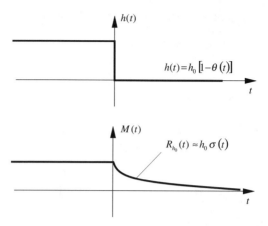

Fig. 6.3. Illustration of the relaxation function or response function in the linear case with $M_0 = 0$

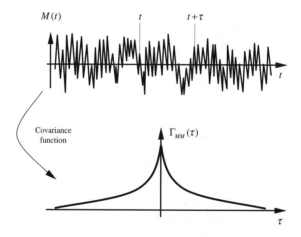

Fig. 6.4. The covariance function characterizes to second order the spontaneous fluctuations in an extensive quantity. In this case, $M_0 = 0$

Stationarity implies that $\langle M(t+\tau) \rangle = \langle M(t) \rangle$. Hence,

$$\Gamma_{MM}(\tau) = \langle M(t)M(t+\tau) \rangle - \langle M(t) \rangle^2 \,.$$

Let $\tilde{P}_t(X)$ be the probability of finding the system in the state X at time $t > 0$ in the absence of any applied field and given that, for negative times $t < 0$, a field of amplitude h_0 was applied. $\tilde{P}_t(X)$ is not therefore representative of equilibrium because the field was suddenly brought to zero at $t = 0$. On the other hand, we have

$$R_{h_0}(t) = \sum_{X \in \Omega} M(X)\tilde{P}_t(X) \, .$$

If $\tilde{P}_{t,0}(X,Y)$ represents the joint probability that the system is in state Y at $t = 0$ and state X at a positive time t, we can then write $\tilde{P}_t(X) = \sum_{Y \in \Omega} \tilde{P}_{t,0}(X,Y)$. At $t = 0$, the system is in thermodynamic equilibrium because we assume that the external field was applied at $t = -\infty$. In practice, this amounts precisely to assuming that the field has been applied long enough for equilibrium to be reached. In the presence of an external applied field, let $P_{\beta,h_0}(Y)$ denote the Gibbs probability, so that

$$P_{\beta,h_0}(Y) = \frac{\exp[-\beta H(Y)]}{Z_\beta} \, ,$$

where

$$H(Y) = H_0(Y) - h_0 M(Y) \, .$$

According to Bayes' relation, we can also write

$$\tilde{P}_{t,0}(X,Y) = \tilde{P}_{t,0}(X \mid Y) P_{\beta,h_0}(Y) \, .$$

Therefore,

$$R_{h_0}(t) = \sum_{X \in \Omega} \sum_{Y \in \Omega} M(X) \tilde{P}_{t,0}(X \mid Y) P_{\beta,h_0}(Y) \, ,$$

and since $\sigma(t) = \partial R_{h_0}/\partial h_0 |_{h_0=0}$, it is easy to see that

$$\sigma(t) = \sum_{Y \in \Omega} M(X) \left[\tilde{P}_{t,0}(X \mid Y) \frac{\partial}{\partial h_0} P_{\beta,h_0}(Y) \Big|_{h_0=0} \right] \, .$$

To obtain a simpler expression, we examine $\partial P_{\beta,h_0}(Y)/\partial h_0 \mid_{h_0=0}$. We have

$$\frac{\partial}{\partial h_0} \exp\left[-\beta H_0(Y) + \beta h_0 M(Y) \right] = \beta M(Y) \exp[-\beta H(Y)] \, ,$$

and hence,

$$\frac{\partial}{\partial h_0} P_{\beta,h_0}(Y) = \beta M(Y) \frac{e^{-\beta H(Y)}}{Z_\beta} - \frac{e^{-\beta H(Y)}}{Z_\beta^2} \left[\sum_{X \in \Omega} \beta M(X) e^{-\beta H(X)} \right] \, .$$

Thus,

$$\frac{\partial}{\partial h_0} P_{\beta,h_0}(Y) \Big|_{h_0=0} = \beta \big[M(Y) - \langle M(X) \rangle \big] P_{\beta,0}(Y) \, .$$

We then see that we can write

$$\sigma(t) = \sum_{X \in \Omega} \sum_{Y \in \Omega} \beta M(X) \big[M(Y) - \langle M(X) \rangle \big] \tilde{P}_{t,0}(X \mid Y) P_{\beta,0}(Y) \, .$$

However, $\tilde{P}_{t,0}(X \mid Y)P_{\beta,0}(Y)$ is the joint probability $P_{\beta,t,0}(X,Y)$ at thermodynamic equilibrium that the system is in state Y at time $t = 0$ and in state X at time t. Indeed, we have $h_0 = 0$ in the last equation, which means that there is no discontinuity in the intensive quantity at $t = 0$. We therefore find that

$$\sigma(t) = \beta \sum_{X \in \Omega} \sum_{Y \in \Omega} \left[M(X)M(Y) - M(X)\langle M(X) \rangle \right] P_{\beta,t,0}(X,Y) .$$

Now

$$\sum_{X \in \Omega} \sum_{Y \in \Omega} M(X)P_{\beta,t,0}(X,Y) = \langle M(X) \rangle ,$$

and hence,

$$\sigma(t) = \beta \left\{ \sum_{X \in \Omega} \sum_{Y \in \Omega} M(X)M(Y)P_{\beta,t,0}(X,Y) - [\langle M(X) \rangle]^2 \right\} .$$

The covariance function of the fluctuations in the extensive quantity is

$$\Gamma_{MM}(t) = \sum_{X \in \Omega} \sum_{Y \in \Omega} M(X)M(Y)P_{\beta,t,0}(X,Y) - [\langle M(X) \rangle]^2 .$$

This relation can simply be written $\sigma(t) = \beta \Gamma_{MM}(t)$. It has been established for positive times t. Let us now consider what happens for negative t. Clearly, $\Gamma_{MM}(t)$ is an even function, whilst $\sigma(t)$ is only defined for positive t. We make the convention that $\sigma(t) = 0$ for negative t. We then obtain the fluctuation–dissipation theorem, which simply says that

$$\sigma(|t|) = \beta \Gamma_{MM}(t) .$$

This relates the relaxation function $\sigma(t)$ and the covariance function $\Gamma_{MM}(t)$ of the fluctuations at thermodynamic equilibrium in the absence of an applied field.

The Wiener–Khinchine theorem says that the Fourier transform of the covariance function of stationary stochastic processes is equal to the spectral density of the fluctuations $\hat{S}_{MM}(\nu)$. The impulse response is related to the relaxation function by $\chi(t) = -d\sigma(t)/dt$, so that in Fourier space, $\hat{\chi}(\nu) = -i2\pi\nu\hat{\sigma}(\nu)$. The susceptibility $\hat{\chi}(\nu)$ is traditionally expressed in physics in terms of its real and imaginary parts, viz., $\hat{\chi}(\nu) = \hat{\chi}'(\nu) - i\hat{\chi}''(\nu)$. Noting that $\sigma_S(t) = \sigma(|t|)$, we immediately obtain $\hat{\chi}''(\nu) = \pi\nu\hat{\sigma}_S(\nu)$. Then recalling that $\beta = 1/k_{\mathrm{B}}T$, where k_{B} is Boltzmann's constant and T the absolute temperature in kelvins, we find that

$$\hat{S}_{MM}(\nu) = k_{\mathrm{B}}T \frac{\hat{\chi}''(\nu)}{\pi\nu} .$$

This relation is the fluctuation–dissipation theorem in Fourier space.

We thus know the spectral density of fluctuations of an extensive quantity at thermodynamic equilibrium. To determine its probability density function, we assume that the spatial correlation length is very small compared with the size of the physical system under investigation. (We assume that variations in the macroscopic quantity can be described by a continuous variable. This is generally a very good approximation.) In this case, we may consider the extensive quantity to be the sum of the values for each cluster which are mutually independent and have second finite moment. The central limit theorem then implies that the probability density function is Gaussian. We have thereby characterized the fluctuations by their probability density function and spectral density.

It is important to emphasize the generality of the result obtained here. It is true at thermodynamic equilibrium for every conjugate pair of intensive and extensive quantities. This generality is what makes the result so powerful and we shall illustrate it below in the case of electric circuits.

The fluctuations in the extensive quantity can be represented as equivalent fluctuations in an applied intensive field. We shall therefore determine the spectral density of the fluctuations in the intensive field $h_\lambda(t)$ that would lead to the fluctuations $M_\lambda(t)$ in the extensive quantity described by the fluctuation–dissipation theorem. We have

$$\hat{S}_{MM}(\nu) = |\hat{\chi}(\nu)|^2 \, \hat{S}_{hh}(\nu) \,,$$

and since $\hat{S}_{MM}(\nu) = k_B T \hat{\chi}''(\nu)/\pi\nu$, we thus see that we should have

$$\hat{S}_{hh}(\nu) = \frac{k_B T \hat{\chi}''(\nu)}{|\hat{\chi}(\nu)|^2 \, \pi\nu} \,.$$

This formulation will prove useful below.

6.7 Noise at the Terminals of an RC Circuit

Current noise is produced by a resistance at nonzero temperature in an electric circuit. This noise can be characterized to second order using the fluctuation–dissipation theorem. We shall consider the case of an RC circuit with the resistor and capacitor connected in series (see Fig. 6.5).

The electric charge Q is the extensive quantity conjugate to the voltage V. To determine the susceptibility, we must consider the deterministic response. We calculate the charge $Q(t)$ on the plates of the capacitor when the applied voltage is $V(t) = V_0\,[1 - \theta(t)]$, with $\theta(t)$ the Heaviside distribution. It is well known from the theory of electric circuits that

$$R\frac{dQ(t)}{dt} + \frac{Q(t)}{C} = V(t) \,.$$

The solution to this differential equation is $Q(t) = CV_0 \exp(-t/RC)$. The response function is therefore $\sigma(t) = C \exp\left(-t/RC\right)$ and the impulse response

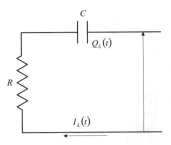

Fig. 6.5. Series RC circuit. $Q_\lambda(t)$ represents the charge fluctuations and $I_\lambda(t)$ the fluctuations in $dQ_\lambda(t)/dt$

is $\chi(t) = (1/R) \exp(-t/RC)$, since we know from Section 3.10 that $\chi(t) = -d\sigma(t)/dt$. It is easy to deduce the a.c. susceptibility by Fourier transform:

$$\hat{\chi}(\nu) = \frac{C}{1 + i2\pi RC\nu} .$$

Let $\hat{\chi}(\nu) = \hat{\chi}'(\nu) - i\hat{\chi}''(\nu)$ with $\hat{\chi}'(\nu)$ and $\hat{\chi}''(\nu)$ real. Then,

$$\hat{\chi}'(\nu) = \frac{C}{1 + (2\pi RC\nu)^2} \quad \text{and} \quad \hat{\chi}''(\nu) = \frac{2\pi RC^2\nu}{1 + (2\pi RC\nu)^2} .$$

To characterize the fluctuations, we determine the covariance function $\Gamma_{QQ}(\tau)$ of the charge. We obtain

$$\Gamma_{QQ}(\tau) = k_B TC \exp\left(-\frac{|\tau|}{RC}\right) ,$$

directly from the fluctuation–dissipation theorem. The total power of the fluctuations is thus $k_B TC$. The spectral density $\hat{S}_{QQ}(\nu)$ is equal to $k_B T \hat{\chi}''(\nu)/(\pi\nu)$ or

$$\hat{S}_{QQ}(\nu) = 2k_B T \frac{RC^2}{1 + (2\pi RC\nu)^2} .$$

Naturally, we could have found this by calculating the Fourier transform of $\Gamma_{QQ}(\tau)$. To determine the spectral density $\hat{S}_{II}(\nu)$ of the current, we simply note that $I(t) = dQ(t)/dt$. Hence, $\hat{S}_{II}(\nu) = 4\pi^2\nu^2 \hat{S}_{QQ}(\nu)$, or

$$\hat{S}_{II}(\nu) = 2k_B T \frac{(2\pi\nu)^2 RC^2}{1 + (2\pi RC\nu)^2} .$$

In electronics, all the power of the fluctuations is often concentrated on the positive frequencies by setting $\hat{P}_{II}(\nu) = 2\hat{S}_{II}(\nu)\theta(\nu)$, whereupon

$$\hat{P}_{II}(\nu) = 4k_B T \frac{(2\pi\nu)^2 RC^2}{1 + (2\pi RC\nu)^2} .$$

If $RC\nu \gg 1$, we obtain $\hat{P}_{II}(\nu) \approx 4k_BT/R$.

Before commenting on the practical significance of this result, let us just point out the theoretical problems it raises. With this approach, we have

$$\int_{-\infty}^{\infty} \hat{S}_{II}(\nu)d\nu = \int_0^{\infty} \hat{P}_{II}(\nu)d\nu = +\infty .$$

On the other hand, $\int_{-\infty}^{\infty} \hat{S}_{QQ}(\nu)d\nu = k_BTC$. In real situations, the presence of unwanted self-inductances limits the spectral band of the fluctuations. In the presence of an inductance L placed in series in the circuit, we would have

$$\hat{S}_{QQ}(\nu) = 2k_BT\frac{RC^2}{(1 - 2\pi\nu^2 LC)^2 + (2\pi RC\nu)^2} ,$$

and hence,

$$\hat{S}_{II}(\nu) = 2k_BT\frac{4\pi^2\nu^2 RC^2}{(1 - 2\pi\nu^2 LC)^2 + (2\pi RC\nu)^2} .$$

We deduce that $\int_{-\infty}^{\infty} \hat{S}_{II}(\nu)d\nu$ is finite. The power dissipated across the resistance R is indeed bounded. In fact it is equal to $R\langle[I_\lambda(t)]^2\rangle$, or $R\int_{-\infty}^{\infty} \hat{S}_{II}(\nu)d\nu$.

This result can be represented in a more memorable way. The charge on the plates of the capacitor fluctuates due to thermal agitation. Let us see whether it is possible to represent these charge fluctuations as being produced by a random voltage generator. We saw in Section 6.6 that the spectral density of the fluctuations in the intensive quantity would then be

$$\hat{S}_{VV}(\nu) = \frac{k_BT\hat{\chi}''(\nu)}{|\hat{\chi}(\nu)|^2 \pi\nu} .$$

Let $\hat{Z}(\nu)$ denote the electrical impedance of the circuit. We have $\hat{u}(\nu) = \hat{Z}(\nu)\hat{j}(\nu)$ and $\hat{j}(\nu) = i2\pi\nu\hat{Q}(\nu)$, where $\hat{j}(\nu)$ is the Fourier transform of the current $j(t)$, and hence $\hat{Q}(\nu) = -i\hat{u}(\nu)/[2\pi\nu\hat{Z}(\nu)]$, or

$$\hat{\chi}(\nu) = -\frac{i\hat{Z}^*(\nu)}{2\pi\nu|\hat{Z}(\nu)|^2} .$$

We thus have

$$|\hat{\chi}(\nu)|^2 = \left[(2\pi\nu)^2|\hat{Z}(\nu)|^2\right]^{-1} \quad \text{and} \quad \hat{\chi}''(\nu) = \frac{\hat{Z}_R(\nu)}{2\pi\nu|\hat{Z}(\nu)|^2} ,$$

where $\hat{Z}_R(\nu)$ is the real part of $\hat{Z}(\nu)$. We deduce that the spectral density of the voltage fluctuations that would produce the same spectral density of current fluctuations as those due to thermal agitation would be given by

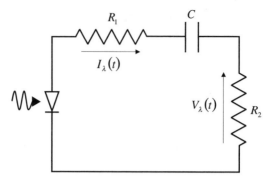

Fig. 6.6. Equivalent circuit for a simplified optical detector

$$\hat{S}_{VV}(\nu) = 2k_\mathrm{B}T\hat{Z}_\mathrm{R}(\nu)\ .$$

This result shows that we can represent the fluctuations due to thermal agitation by a random voltage generator with spectral density given by $\hat{S}_{VV}(\nu) = 2k_\mathrm{B}T\hat{Z}_\mathrm{R}(\nu)$. The real part of the electrical impedance RLC of the circuit is equal to the resistance R and we therefore find that

$$\hat{S}_{VV}(\nu) = 2k_\mathrm{B}TR\ ,$$

or equivalently,

$$\hat{P}_{VV}(\nu) = 4k_\mathrm{B}TR\ .$$

We now illustrate this result in the case of an optical detector. We assume that we may treat this detector as a perfect current generator placed in series with a resistance R_1 and a capacitance C, as shown in Fig. 6.6. We will also assume that the detector is mounted on a resistance R_2.

We begin by finding the potential difference $u^{(2)}(t)$ across the resistance R_2 when a current $i(t)$ is generated by the detector. The result is simply $u^{(2)}(t) = R_2 i(t)$.

Now consider the fluctuations in the potential difference $V_\lambda(t)$ across the resistance R_2. We have $\hat{P}_{II}(\nu) = 4k_\mathrm{B}T(2\pi\nu)^2 RC^2/\left[1 + (2\pi RC\nu)^2\right]$, where $R = R_1 + R_2$. We thus deduce that

$$\hat{P}_{VV}(\nu) = 4k_\mathrm{B}TR_2^2\frac{(2\pi\nu)^2(R_1 + R_2)C^2}{1 + \left[2\pi(R_1 + R_2)C\nu\right]^2}\ .$$

This calculation is commonly carried out in a different way in electronics. When there is no current in the detector, the noise is calculated by associating a random voltage generator with each resistor: $u_\lambda^{(1)}(t)$ for R_1 and $u_\lambda^{(2)}(t)$ for R_2. The spectral densities are $\hat{P}_{11}(\nu) = 4k_\mathrm{B}TR_1$ and $\hat{P}_{22}(\nu) = 4k_\mathrm{B}TR_2$ (see Fig. 6.7).

The spectral density of the random current is then

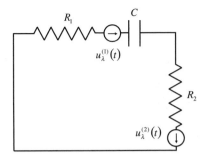

Fig. 6.7. Equivalent circuit with voltage generators associated with each resistor

$$\hat{P}_{II}(\nu) = \frac{\hat{P}_{11}(\nu) + \hat{P}_{22}(\nu)}{|\hat{Z}(\nu)|^2} .$$

Putting $\hat{Z}(\nu) = R_1 + R_2 + 1/(\mathrm{i}2\pi\nu C)$, $\hat{P}_{11}(\nu) = 4k_\mathrm{B}TR_1$, and $\hat{P}_{22}(\nu) = 4k_\mathrm{B}TR_2$, we find that

$$\hat{P}_{II}(\nu) = 4k_\mathrm{B}T(R_1 + R_2)\frac{(2\pi\nu C)^2}{1 + \left[2\pi\nu C(R_1 + R_2)\right]^2} ,$$

which is equivalent to the result obtained above.

In the high frequency regime, i.e., when $RC\nu \gg 1$, we have simply

$$\hat{P}_{VV}(\nu) \approx 4k_\mathrm{B}T\frac{R_2^2}{R_1 + R_2} .$$

In a frequency band of width $\delta\nu$, the noise power will therefore be

$$P_\mathrm{N} \approx \frac{4k_\mathrm{B}TR_2^2\delta\nu}{R_1 + R_2} ,$$

whilst the signal power will be $P_\mathrm{S} \approx R_2^2|\hat{\imath}(\nu)|^2\delta\nu$, where $\hat{\imath}(\nu)$ is the Fourier transform of $i(t)$. The signal-to-noise ratio will thus be

$$\frac{P_\mathrm{S}}{P_\mathrm{N}} \approx \frac{R_1 + R_2}{4k_\mathrm{B}T}|\hat{\imath}(\nu)|^2 .$$

The signal-to-noise ratio will therefore grow with increasing $R_1 + R_2$ and decreasing T.

Note that the result would be different if the detector was treated as a voltage generator $u(t)$. In this case, we would have (still assuming that $RC\nu \gg 1$) $\hat{\imath}(\nu) = \hat{u}(\nu)/(R_1 + R_2)$, $\hat{P}_{VV}(\nu) = |\hat{u}(\nu)|^2 R_2^2/(R_1 + R_2)^2$, and hence,

$$\frac{P_\mathrm{S}}{P_\mathrm{N}} \approx \frac{|\hat{u}(\nu)|^2}{4k_\mathrm{B}T(R_1 + R_2)} .$$

6.8 Phase Transitions

A phase transition corresponds to the appearance of a singularity in the free energy or one of its temperature derivatives. A sudden change is then observed in the thermodynamic properties of the physical system.[2] There are many well known examples of phase transitions. Melting corresponds to a solid–liquid transition, and evaporation to the liquid–gas transition. Different materials can manifest transitions with respect to various physical properties. Examples are the conductor–insulator transition and the paramagnetic–ferromagnetic transition.

Certain phase transitions are accompanied by a discontinuity in the entropy, leading to the existence of latent heat. This happens in the solid–liquid transition of water, wherein we may observe a change in density and the coexistence of the liquid and solid phases. These phase transitions are first order transitions, in the sense that the first derivatives of the free energy are discontinuous.

There is another type of phase transition which occurs without latent heat or discontinuities in the density. These are the second order phase transitions for which the discontinuities appear in the second derivatives of the free energy, i.e., the susceptibility, specific heat, and so on. An example is the paramagnetic–ferromagnetic phase transition. We shall consider this classic example here as an illustration. However, although we are using the language of magnetizm, it should be remembered that all the ideas discussed here can be generalised to other types of second order transition. This study, which has fascinated many physicists, has greatly stimulated the development of physics, especially the branch known as condensed matter physics. Here we shall be concerned only with those aspects which touch upon the fluctuations in extensive quantities that are characteriztic of the transition. Furthermore, we shall limit the discussion to a few very simple ideas, although nonetheless fundamental and illustrative.

To give a more precise description of the various types of phase transition, Landau introduced the notion of order parameter. This parameter is associated with the change of symmetry which accompanies the change of state between the two phases. It must be zero in the phase having the greater symmetry and nonzero in the other phase. The idea of symmetry must generally be understood in the wide sense. For the paramagnetic–ferromagnetic transition, it is related to the magnetization. When there is no applied magnetic field, the magnetization is zero in the paramagnetic phase but nonzero in the ferromagnetic phase. The situation is simplified if we consider anisotropic materials for which the magnetization is parallel to a favored direction in space. Once we have chosen a direction parallel to this favored direction, then below the critical temperature T_C and in the absence of any applied field, there are two possible situations: the magnetization can be positive or negative (see

---

[2] We are following the approach used in [1].

Fig. 6.8). Still in the absence of an applied field, cooling the material from a temperature above T_C down to a temperature below T_C, the symmetry is broken since either a positive or a negative magnetization will be obtained. The choice of sign occurs at the critical temperature, a temperature where the thermodynamic properties of the system are singular.

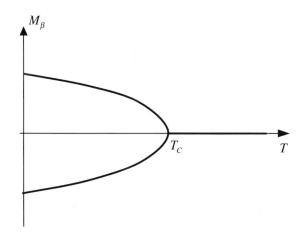

Fig. 6.8. Change in magnetization with temperature for the paramagnetic–ferromagnetic phase transition

The order parameter is then simply proportional to the magnetization since it is zero if $T > T_C$ and nonzero if $T < T_C$, and it does indeed describe the change in symmetry.

Landau's idea was to classify phase transitions in terms of variations in the order parameter. A transition is first order if the order parameter changes in a discontinuous manner at T_C, whilst it is second order if the order parameter changes continuously at T_C. The order parameter of the paramagnetic–ferromagnetic transition is proportional to the magnetization. Since it varies continuously near T_C, the transition is second order. We shall focus on this type of phase transition. The free energy, internal energy and entropy are all continuous at T_C. In contrast, second derivatives such as the specific heat and the susceptibility diverge. We have seen, for example, that $\chi_\beta = -\partial^2 F_\beta / \partial h^2$ diverges, as shown in Fig. 6.9.

The symmetry breaking which occurs at the critical temperature reveals itself through a break in the ergodicity. Indeed, although the system is effectively ergodic at temperatures above T_C, this is no longer so for lower temperatures. The configuration space Ω separates into two disjoint subspaces Ω_+ and Ω_- such that $\Omega = \Omega_+ \cup \Omega_-$. We define Ω_+ as the set of states X in which the magnetization $M(X)$ is positive, whilst Ω_- corresponds to states with negative magnetization.

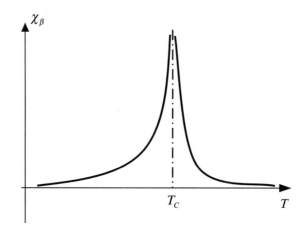

Fig. 6.9. Divergence in the magnetic susceptibility at the critical temperature T_C during a second order phase transition of paramagnetic–ferromagnetic type

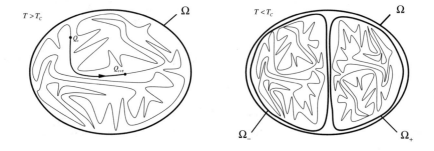

Fig. 6.10. Ergodicity breaking which occurs during the paramagnetic–ferromagnetic transition at the critical temperature T_C

In the low temperature phase, states evolve in such a way that the magnetization keeps the same sign. Let X^t be the state of the system at a given time t. For $T < T_C$, we observe that $M(X^t)$ remains strictly positive or strictly negative as time goes by. Hence if X^t is in Ω_+, it will stay in Ω_+ as the system evolves with time. Likewise if X^t is in Ω_-, it will stay in Ω_-. The two subspaces Ω_+ and Ω_- are invariant under the action of the evolution operator and they characterize the ergodicity breaking (see Fig. 6.10).

In practice, because of effects due to the finite size of the thermodynamic system, this ergodicity breaking is never perfect. The magnetization can then change sign. We will not analyze this aspect here and will simply assume that the ergodicity breaking is total.

We mentioned earlier that the susceptibility diverges at the critical temperature. The study of this divergence led to significant progress in understanding second order phase transitions. One usually defines the reduced temperature $t_R = (T - T_C)/T_C$. Then β is a function of t_R and T_C. Divergences are characterized by their asymptotic behavior which can often be written in the form of a power law. The susceptibility is then written

$$\chi_\beta \propto |t_R|^{-\gamma} \, ,$$

where the proportionality sign $\propto$ indicates that we are interested in the asymptotic behavior and will not describe any multiplicative factors. We may proceed in the same way with the other quantities. The specific heat varies as $C_V \propto |t_R|^{-\alpha}$ and the free energy as $F_\beta \propto |t_R|^\lambda$. The parameters α, λ and γ are the critical exponents.

6.9 Critical Fluctuations

The symmetry breaking which occurs at the critical temperature is reflected in the choice of a positive or negative magnetization at temperatures below the critical temperature. This is not without consequence for the behavior of the system in the vicinity of this temperature. Indeed it turns out that the power of the fluctuations in the total magnetization $M(X)$ diverges at T_C.

Consider temperatures slightly above T_C, so that we may assume the system to be ergodic. We have seen that

$$\langle [M(X)]^2 \rangle - [\langle M(X) \rangle]^2 = k_B T \chi_\beta \, .$$

As T approaches T_C, the susceptibility χ_β diverges and the same is therefore true for the power of the magnetization fluctuations $\langle [M(X)]^2 \rangle - [\langle M(X) \rangle]^2$. This is quite characteriztic of second order phase transitions. Near the critical temperature, the value of the susceptibility χ_β increases and so the response of the system to an external excitation is greater. However, this leads to a growth in the fluctuations in the macroscopic quantity since their power diverges (see Fig. 6.11).

From the standpoint of an engineer, we could say that if the signal increases (the signal being the response to some external excitation here), the same is true of the noise (i.e., fluctuations in the quantity). From a more fundamental standpoint, the divergence of the fluctuations lies at the origin of several fascinating types of thermodynamic behavior that can be observed close to the transition temperature, especially with regard to the critical exponents. Indeed, these parameters manifest universality properties, in the sense that their values do not depend on the details of interparticle interactions, but only on the dimension[3] of the system under investigation and symmetry properties of the order parameter. We shall briefly discuss this in the following.

---

[3] Two for a surface material and three for a bulk material.

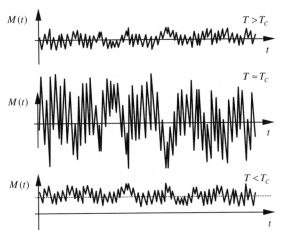

Fig. 6.11. Divergence in the power of the magnetization fluctuations at the critical temperature T_C for a second order transition of paramagnetic–ferromagnetic type

Let us consider the local magnetization $m(\boldsymbol{r},t)$ at time t of the atom located at the site $\boldsymbol{r}$. The total magnetization at time t is then simply $M(t) = \sum_{r \in L} m(\boldsymbol{r}, t)$, where L is the lattice on which the magnetic atoms are located. We have already defined the total covariance function by

$$\Gamma_{mm}(\boldsymbol{r}_1, t_1, \boldsymbol{r}_2, t_2) = \langle m(\boldsymbol{r}_1, t_1) m(\boldsymbol{r}_2, t_2) \rangle - \langle m(\boldsymbol{r}_1, t_1) \rangle \langle m(\boldsymbol{r}_2, t_2) \rangle \;.$$

For homogeneous, stationary systems, $\Gamma_{mm}(\boldsymbol{r}, t, \boldsymbol{r} + \boldsymbol{d}, t + \tau)$ does not depend only on $\boldsymbol{d}$ and τ, and we can thus simplify this expression:

$$\Gamma_{mm}(\boldsymbol{r}, t, \boldsymbol{r} + \boldsymbol{d}, t + \tau) = \Gamma_{mm}(\boldsymbol{d}, \tau) \;.$$

The spatial covariance function is simply

$$\Gamma_{mm}(\boldsymbol{d}) = \langle m(\boldsymbol{r}) m(\boldsymbol{r} + \boldsymbol{d}) \rangle - \langle m(\boldsymbol{r}) \rangle^2 \;.$$

It is obtained from the spatio-temporal covariance $\Gamma_{mm}(\boldsymbol{d}, \tau)$ by noting that $\Gamma_{mm}(\boldsymbol{d}) = \Gamma_{mm}(\boldsymbol{d}, 0)$. Well outside the neighbourhood of the critical point, the correlation function is generally well described by an exponential correlation function: $\Gamma_{mm}(\boldsymbol{d}) = \Gamma_0 \exp(-\|\boldsymbol{d}\|/\xi)$, where ξ is the correlation length.

As already mentioned, at the critical temperature, the total variance of the magnetization diverges:

$$\langle [M(X) - \langle M(X) \rangle]^2 \rangle \longrightarrow \infty \;.$$

We also note that, for $T \approx T_C$ and $T \geqslant T_C$, the correlation length ξ diverges as a power law in the reduced temperature: $\xi \propto |t_R|^{-\nu}$, where ν can be considered as a new critical exponent.

At the critical temperature, the spatial correlation function is well described by a power law: $\Gamma_{mm}(\boldsymbol{r}) \propto 1/\|\boldsymbol{r}\|^{\varsigma}$. This type of correlation function is invariant under transformations of the type $\boldsymbol{r} \to \boldsymbol{r}/a$, and this reflects an invariance under change of scale in the physical system. This property, known as scale invariance, lies at the heart of the so-called renormalization group techniques which have done much to improve our understanding of second order phase transitions.

Let us just note that the divergence of the fluctuation power at the critical temperature plays a fundamental role. Indeed, there are then clusters of all sizes which react collectively, and this tends to erase certain features related to the peculiarities of the individual microscopic entities. This is manifested through the universality properties of the critical exponents.

Exercises

Exercise 6.1. Introduction

Consider a system of N particles, each of which has two energy levels E_1 and E_2, with $E_1 < E_2$. Let q be the probability that a particle is in the state e_1 with energy E_1 and p the probability that it is in the state e_2 with energy E_2. We then have $p + q = 1$.

(1) What is the average energy U of the system expressed in terms of E_1, E_2 and p?
(2) What is the variance σ_U^2 of the total energy expressed in terms of E_1, E_2 and p?
(3) For what value of p is σ_U^2 maximal?
(4) Express the entropy S of the system in terms of p.
(5) For what value of p is the entropy maximal?

Assume now that the system is in thermodynamic equilibrium so that

$$p = \frac{\exp(-\beta E_2)}{\exp(-\beta E_2) + \exp(-\beta E_1)} \ .$$

(6) What is the value of σ_U^2 when the temperature tends to 0?
(7) What is the value of σ_U^2 when the temperature tends to $+\infty$?
(8) How does the entropy S evolve with temperature?

Let $E(n)$ be the energy of particle n, where $E(n) = E_1$ or $E(n) = E_2$. The energy $H(X)$ for a given configuration X is thus

$$H(X) = \sum_{n=1}^{N} E(n) \ .$$

(9) What is the probability distribution of the state X and what happens when the temperature tends to 0?

Exercise 6.2. Magnetic Field I

Consider a system of N independent particles, each of which has two states. In the presence of an applied magnetic field, the energy of the first state becomes $\epsilon_1 = E_0 - \alpha h$ whilst that of the second becomes $\epsilon_2 = E_0 + \alpha h$.

(1) Calculate the partition function for the system.
(2) Calculate the average energy.
(3) Study its behavior as a function of βh. Compare the value of the average energy with E_0.
(4) Interpret this result.

Exercise 6.3. Magnetic Field II

A system at thermodynamic equilibrium contains N non-interacting particles, each of which has $2n+1$ states with the same energy E_0 but different possible magnetic moments m_ℓ, where ℓ varies from $-n$ to n. In the presence of an applied magnetic field h, the energy of the state labeled ℓ becomes $E_\ell^h = E_0 - hm_\ell$, where m_ℓ is the magnetization of the state.

(1) Define the mean magnetization per particle by $\langle m \rangle = \sum_{\ell=-n}^{n} m_\ell P_\ell$, where P_ℓ is the probability of finding a given particle in the state ℓ. Show that the mean magnetization per particle can be written

$$\langle m \rangle = \frac{1}{N\beta} \frac{\partial \ln Z_{\beta,h}}{\partial h} .$$

(2) Determine $(1/N\beta^2)(\partial^2 \ln Z_{\beta,h}/\partial h^2)$. What is the physical meaning of this quantity?

Exercise 6.4. Gravity

Consider a physical system made up of N microscopic particles of mass m in suspension in a liquid placed in a reservoir. This reservoir is of infinite height and the base is located arbitrarily at the coordinate origin. We neglect the kinetic energy. To simplify the analysis, assume that the Ox axis (height) is quantized into discrete states:

$$x = np ,$$

where n is a positive or zero integer and p a positive real number. The potential energy of a particle can then be written

$$h(n) = mgnp , \quad n \geq 0 .$$

The total energy is thus

$$H(n_1, n_2, \ldots, n_N) = \sum_{j=1}^{N} mgn_j p \quad \text{with} \quad \forall j, \ n_j \geq 0 .$$

(1) Calculate the entropy of the system.
(2) Give the limiting value of the entropy when the temperature T goes to zero.
(3) Give the limiting value of the entropy when mgp is very small compared with k_BT, where k_B is Boltzmann's constant and T the temperature in kelvins.
(4) What happens when $p \longrightarrow 0$? Can you interpret?

Exercise 6.5. Two-State System

Consider a surface in contact with a mixture of gases, containing two sorts of atoms denoted A and B. We shall make a highly simplified model of the adsorption of atoms onto the surface. To this end, imagine that the surface comprises N sites at which an atom of type A or B can be adsorbed. Let E_A and E_B be the energies of sites occupied by atoms of type A and B, respectively. Let $C_A(T)$ and $C_B(T)$ be the concentrations at equilibrium of atoms of types A and B adsorbed onto the surface, where T is the temperature.

Assuming that we have experimental results representing the natural logarithm of the concentration ratio in terms of the reciprocal temperature, i.e., $\ln\left[C_A(T)/C_B(T)\right]$ as a function of $1/k_BT$, where k_B is Boltzmann's constant, suggest a way of deducing the energy difference $E_A - E_B$ from these results.

Exercise 6.6. Harmonic Oscillator

Consider a simplified thermodynamic system made up of N atoms which may oscillate independently of each other. The quantum harmonic oscillator is characterized as having an infinite number of equally spaced energy levels $E_n = nE_0$, where n is a non-negative integer. (We neglect the background energy $E_0/2$.)

(1) Calculate the partition function for the system.
(2) Deduce the free energy F_β, the average energy U_β, and the entropy S_β of the system.
(3) What is the entropy in the high and low temperature limits?

Exercise 6.7. Energy Fluctuations

Show that the variance of the energy fluctuations in a system can be written

$$\sigma_U^2 = a\frac{\partial}{\partial\beta}U_\beta ,$$

and determine a.

Exercise 6.8. Fluctuation–Dissipation Theorem

Suppose that the covariance function of the magnetization fluctuations in a system at equilibrium is

$$\Gamma_{MM}(t) = A \exp\left(-\frac{t^2}{2\tau^2}\right) .$$

(1) Express A in terms of the total susceptibility χ_0 and temperature T.
(2) Express the impulse response $\chi(t)$ of the magnetization of the system at equilibrium.

Exercise 6.9. Noise in an RC Circuit

We have obtained two theorems which describe fluctuations at thermodynamic equilibrium. The first, in Section 6.4, can be used to determine the total power of the fluctuations. The second, in Section 6.6, is the true fluctuation–dissipation theorem. Let us now analyze the consequences of these theorems for thermodynamic fluctuations in the charge on the plates of a capacitor in a series RC circuit (see Section 6.7) when the resistance R tends to 0.

(1) State the expression for the spectral density of the charges on the capacitor plates of an RC circuit.
(2) Give the limit of this spectral density when R tends to 0.
(3) Determine the total power of the fluctuations in the charge.
(4) Comparing the results, what can you deduce?

7

Statistical Estimation

Statistical estimation is a very important tool in engineering and experimental physics. The problem of estimation often arises so "naturally" that the very question of statistical estimation seems superfluous in the first instance. However, every time we need to determine a parameter from experimental measurements, the task can be analyzed as a problem of statistical estimation. In this context, estimation involves making the best possible use of the measurements that have been recorded. In this chapter, we begin by illustrating with simple examples. We then define quality criteria for estimation and discuss maximum likelihood techniques and their more pertinent features.

7.1 The Example of Poisson Noise

Imagine that we have measured the number of particles detected during a time interval τ. In Section 4.9, we showed that the Poisson distribution provides a simple model for describing this number of particles. If $N(t, t + dt)$ represents the mean number (in the sense of expectation values) of particles detected over the time interval $[t, t + dt]$, the flux at time t is

$$\Phi(t) = \lim_{dt \to 0} \frac{dN(t, t + dt)}{dt} .$$

The flux is stationary if $\Phi(t)$ is independent of t and will then be denoted by Φ. The probability $P_{\Phi,\tau}(n)$ of detecting n particles during a time interval of length τ is then

$$P_{\Phi,\tau}(n) = e^{-\Phi\tau} \frac{(\Phi\tau)^n}{n!} .$$

The mean $\langle n \rangle$ and variance $\langle n^2 \rangle - \langle n \rangle^2$ of the Poisson distribution defined by $P(n) = \exp(-\mu)\mu^n/n!$ are equal to μ, and we thus deduce that

$$\langle n \rangle = \Phi\tau \quad \text{and} \quad \langle n^2 \rangle - \langle n \rangle^2 = \Phi\tau .$$

Let n_j be the number of particles measured between times $j\tau$ and $(j+1)\tau$. Then n_j is indeed a realization of a random variable n_λ distributed according to the probability law $P_{\Phi,\tau}(n)$. Suppose we have carried out P measurements so that we obtain P values $n_1, n_2, \ldots, n_P$. This set of P values corresponds to the sample obtained from the experiment and we shall denote it by χ, i.e., $\chi = \{n_1, n_2, \ldots, n_P\}$. Given τ, we may ask the following practical question: How can we determine Φ in the best possible way from χ? In other words, is there an "optimal" method for determining the flux from measurements.

One "natural" method would be to identify the true statistical moments with the empirical means estimated from the measurements. In this case, we would set

$$m_P = \frac{1}{P} \sum_{j=1}^{P} n_j \,,$$

and we would say $\langle n \rangle \approx m_P$. We then obtain $\hat{\Phi}\tau = (1/P) \sum_{j=1}^{P} n_j$. To put it more precisely, we say that we consider $\hat{\Phi}$ defined by

$$\hat{\Phi} = \frac{1}{P\tau} \sum_{j=1}^{P} n_j$$

to be an estimator of Φ. At this stage, we should note that this choice is totally arbitrary. Indeed, it is easy to propose another solution which seems just as natural, namely, we can estimate the variance of the set χ. As we said earlier, the variance of the Poisson distribution is also equal to $\Phi\tau$. A simple estimator for the variance can therefore be obtained with

$$\sigma_P^2 = \frac{1}{P} \sum_{j=1}^{P} n_j^2 - \left(\frac{1}{P} \sum_{j=1}^{P} n_j \right)^2 .$$

We thus discover that we could define a new estimator $\hat{\Phi}'$ of Φ by writing $\langle n^2 \rangle - \langle n \rangle^2 \approx \sigma_P^2$. We would then obtain

$$\hat{\Phi}'\tau = \frac{1}{P} \sum_{j=1}^{P} n_j^2 - \left(\frac{1}{P} \sum_{j=1}^{P} n_j \right)^2 .$$

We could then consider the following estimator $\hat{\Phi}'$ of Φ:

$$\hat{\Phi}' = \frac{1}{\tau} \left[\frac{1}{P} \sum_{j=1}^{P} n_j^2 - \left(\frac{1}{P} \sum_{j=1}^{P} n_j \right)^2 \right] .$$

We must therefore ask whether it is better to use $\hat{\Phi}$ or $\hat{\Phi}'$. In order to obtain a precise answer, we must set up criteria for choosing amongst such possibilities. This will become possible when we have analyzed the main features of estimators in the next section.

7.2 The Language of Statistics

Consider a real stochastic vector X_λ with n components. The simplest case corresponds to $n = 1$, when X_λ is a simple random variable. When we gather P experimental results, we obtain P realizations $(x_1, x_2, \ldots, x_P)$ of X_λ. We shall group these P vectors together into a set $\chi = \{x_1, x_2, \ldots, x_P\}$. Using the vocabulary of statistics, the set χ will be called the sample.

Note that when we consider a particular experiment, which has thus given rise to P vectors, these P vectors $x_1, x_2, \ldots, x_P$ are fixed. In the last example, we had $\chi = \{n_1, n_2, \ldots, n_P\}$, where the values $n_1, n_2, \ldots, n_P$ represent the P measurements. However, if we consider a potential experiment, each result is itself a stochastic vector and we shall then denote the P potential observations by $X_{\lambda(1)}, X_{\lambda(2)}, \ldots, X_{\lambda(P)}$. In this case χ is itself a set of stochastic vectors that we could denote by $\chi_{\lambda_T} = \{X_{\lambda(1)}, X_{\lambda(2)}, \ldots, X_{\lambda(P)}\}$, where $\lambda_T = \{\lambda(1), \lambda(2), \ldots, \lambda(P)\}$.

Any function $T(\chi) = t(x_1, x_2, \ldots, x_P)$ of the sample $\chi = \{x_1, x_2, \ldots, x_P\}$ is called a statistic of χ. Although this may sometimes be ambiguous, it is common usage and we shall follow it. In the last section, we introduced the statistics $\hat{\Phi}$ and $\hat{\Phi}'$, which we should now write

$$\hat{\Phi}(\chi) = \frac{1}{P\tau} \sum_{j=1}^{P} n_j \ \text{ and } \ \hat{\Phi}'(\chi) = \frac{1}{\tau} \left[\frac{1}{P} \sum_{j=1}^{P} n_j^2 - \left(\frac{1}{P} \sum_{j=1}^{P} n_j \right)^2 \right].$$

In order to approach the problem from the standpoint of statistical estimation, let us suppose that X_λ is a stochastic vector with probability density function $P_\theta(x)$, where θ is an *a priori* unknown parameter. Note that θ may be a scalar or a vector of some known dimension. In the context of the Poisson noise example, we assumed that $P_{\Phi,\tau}(n) = \mathrm{e}^{-\Phi\tau}(\Phi\tau)^n/n!$ and the parameter θ is then simply Φ. The problem of estimation then involves somehow "quessing" the value of θ from the set χ. Both $\hat{\Phi}(\chi)$ and $\hat{\Phi}'(\chi)$ are statistics we can use to estimate Φ.

We noted earlier that, when we consider P potential measurements $X_{\lambda(1)}$, $X_{\lambda(2)}, \ldots, X_{\lambda(P)}$, χ is itself a set of random variables. In this case, we can say that $\hat{\Phi}(\chi_{\lambda_T})$ and $\hat{\Phi}'(\chi_{\lambda_T})$ are two possible estimators for Φ. Moreover, $\hat{\Phi}(\chi_{\lambda_T})$ and $\hat{\Phi}'(\chi_{\lambda_T})$ are two random variables, whilst Φ is a value that can be considered as deterministic but unknown.

7.3 Characterizing an Estimator

In the last section, we saw that there can be different estimators for the same parameter θ. The obvious question arises as to whether there are scientific criteria for favoring one estimator over another. We shall not discuss the trivial although otherwise worthy case where there exist extrinsic arguments, unrelated to the quality of the estimate itself. Such arguments may concern such features as computation time or the possibility of making the estimate using analog techniques, which are sometimes decisive factors when choosing estimation methods. We may then say that the processing structure prevails. On the other hand, when it is the quality of the estimate alone which is taken into acount, we must specify how it can be quantified. This is what we intend to examine here.

We have made considerable use of the notion of expectation value in previous chapters and we shall consider this type of average once again here, still with the notation $\langle\ \rangle$. We can thereby define the expectation value (also known as the statistical mean) of any statistic $T(\chi_\lambda)$. To simplify the notation, we have written $\lambda_T = \lambda$, although we shall keep the explicit mention of λ to emphasize the fact that we are now considering P potential measurements and hence that χ_λ is indeed a set of random variables. To determine this expectation value, we must define the probability $L_\theta(x_1, x_2, \ldots, x_P)$ of observing the sample χ. In the problem situation specific to estimation, the true value θ_0 of the parameter θ is unknown. The probability $L_\theta(x_1, x_2, \ldots, x_P)$ should therefore be considered as a function of θ. It is often called the likelihood of the hypothesis which attributes the value θ to the unknown parameter when the observed sample is $\chi = \{x_1, x_2, \ldots, x_P\}$, and this explains why the symbol L is used to denote it. The expected value of the statistic $T(\chi_\lambda)$ is then

$$\langle T(\chi_\lambda)\rangle = \int \ldots \int T(x_1, x_2, \ldots, x_P)L_\theta(x_1, x_2, \ldots, x_P)\mathrm{d}x_1\mathrm{d}x_2\ldots\mathrm{d}x_P\ .$$

To begin with, and in order to simplify the discussion, we consider the case where θ is a scalar parameter. If $T(\chi_\lambda)$ is an estimator of θ_0, we clearly hope that $T(\chi_\lambda)$ will be as close as possible to θ_0. To make this idea more precise, we now present the main features of a statistical estimator.

The bias of an estimator $T(\chi_\lambda)$ for the parameter θ_0 is defined to be the difference between the expected value of $T(\chi_\lambda)$ and the true value θ_0. More precisely, the bias b_T of the estimator $T(\chi_\lambda)$ for θ_0 is defined as

$$b_T = \langle T(\chi_\lambda)\rangle - \theta_0\ .$$

Let us apply this definition to the example in Section 7.1. To simplify, we will consider $\theta_0 = \Phi_T$ and thus $P_{\theta_0}(n) = \mathrm{e}^{-\theta_0}\theta_0^n/n!$. We can study the bias of the estimator $\hat{\theta}(\chi_\lambda)$ defined by

$$\hat{\theta}(\chi_\lambda) = \frac{1}{P}\sum_{j=1}^{P}N_{\lambda(j)}\ .$$

Then,

$$\langle\hat{\theta}(\chi_\lambda)\rangle = \left\langle\frac{1}{P}\sum_{j=1}^{P}N_{\lambda(j)}\right\rangle\ ,$$

or

$$\langle \hat{\theta}(\chi_\lambda) \rangle = \frac{1}{P} \sum_{j=1}^{P} \langle N_{\lambda(j)} \rangle \ .$$

Now,

$$\langle N_{\lambda(j)} \rangle = \theta_0 \ ,$$

and hence,

$$\langle \hat{\theta}(\chi_\lambda) \rangle = \theta_0 \ .$$

The bias is zero and we thus say that the estimator is unbiased.

Consider now the estimator

$$\hat{\theta}'(\chi_\lambda) = \frac{1}{P} \sum_{j=1}^{P} N_{\lambda(j)}^2 - \left(\frac{1}{P} \sum_{j=1}^{P} N_{\lambda(j)} \right)^2 \ .$$

We have $\langle N_{\lambda(j)} \rangle = \theta_0$ and $\langle N_{\lambda(j)}^2 \rangle = \theta_0 + \theta_0^2$, so that

$$\langle N_{\lambda(j)} N_{\lambda(k)} \rangle = (\theta_0 + \theta_0^2) \delta_{j-k} + \theta_0^2 (1 - \delta_{j-k}) \ ,$$

where δ_n is the Kronecker delta. Given that

$$\langle \hat{\theta}'(\chi_\lambda) \rangle = \frac{1}{P} \sum_{j=1}^{P} \langle N_{\lambda(j)}^2 \rangle - \frac{1}{P^2} \sum_{j=1}^{P} \sum_{k=1}^{P} \langle N_{\lambda(j)} N_{\lambda(k)} \rangle \ ,$$

we deduce that $\langle \hat{\theta}'(\chi_\lambda) \rangle = \theta_0 + \theta_0^2 - \theta_0^2 - \theta_0/P$, whereupon

$$\langle \hat{\theta}'(\chi_\lambda) \rangle = \frac{P-1}{P} \theta_0 \ .$$

This result shows that, unlike $\hat{\theta}(\chi_\lambda)$, $\hat{\theta}'(\chi_\lambda)$ is a biased estimator of θ_0.

In the last example, we observe that the bias of the estimator is due to the finiteness of the number of measurements in the sample χ. To be precise, when P tends to infinity, $\hat{\theta}'(\chi_\lambda)$ becomes an unbiased estimator of θ. More generally, we say that an estimator $T(\chi_\lambda)$ of a parameter θ is asymptotically umbiaised if

$$\lim_{P \to \infty} \langle T(\chi_\lambda) \rangle = \theta_0 \ .$$

$\hat{\theta}'(\chi_\lambda)$ is thus a biased but asymptotically umbiaised estimator of θ.

It is useful to obtain a better characterization of the behavior of an estimator $T(\chi)$ when the size P of the sample is finite, as we have just done. For this purpose, suppose that we carry out N experiments and that each one produces a sample χ_j. For each sample, we can determine the value of the statistic $T(\chi_j)$ and hence plot the histogram of the values obtained, as shown in Fig. 7.1.

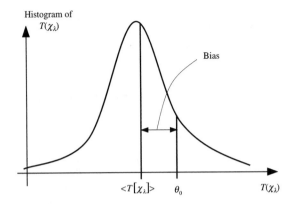

Fig. 7.1. Schematic representation of the histogram of an estimator

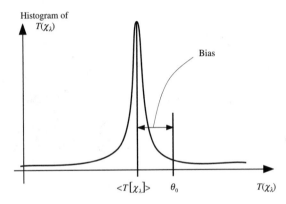

Fig. 7.2. Schematic representation of the histogram of an estimator with lower variance

We can make two observations with regard to this diagram. It is of course important that the bias of an estimator should be small. It is, however, equally important that the fluctuations in the estimator relative to its mean value should also be small. Figure 7.2 shows another example of an estimator of θ which has the same bias as the one in Fig. 7.1, but with smaller fluctuations.

The variance provides a standard way of characterizing the fluctuations in a random variable. We will thus consider that the smaller the variance of an estimator, the better it is. We define the variance σ_T^2 of an estimator $T(\chi)$ by

$$\sigma_T^2 = \left\langle [T(\chi_\lambda) - \langle T(\chi_\lambda)\rangle]^2 \right\rangle .$$

A simple illustration of this idea is shown in Fig. 7.3.

It is quite clear that, amongst all the statistics which are potential estimators of a parameter θ, those without bias which also lead to the lowest possible variance are going to be particularly interesting. These unbiased estimators

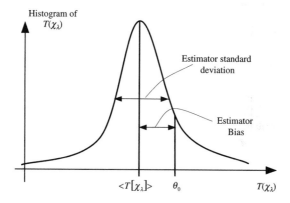

Fig. 7.3. Schematic illustration of the standard deviation (or equivalently, the variance) and the bias of an estimator

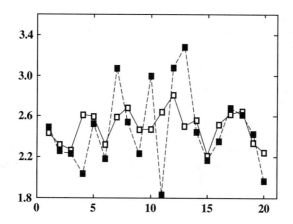

Fig. 7.4. Values of the estimators obtained with $\hat{\theta}(\chi)$ (*white squares*) and $\hat{\theta}'(\chi)$ (*black squares*)

with minimal variance do indeed play a key role in the context of estimation theory.

Let us return to the example of Poisson noise. We have generated several samples χ_j containing 100 Poisson variables with $\theta_0 = 2.5$. For each sample χ_j, we have determined the values of the two estimators $\hat{\theta}(\chi_j)$ and $\hat{\theta}'(\chi_j)$. The values of $\hat{\theta}(\chi_j)$ and $\hat{\theta}'(\chi_j)$ are shown in Fig. 7.4 for 20 samples ($j = 1, \ldots, 20$). They thus correspond to the mean and variance of each set of independent realizations of 100 Poisson variables with parameter $\theta_0 = 2.5$.

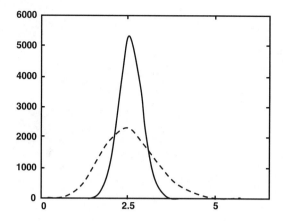

Fig. 7.5. Histograms of $\hat{\theta}(\chi)$ (*continuous line*) and $\hat{\theta}'(\chi)$ (*dashed line*) determined from 20 000 samples of 100 Poisson variables each with $\theta_0 = 2.5$

Figure 7.5 shows the histograms of $\hat{\theta}(\chi)$ and $\hat{\theta}'(\chi)$ determined from 20 000 samples containing 100 Poisson variables each. The means obtained with the estimators $\hat{\theta}(\chi)$ and $\hat{\theta}'(\chi)$ are 2.5002 and 2.4979, respectively. The variances differ much more in the two cases:

- 0.0254 for $\hat{\theta}(\chi)$, i.e., a standard deviation of 0.1595,
- 0.1510 for $\hat{\theta}'(\chi)$, i.e., a standard deviation of 0.3886.

We thus observe that, in this particular case, the estimator $\hat{\theta}(\chi)$ is better than $\hat{\theta}'(\chi)$. This raises the question as to whether $\hat{\theta}(\chi)$ is actually the best possible estimator, or more precisely, whether it is the estimator with the lowest possible variance amongst all unbiased estimators.

7.4 Maximum Likelihood Estimator

The notion of likelihood plays a very important role in the field of statistical estimation. We begin by illustrating this with the Poisson distribution. Suppose that the P realizations of the sample $\chi = \{n_1, n_2, \ldots, n_P\}$ are independent and identically distributed. The likelihood is then

$$L(\chi|\theta) = \prod_{i=1}^{P} \left(e^{-\theta} \frac{\theta^{n_i}}{n_i!} \right) \ .$$

In many circumstances it is the natural logarithm of this quantity that arises and it is common parlance to speak of the log-likelihood, viz.,

$$\ell(\chi|\theta) = \sum_{i=1}^{P} \big[-\theta + n_i \ln \theta - \ln(n_i!) \big] \ .$$

Quite generally, when the realizations are independent and identically distributed, we have

$$\ell(\chi|\theta) = \sum_{i=1}^{P} \ln P_\theta(x_i) \ ,$$

where x_i is the value of the ith realization.

In Section 5.1, we saw that $-\ln P_\theta(x_i)$ is the information content of the realization of x_i. We thus see that the log-likelihood $\ell(\chi|\theta)$ is the opposite of the information contained in the realization of χ. We also note that the mean value of the log-likelihood is proportional to the negative of the entropy of the distribution. Indeed,

$$\langle \ell(\chi|\theta) \rangle = \sum_{i=1}^{P} \langle \ln P_\theta(x_i) \rangle \ ,$$

or

$$\langle \ell(\chi|\theta) \rangle = P \langle \ln P_\theta(x) \rangle = P \sum_{x} P_\theta(x) \ln P_\theta(x) \ .$$

The value of the parameter θ which maximizes the likelihood is the maximum likelihood estimator. As the logarithm function is an increasing function, maximizing the likelihood is equivalent to maximizing the log-likelihood. This in turn amounts to seeking the value of θ that minimizes the information contained in the realization of χ.

Let us examine the result obtained for the Poisson distribution. To find the maximum of $\ell(\chi|\theta)$, we require the value of θ that makes $\partial \ell(\chi|\theta)/\partial\theta$ zero:

$$\frac{\partial}{\partial \theta} \ell(\chi|\theta) = \sum_{i=1}^{P} \left(-1 + \frac{n_i}{\theta} \right) = 0 \ .$$

We then obtain

$$\hat{\theta} = \frac{1}{P} \sum_{i=1}^{P} n_i \ ,$$

and this is indeed a maximum since

$$\frac{\partial^2 \ell(\chi|\theta)}{\partial \theta^2} = - \sum_{i=1}^{P} \frac{n_i}{\theta^2} < 0 \ .$$

In other words, $T(\chi) = (1/P) \sum_{i=1}^{P} n_i$ is the statistic which corresponds to the maximum likelihood estimator for θ. We see that in the particular case of the Poisson distribution, the maximum likelihood estimator corresponds to identifying the first moment. We have already seen that this estimator has lower variance than the one obtained by identifying the second moment.

The maximum likelihood estimator is invariant under reparametrization of the probability distribution. This is a useful property because it leads to a result that is independent of the arbitrariness involved in choosing a method of parametrization. Consider a family of probability laws or probability density functions $P_\theta(x)$. Suppose now that we reparametrize this family by making the change $\mu = g(\theta)$, where g is a bijective function. We then obtain the family of probability laws $\tilde{P}_\mu(x)$ such that $\tilde{P}_{\mu=g(\theta)}(x) = P_\theta(x)$. For example, the family of exponential laws is defined by $P_\theta(x) = (1/\theta)\exp(-x/\theta)$ with $\theta > 0$. We may consider the reparametrization $\mu = 1/\theta$, which gives $P_\mu(x) = \mu\exp(-\mu x)$.

It is a straightforward matter to show that if $T(\chi)$ is the maximum likelihood estimator for θ, then $g(T(\chi))$ is the maximum likelihood estimator for μ. Indeed, setting $\theta_{\mathrm{ML}} = T(\chi)$. We have $\forall\,\theta$, $\ell(\chi|\theta_{\mathrm{ML}}) \geqslant \ell(\chi|\theta)$, whilst

$$\sum_{i=1}^{P} \ln P_\theta(x_i) = \sum_{i=1}^{P} \ln\left[\tilde{P}_{\mu=g(\theta)}(x_i)\right] ,$$

and hence $\ell(\chi|\theta) = \tilde{\ell}(\chi|\mu = g(\theta))$, where

$$\tilde{\ell}(\chi|\mu) = \sum_{i=1}^{P} \ln\left[\tilde{P}_\mu(x_i)\right] .$$

Hence, $\ell(\chi|\theta_{\mathrm{ML}}) \geqslant \ell(\chi|\theta)$ implies that $\tilde{\ell}(\chi|g(\theta_{\mathrm{ML}})) \geqslant \tilde{\ell}(\chi|g(\theta))$ or $\tilde{\ell}(\chi|\mu_{\mathrm{ML}}) \geqslant \tilde{\ell}(\chi|\mu)$, $\forall\,\mu$ with $\mu_{\mathrm{ML}} = g(\theta_{\mathrm{ML}})$.

Let us illustrate this result in the case of the exponential distribution. We have

$$\ell(\chi|\theta) = \sum_{i=1}^{P}\left(-\frac{x_i}{\theta} - \ln\theta\right) ,$$

and the maximum likelihood estimator is thus obtained from $\partial\ell(\chi|\theta)/\partial\theta = 0$, or $\theta_{\mathrm{ML}} = T(\chi) = (1/P)\sum_{i=1}^{P} x_i$. We also have $\tilde{\ell}(\chi|\mu) = \sum_{i=1}^{P}(-\mu x_i + \ln\mu)$, which leads to

$$\mu_{\mathrm{ML}} = \tilde{T}(\chi) = \left(\frac{1}{P}\sum_{i=1}^{P} x_i\right)^{-1} .$$

We do indeed obtain $\mu_{\mathrm{ML}} = 1/\theta_{\mathrm{ML}}$.

Many estimation methods do not possess this invariance property, and this is considered to be a weak point in such techniques.

Unlike the method of identifying moments, the maximum likelihood method leads to a unique estimator. This is a clear advantage. The question then arises as to whether this estimator always displays properties as good as those obtained by identifying moments. More generally, we may wonder whether this method guarantees us the best estimator. The answer is unfortunately negative. We do not always obtain the best estimator, i.e., the one with minimal variance. Later we shall examine some results which will elucidate these questions.

7.5 Cramer–Rao Bound in the Scalar Case

In this section, we shall be concerned with the minimal value that can be attained by the variance of an estimator for a scalar parameter θ. Later we shall consider the general case where the parameter θ to be estimated is a vector, but for simplicity, we prefer to approach the problem in two stages.

In order to emphasize the physical meaning of likelihood, which corresponds to the probability of observing a sample χ under the assumption that the parameter of the law is θ, we write $L(\chi|\theta) = L_\theta(x_1, x_2, \ldots, x_P)$. As already mentioned, the true value of the parameter θ_0 remains unknown and the problem is to estimate it from the sample χ. Note first that, when we consider the expectation value of the estimator, the result corresponds to the one we would obtain if we were to carry out an infinite number of independent experiments with different samples. This is the mean we considered in Chapter 6 in the context of classical statistical physics. In this type of experiment, each sample $\chi_\lambda = \{X_{\lambda(1)}, X_{\lambda(2)}, \ldots, X_{\lambda(P)}\}$ is generated with the probability law $L(\chi|\theta_0) = L_{\theta_0}(x_1, x_2, \ldots, x_P) = L(x_1, x_2, \ldots, x_P|\theta_0)$. We thus naturally obtain

$$\langle T(\chi_\lambda) \rangle = \int \ldots \int T(x_1, x_2, \ldots, x_P) L(x_1, x_2, \ldots, x_P|\theta_0) \mathrm{d}x_1 \mathrm{d}x_2 \ldots \mathrm{d}x_P \ .$$

Note, however, that we can also consider the expectation value as a mathematical operator. The idea is simply to calculate the mean of $T(x_1, x_2, \ldots, x_P)$ with a probability law

$$L(x_1, x_2, \ldots, x_P|\theta) \ .$$

To emphasize the dependence on θ, we will write

$$\langle T(\chi_\lambda) \rangle_\theta = \int \ldots \int T(x_1, x_2, \ldots, x_P) L(x_1, x_2, \ldots, x_P|\theta) \mathrm{d}x_1 \mathrm{d}x_2 \ldots \mathrm{d}x_P \ .$$

In this case, $\langle T(\chi_\lambda) \rangle_\theta$ is a function of θ which we shall also write

$$\langle T(\chi_\lambda) \rangle_\theta = h(\theta) \ .$$

This is the mean of the statistic $T(\chi_\lambda)$ which we would obtain for random samples χ_λ that would be generated with the probability law $L(\chi|\theta)$. To simplify the formulas, we use the notation

$$\int T(\chi) L(\chi|\theta) \mathrm{d}\chi$$

$$= \int \ldots \int T(x_1, x_2, \ldots, x_P) L(x_1, x_2, \ldots, x_P|\theta) \mathrm{d}x_1 \mathrm{d}x_2 \ldots \mathrm{d}x_P \ .$$

Let us examine in detail the case where χ takes continuous values, which will justify writing the above relations in integral form. When χ takes discrete values, the integrals are simply replaced by discrete sums.

We can define the variance of the statistic $T(\chi_\lambda)$ for any θ:

$$\sigma_T^2(\theta) = \int \left[T(\chi) - \langle T(\chi_\lambda) \rangle_\theta \right]^2 L(\chi|\theta) \mathrm{d}\chi .$$

In Section 7.12, we show that if the domain of definition of X_λ does not depend on θ, the variance of the statistic $T(\chi_\lambda)$ cannot be less than a certain limiting value:

$$\sigma_T^2(\theta) \geqslant \frac{-\left| \frac{\partial}{\partial \theta} h(\theta) \right|^2}{\int \frac{\partial^2 \ln L(\chi|\theta)}{\partial \theta^2} L(\chi|\theta) \mathrm{d}\chi} .$$

Naturally, we must also assume that the logarithm of the likelihood does have a second derivative. This inequality, which provides a lower bound for the variance $\sigma_T^2(\theta)$ of the estimator $T(\chi_\lambda)$, is a classical result in statistics, known as the Cramer–Rao bound. We note immediately that this inequality holds whatever value we take for θ.

If the estimator is unbiased, the expression for the Cramer–Rao bound simplifies. Indeed, in this case, we have $h(\theta) = \langle T(\chi_\lambda) \rangle_\theta = \theta$ and hence $\partial h(\theta)/\partial \theta = 1$, which implies

$$\sigma_T^2(\theta) \geqslant \frac{1}{I_F(\theta)} ,$$

where

$$I_F(\theta) = - \int \frac{\partial^2 \ln L(\chi|\theta)}{\partial \theta^2} L(\chi|\theta) \mathrm{d}\chi .$$

We see from this expression that the variance of an unbiased estimator cannot be less than a certain lower bound. In the case of unbiased estimators, this bound does not depend on the estimator chosen. It only depends on the mean value of the curvature of the logarithm of the likelihood (see Fig. 7.6).

In the neighborhood of a maximum, the second derivative of the likelihood is negative, and the first derivative is decreasing. Its absolute value is all the greater as the curvature is large. In other words, the more sharply peaked the likelihood is as a function of the parameter we wish to estimate, the more precisely we may hope to estimate that parameter (see Fig. 7.7).

Note, however, that it is the expectation value of the second derivative of the log-likelihood which comes into the expression for the Cramer–Rao bound, since

$$\left\langle \frac{\partial^2}{\partial \theta^2} \ln L(\chi|\theta) \right\rangle = \int L(\chi|\theta) \frac{\partial^2}{\partial \theta^2} \ln L(\chi|\theta) \mathrm{d}\chi .$$

The quantity

$$I_F(\theta) = - \left\langle \frac{\partial^2}{\partial \theta^2} \ln L(\chi|\theta) \right\rangle$$

is also known in statistics as the Fisher information.

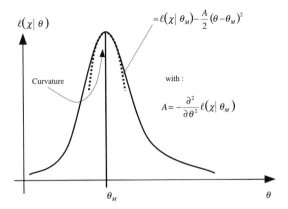

Fig. 7.6. The role played by the curvature in the Cramer–Rao bound, where θ_M is simply the value of θ maximizing the log-likelihood $\ell(\chi|\theta) = \ln L(\chi|\theta)$

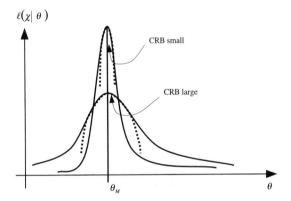

Fig. 7.7. Relation between the Cramer–Rao bound (CRB) and the shape of the log-likelihood

7.6 Exponential Family

The Cramer–Rao bound represents the minimal value that can be attained by the variance of an estimator, and hence of a statistic. If the variance of some estimator reaches this bound, it is referred to as an efficient estimator. It is interesting to know for what types of probability density function an efficient estimator actually exists. In Section 7.13 of this chapter, we show that they can all be written in the form

$$P_\theta(x) = \frac{\exp\left[a(\theta)t(x) + f(x)\right]}{Z(\theta)} \ ,$$

where $Z(\theta) = \int \exp\left[a(\theta)t(x) + f(x)\right]dx$. Statistics which attain the Cramer–Rao bound when we observe independent and identically distributed realizations are then proportional to

$$T(\chi) = \sum_{i=1}^{P} t(x_i) \; .$$

Note that, generally, when the statistic $T(\chi)$ is an estimator, we consider instead

$$T'(\chi) = (1/P) \sum_{i=1}^{P} t(x_i).$$

The variance of the estimator $T'(\chi)$ is then simply the variance of $T(x)$ divided by P^2. The above probability density functions define the family of exponential probability densities. We also speak more simply of the exponential family. Note that the log-likelihood of χ is simply

$$\ell(\chi|\theta) = \left[a(\theta)T(\chi) + \sum_{i=1}^{P} f(x_i) \right] - P \ln Z(\theta) \; .$$

In Section 7.13, we show that the variance of $T(\chi)$ is given by

$$\sigma_T^2(\theta) = \left| \frac{\frac{\partial}{\partial \theta} h(\theta)}{\alpha_0(\theta)} \right| \; ,$$

where $h(\theta) = \langle T(\chi) \rangle_\theta$ and $\alpha_0(\theta) = \partial a(\theta)/\partial \theta$.

When the probability density function of a probability law in the exponential family can be written in the form

$$P_\theta(x) = \frac{\exp \left[\theta t(x) + f(x) \right]}{Z(\theta)} \; ,$$

we say that it is in the canonical or natural form and θ is then the canonical or natural parameter of the law. Let us examine the Cramer–Rao bound when a law is written in the canonical form. We then have $a(\theta) = \theta$ and hence $\alpha_0(\theta) = \partial a(\theta)/\partial \theta = 1$. Moreover, $T(\chi) = \sum_{i=1}^{P} t(x_i)$ and we have $\langle T(\chi) \rangle_\theta = P \langle t(x) \rangle_\theta$. We thus obtain

$$\sigma_T^2(\theta) = P^2 \sigma_t^2(\theta) = P \left| \frac{\partial}{\partial \theta} \langle t(x) \rangle_\theta \right| \; ,$$

where $\sigma_t^2(\theta) = \langle [t(x)]^2 \rangle_\theta - [\langle t(x) \rangle_\theta]^2$.

If, furthermore, we consider $T'(\chi) = (1/P)T(\chi)$ and $T'(\chi)$ is an unbiased estimator of θ, we have $\sigma_{T'}^2(\theta) = (1/P^2)\sigma_T^2(\theta) = \sigma_t^2(\theta)$ and $\langle t(x) \rangle_\theta = \theta$, so that $\sigma_{T'}^2(\theta) = 1/P$. This relation only holds for unbiased estimators deduced from the canonical form.

The exponential family plays an important role because, as we shall see later, its probability laws have simple optimality properties.

7.7 Example Applications

It is easy to see that the Poisson distribution belongs to the exponential family. Indeed, we have $P(x) = e^{-\theta}\theta^x/x!$ and we can write

$$P_\theta(x) = \frac{\exp\left[a(\theta)t(x) + f(x)\right]}{Z(\theta)} ,$$

where $Z(\theta) = e^\theta$, $f(x) = -\ln(x!)$, $t(x) = x$ and $a(\theta) = \ln(\theta)$. Let us determine the variance of the estimator $m(\chi_\lambda) = (1/P)\sum_{j=1}^{P} x_{\lambda(j)}$ which is unbiased. According to the last section this statistic reaches the Cramer–Rao bound. We observe that $T(\chi_\lambda) = \sum_{j=1}^{P} x_{\lambda(j)}$. It is easy to see that $h(\theta) = \int T(\chi)L(\chi|\theta)d\chi = P\theta$, and hence $\partial h(\theta)/\partial\theta = P$. Moreover, since $a(\theta) = \ln\theta$, we have $\alpha_0(\theta) = \partial a(\theta)/\partial\theta = 1/\theta$ and hence $\sigma_T^2(\theta) = P\theta$. The variance of the estimator $m(\chi_\lambda)$ of θ is therefore

$$\sigma_m^2(\theta) = \frac{\theta}{P} .$$

We estimated the variance of this estimator in Section 7.3. We had $P = 100$ and $\theta = 2.5$ and it was found that $\sigma_T^2(\theta) = 0.0254$, which is indeed of the order of θ/P. There is no surprise here, since $m(\chi_\lambda)$ is an efficient estimator of θ.

The Gaussian case is particularly interesting because it is often a good model when measurements are perturbed by additive noise. Suppose we carry out P measurements corresponding to the model

$$x_i = \theta + y_i ,$$

where $i \in [1, P]$ and y_i is a random variable with zero mean and variance b^2. If we seek to estimate θ from the P measurements x_i with $i \in [1, P]$, we can consider the estimator $m(\chi_\lambda) = (1/P)\sum_{j=1}^{P} x_{\lambda(j)}$ which is unbiased and set $T(\chi_\lambda) = \sum_{j=1}^{P} x_{\lambda(j)}$. The variance of this estimator is easily determined. Note first that x_i belongs to the exponential family. Indeed,

$$P_\theta(x) = \frac{\exp\left[(2\theta x - x^2)/2b^2\right]}{\sqrt{2\pi}bZ(\theta)} ,$$

where $Z(\theta) = \exp(\theta^2/2b^2)$. It is easy to see that $h(\theta) = \int T(\chi)L(\chi|\theta)d\chi = P\theta$, $a(\theta) = \theta/b^2$ and $\alpha_0(\theta) = 1/b^2$, and hence $\sigma_T^2(\theta) = Pb^2$. The variance of the estimator $m(\chi_\lambda)$ of θ is therefore

$$\sigma_m^2(\theta) = \frac{b^2}{P} .$$

It is easy to show that the probability laws in Table 7.1 belong to the exponential family. We leave it to the reader to reformulate these laws in order to show that they belong to the exponential family.

Table 7.1. Some probability laws in the exponential family

Name	Probability density function	Parameters
Bernoulli	$(1 - q)\delta(x) + q\delta(x - 1)$	q
Poisson	$\displaystyle\sum_{n=0}^{\infty} \exp(-\mu)\delta(x - n)\mu^n/n!$	μ
Gamma	$\begin{cases} \dfrac{\beta^\alpha x^{\alpha-1}}{\Gamma(\alpha)} \exp(-\beta x) & \text{if } x \geqslant 0 \\ 0 & \text{otherwise} \end{cases}$	α and β
Gaussian	$\dfrac{1}{\sqrt{2\pi}\sigma} \exp\left[-\dfrac{(x - m)^2}{2\sigma^2}\right]$	m and σ

7.8 Cramer–Rao Bound in the Vectorial Case

The Cramer–Rao relation is more complex in the vectorial case than in the scalar case. The parameter to be estimated is now a vector in $\mathbb{R}^n$, which we denote by $\boldsymbol{\theta} = (\theta_1, \theta_2, \ldots, \theta_n)^{\mathrm{T}}$. We consider a vectorial statistic $\boldsymbol{T}(\chi_\lambda)$ of the same dimension as $\boldsymbol{\theta}$. It is useful to introduce the Fisher information matrix $\overline{\overline{J}}$ with entries

$$J_{ij} = -\left\langle \frac{\partial^2}{\partial\theta_i \partial\theta_j} \ell(\chi|\boldsymbol{\theta}) \right\rangle_{\boldsymbol{\theta}} ,$$

where $\ell(\chi|\boldsymbol{\theta}) = \ln L(\chi|\boldsymbol{\theta})$. As the statistic $\boldsymbol{T}(\chi_\lambda)$ is a vector, its fluctuations are characterized by its covariance matrix $\overline{\overline{\Gamma}}$:

$$\Gamma_{ij} = \langle \delta T_i(\chi)\delta T_j(\chi)\rangle_{\boldsymbol{\theta}} ,$$

where $\delta T_i(\chi) = T_i(\chi) - \langle T_i(\chi)\rangle_{\boldsymbol{\theta}}$. In the case where $\boldsymbol{T}(\chi_\lambda)$ is an unbiased estimator of $\boldsymbol{\theta}$, whatever the complex vector $\boldsymbol{u} \in \mathbb{C}^n$, we have

$$\boldsymbol{u}^\dagger \overline{\overline{\Gamma}} \boldsymbol{u} \geqslant \boldsymbol{u}^\dagger \overline{\overline{J}}^{-1} \boldsymbol{u} ,$$

where $\boldsymbol{u}^\dagger$ is the transposed complex conjugate of $\boldsymbol{u}$ and $\overline{\overline{J}}^{-1}$ is the matrix inverse to $\overline{\overline{J}}$. This is proved in Section 7.14 at the end of this chapter.

Let us illustrate this result in the case where we wish to estimate the mean of two-dimensional Gaussian vectors. We have

$$P_{m_1,m_2}(\boldsymbol{x}) = \frac{1}{2\pi\sqrt{|\overline{\overline{C}}|}} \exp\left[-\frac{1}{2}(\boldsymbol{x} - \boldsymbol{m})^{\mathrm{T}}\overline{\overline{C}}^{-1}(\boldsymbol{x} - \boldsymbol{m})\right] ,$$

where $\boldsymbol{m} = (m_1, m_2)^{\mathrm{T}}$ and $|\overline{\overline{C}}|$ is the determinant of the covariance matrix $\overline{\overline{C}}$. The log-likelihood is thus

$$\ell(\chi|\boldsymbol{m}) = -P\ln\left(2\pi\sqrt{|\overline{\overline{C}}|}\right) - \frac{1}{2}\sum_{i=1}^{P}\left[(\boldsymbol{x}_i - \boldsymbol{m})^\mathrm{T}\overline{\overline{C}}^{-1}(\boldsymbol{x}_i - \boldsymbol{m})\right],$$

where $\chi = \{\boldsymbol{x}_1, \boldsymbol{x}_2, \dots, \boldsymbol{x}_P\}$. The Fisher matrix is obtained from

$$\frac{\partial^2}{\partial m_1^2}\ell(\chi|\boldsymbol{m}), \quad \frac{\partial^2}{\partial m_2^2}\ell(\chi|\boldsymbol{m}), \quad \text{and} \quad \frac{\partial^2}{\partial m_1 \partial m_2}\ell(\chi|\boldsymbol{m}),$$

and is therefore

$$\overline{\overline{J}} = P\overline{\overline{C}}^{-1}.$$

We thus obtain

$$\overline{\overline{J}}^{-1} = \frac{1}{P}\overline{\overline{C}},$$

and the Cramer–Rao bound is therefore

$$u_1^2\Gamma_{11} + u_2^2\Gamma_{22} + 2u_1u_2\Gamma_{12} \geqslant \left(u_1^2C_{11} + u_2^2C_{22} + 2u_1u_2C_{12}\right)/P.$$

In particular, we can have $\Gamma_{12} \neq 0$, which implies that, if the covariance matrix of the fluctuations is not diagonalized, there may be correlations between the joint estimation errors of m_1 and m_2.

It is shown in Section 7.14 that probability laws with statistics which attain the Cramer–Rao bound all belong to the exponential family. In the vectorial case, these laws have the form

$$P(\boldsymbol{x}|\boldsymbol{\theta}) = \exp\left[\sum_{j=1}^{n} a_j(\boldsymbol{\theta})t_j(\boldsymbol{x}) + b(\boldsymbol{\theta}) + f(\boldsymbol{x})\right],$$

which can also be written

$$P(\boldsymbol{x}|\boldsymbol{\theta}) = \frac{\exp\left[\sum_{j=1}^{n} a_j(\boldsymbol{\theta})t_j(\boldsymbol{x}) + f(\boldsymbol{x})\right]}{Z(\boldsymbol{\theta})}.$$

In the case of unbiased estimators, if they attain the Cramer–Rao bound, this implies that $\boldsymbol{u}^\dagger \overline{\overline{\Gamma}} \boldsymbol{u} = (1/P)\boldsymbol{u}^\dagger \overline{\overline{C}}^{-1}\boldsymbol{u}$. In the Gaussian example discussed above, we thus have $\Gamma_{12} = C_{12}/P$. There is indeed a coupling between the estimation errors for m_1 and m_2.

7.9 Likelihood and the Exponential Family

We have seen in the last few sections that the Cramer–Rao bound fixes the minimal value that can be reached by the variance of any statistic. Moreover, for statistics corresponding to unbiased estimators, this bound is independent of the statistic under consideration. If there is to exist an efficient statistic, i.e., one which reaches the Cramer–Rao bound, the probability law of the random variable must belong to the exponential family. In this section, we shall examine these properties in detail and elucidate the conditions under which the maximum likelihood estimator will be efficient.

For reasons of simplicity, we shall consider the scalar case. In the exponential family, the probability density function is

$$P_\theta(x) = \frac{\exp\left[a(\theta)t(x) + f(x)\right]}{Z(\theta)} \, ,$$

and the likelihood of a P-samples χ corresponding to the realization of independent random variables is given by

$$L(\chi|\theta) = \frac{\exp\left[a(\theta)T(\chi) + F(\chi)\right]}{Z_P(\theta)} \, ,$$

with $T(\chi) = \sum_{n=1}^{P} t(x_n)$, $F(\chi) = \sum_{n=1}^{P} f(x_n)$ and $Z_P(\theta) = [Z(\theta)]^P$.

It is interesting to observe to begin with that the likelihood can be written in the form

$$L(\chi|\theta) = g\big(T(\chi)|\theta\big)h(\chi) \, .$$

To see this, we set $g\big(T(\chi)|\theta\big) = \exp\left[a(\theta)T(\chi)\right]/Z_P(\theta)$ and $h(\chi) = \exp[F(\chi)]$. If, for a given probability law, the likelihood can be decomposed into a product $L(\chi|\theta) = g\big(T(\chi)|\theta\big)h(\chi)$, we say that $T(\chi)$ is a sufficient statistic of the law for θ. Although this concept is very important in statistics, we shall limit ourselves to a few practical results in the present context.

First of all, the conditional probability of observing χ given $T(\chi)$ is independent of θ. To show this, consider the case where $T(\chi)$ has discrete values. We have

$$P_\theta\big(\chi|T(\chi)\big) = \frac{P_\theta\big(\chi, T(\chi)\big)}{P_\theta\big(T(\chi)\big)} \, .$$

Now when we know χ, we automatically know $T(\chi)$ and therefore $P_\theta\big(\chi, T(\chi)\big) = P_\theta(\chi) = L(\chi|\theta)$. Moreover, $P_\theta\big(T(\chi)\big)$ is obtained by summing the probability $P_\theta(\chi)$ over all samples χ which have the same value for the statistic $T(\chi)$. Therefore,

$$P_\theta\big(T(\chi) = T\big) = \sum_{\chi|T(\chi)=T} P_\theta(\chi) \, .$$

In the case where $T(\chi)$ is a sufficient statistic, we have

$$L(\chi|\theta) = g\big(T(\chi)|\theta\big)h(\chi) \, ,$$

and hence,

$$P_\theta\big(T(\chi) = T\big) = g(T|\theta) \sum_{\chi|T(\chi)=T} h(\chi) \, .$$

Defining

$$H(T) = \sum_{\chi|T(\chi)=T} h(\chi) \, ,$$

we have $P_\theta(T(\chi)) = g(T(\chi)|\theta)H(T(\chi))$ and, consequently,

$$P_\theta(\chi|T(\chi)) = \frac{L(\chi|\theta)}{g(T(\chi)|\theta)H(T(\chi))} \, ,$$

or

$$P_\theta(\chi|T(\chi)) = \frac{g(T(\chi)|\theta)h(\chi)}{g(T(\chi)|\theta)H(T(\chi))} = \frac{h(\chi)}{H(T(\chi))} \, ,$$

which proves the above claim.

This property implies that, once $T(\chi)$ has been given, the sample χ contains no more useful information for estimation of θ. We say that $T(\chi)$ is a sufficient statistic for the estimation of θ. It can be shown that the existence of a sufficient statistic is closely linked to membership of the exponential family, but we shall not examine this feature in any more detail here.

Note that if a sufficient statistic exists, the maximum likelihood estimator will only depend on the sample via this statistic. Indeed, as the likelihood is equal to $L(\chi|\theta) = g(T(\chi)|\theta)h(\chi)$, the log-likelihood is

$$\ell(\chi|\theta) = \ln\left[g(T(\chi)|\theta)\right] + \ln\left[h(\chi)\right] \, .$$

The maximum likelihood estimator $\hat{\theta}_{\mathrm{ML}}(\chi)$ is the value of θ that maximizes $\ell(\chi|\theta)$. As this is equivalent to maximizing $\ln\left[g(T(\chi)|\theta)\right]$, it follows that $\hat{\theta}_{\mathrm{ML}}(\chi)$ can only be a function of $T(\chi)$. It can be shown that in the case of the exponential family, if there is an unbiased estimator which only depends on the sufficient statistic $T(\chi)$, then it must have the minimal variance. (In particular, one can appeal to the more general Lehmann–Scheffé theorem. However, to simplify the discussion, we only consider probability distributions in the exponential family here. The results are then simpler and easier to use.) We thus see that, in the exponential family, if the maximum likelihood estimator is unbiased, it will have minimal variance. This result often justifies the use of the maximum likelihood technique to estimate in the exponential family. It can also be generalized to the vectorial case.

Let us go further in our analysis of the maximum likelihood estimator for the case of the exponential family. We have $\ell(\chi|\theta) = a(\theta)T(\chi) + F(\chi) - P\ln Z(\theta)$ and introduce the notation $b(\theta) = -\ln Z(\theta)$. We can then write $\ell(\chi|\theta) = a(\theta)T(\chi) + F(\chi) + Pb(\theta)$. The maximum in θ is obtained when

$$\frac{\partial\ell(\chi|\theta)}{\partial\theta} = a'(\theta)T(\chi) + Pb'(\theta) = 0 \, ,$$

where we have set $\partial a(\theta)/\partial\theta = a'(\theta)$ and $\partial b(\theta)/\partial\theta = b'(\theta)$. Finally, we obtain $-b'(\theta)/a'(\theta) = T(\chi)/P$. $\hat{\theta}_{\mathrm{ML}}(\chi)$ is obtained by inverting the equation:

$$-\frac{b'\big(\hat{\theta}_{\mathrm{ML}}(\chi)\big)}{a'\big(\hat{\theta}_{\mathrm{ML}}(\chi)\big)} = T(\chi)/P \; .$$

In the case of a canonical parametrization, $a'(\theta) = 1$ and the maximum likelihood estimator then simplifies to $b'(\theta) = -T(\chi)/P$. If it is unbiased, it will have minimal variance. Note, however, that since only $T(\chi)/P$ is efficient, i.e., only $T(\chi)/P$ attains the Cramer–Rao bound, $-b'(\theta)/a'(\theta)$ is the only function of θ that can be efficiently estimated.

The maximum likelihood estimator corresponds to the equality

$$T(\chi) = \langle T(\chi)\rangle_{\hat{\theta}_{\mathrm{ML}}(\chi)} \; .$$

Indeed,

$$J(\theta) = \int \exp\big[a(\theta)T(\chi) + F(\chi) + Pb(\theta)\big]\mathrm{d}\chi = 1 \; ,$$

and hence $\mathrm{d}J(\theta)/\mathrm{d}\theta = 0$, so that

$$\int \big[a'(\theta)T(\chi) + Pb'(\theta)\big] \exp\big[a(\theta)T(\chi) + F(\chi) + Pb(\theta)\big]\mathrm{d}\chi = 0 \; .$$

Now $\langle T(\chi)\rangle_\theta = \int T(\chi)L(\chi|\theta)\mathrm{d}\chi$, so that $a'(\theta)\langle T(\chi)\rangle_\theta + Pb'(\theta) = 0$, and hence finally $\langle T(\chi)\rangle_\theta = -Pb'(\theta)/a'(\theta)$. Given that for $\theta = \hat{\theta}_{\mathrm{ML}}(\chi)$ we have

$$-b'\big(\hat{\theta}_{\mathrm{ML}}(\chi)\big)/a'\big(\hat{\theta}_{\mathrm{ML}}(\chi)\big) = T(\chi)/P,$$

we do indeed obtain

$$T(\chi) = \langle T(\chi)\rangle_{\hat{\theta}_{\mathrm{ML}}(\chi)} \; .$$

For independent realizations, we have $T(\chi) = \sum_{n=1}^{P} t(x_n)$. We deduce that $\langle T(\chi)\rangle_\theta = P\langle t(x)\rangle_\theta$ and hence,

$$\frac{1}{P}\sum_{n=1}^{P} t(x_n) = \langle t(x)\rangle_{\hat{\theta}_{\mathrm{ML}}(\chi)} \; ,$$

which is the analogue of a moment method since it amounts to identifying the mean of $t(x)$.

7.10 Examples in the Exponential Family

In this section, we illustrate the results of the last few sections with five examples from the exponential family. We will consider the Poisson distribution, the Gamma distribution, two examples of the Gaussian distribution, and the Weibull distribution. We use the notation of the last section and we assume that the P-sample χ corresponds to independent realizations.

7.10.1 Estimating the Parameter in the Poisson Distribution

The probability distribution is

$$P_N(n) = \frac{\exp(-\theta)\theta^n}{n!} \ ,$$

where θ is the parameter to be estimated. When we observe a P-sample $\chi = \{n_1, n_2, \ldots, n_P\}$, the log-likelihood is

$$\ell(\chi) = -P\theta + T(\chi)\ln\theta - \sum_{i=1}^{P} \ln(n_i!) \ ,$$

where the sufficient statistic $T(\chi)$ is simply $T(\chi) = \sum_{i=1}^{P} n_i$. We thus have

$$\left. \begin{array}{l} a(\theta) = \ln\theta \ , \ a'(\theta) = 1/\theta \\ b(\theta) = -\theta \ , \ b'(\theta) = -1 \end{array} \right\} \implies -\frac{b'(\theta)}{a'(\theta)} = \theta \ .$$

We see that the maximum likelihood estimator of θ leads to

$$\hat{\theta}_{\mathrm{ML}}(\chi) = \frac{1}{P}T(\chi) = \frac{1}{P}\sum_{i=1}^{P} n_i \ .$$

According to the results of the last section, this estimator is therefore efficient. Let us return for a moment to the example discussed at the beginning of this chapter. We see that we now obtain an unambiguous answer concerning the best way to estimate the parameter in the Poisson distribution, and hence the particle flux, if the relevant criterion is the variance of the estimator when there is no bias.

7.10.2 Estimating the Mean of the Gamma Distribution

The probability distribution is given by

$$P_X(x) = \frac{x^{\alpha-1}}{\theta^\alpha \Gamma(\alpha)} \exp\left(-\frac{x}{\theta}\right) \ ,$$

where θ is the parameter to be estimated and we assume that α is given. When we observe a P-sample $\chi = \{n_1, n_2, \ldots, n_P\}$, the log-likelihood is

$$\ell(\chi) = -P\alpha\ln\theta - \frac{1}{\theta}T(\chi) + (\alpha-1)\sum_{i=1}^{P}\ln x_i - P\ln\Gamma(\alpha) \ ,$$

where the sufficient statistic $T(\chi)$ is simply $T(\chi) = \sum_{i=1}^{P} x_i$. We thus have

$$a(\theta) = -1/\theta , \quad a'(\theta) = 1/\theta^2 \left. \right\} \implies -\frac{b'(\theta)}{a'(\theta)} = \alpha\theta .$$
$$b(\theta) = -\alpha \ln\theta , \, b'(\theta) = -\alpha/\theta$$

We see that the maximum likelihood estimator of θ leads to

$$\hat{\theta}_{\text{ML}}(\chi) = \frac{1}{\alpha P} T(\chi) = \frac{1}{\alpha P} \sum_{i=1}^{P} x_i .$$

This estimator is therefore efficient, since it is proportional to the sufficient statistic $T(\chi)$.

Let us analyze the result we would have obtained if we had used the notation

$$P_X(x) = \frac{x^{\alpha-1}\theta^\alpha}{\Gamma(\alpha)} \exp(-\theta x) .$$

The log-likelihood would then have been

$$\ell(\chi) = P\alpha \ln\theta - \theta T(\chi) + (\alpha - 1) \sum_{i=1}^{P} \ln x_i - P \ln \Gamma(\alpha) .$$

We would thus have found

$$a(\theta) = -\theta , \quad a'(\theta) = -1 \left. \right\} \implies -\frac{b'(\theta)}{a'(\theta)} = \frac{\alpha}{\theta} .$$
$$b(\theta) = \alpha \ln\theta , \, b'(\theta) = \alpha/\theta$$

We see that the maximum likelihood estimator of θ leads to

$$\hat{\theta}_{\text{ML}}(\chi) = \frac{\alpha P}{T(\chi)} .$$

We cannot now deduce that this estimator is efficient.

7.10.3 Estimating the Mean of the Gaussian Distribution

The probability distribution is given by

$$P_X(x) = \frac{1}{\sqrt{2\pi}\sigma} \exp\left[-\frac{(x-\theta)^2}{2\sigma^2}\right] ,$$

where θ is the parameter to be estimated and we assume that σ^2 is given. When we observe a P-sample $\chi = \{n_1, n_2, \ldots, n_P\}$, the log-likelihood is

$$\ell(\chi) = -\frac{1}{2\sigma^2}\left[\sum_{i=1}^{P} x_i^2 - 2\theta T(\chi) + P\theta^2\right] - P\ln\sigma - P\ln\sqrt{2\pi} ,$$

where the sufficient statistic $T(\chi)$ is still simply $T(\chi) = \sum_{i=1}^{P} x_i$. We thus have

$$\left.\begin{array}{ll} a(\theta) = \theta/\sigma^2 \,, & a'(\theta) = 1/\sigma^2 \\ b(\theta) = -\theta^2/(2\sigma^2) \,, & b'(\theta) = -\theta/\sigma^2 \end{array}\right\} \implies -\frac{b'(\theta)}{a'(\theta)} = \theta \,.$$

We see that the maximum likelihood estimator of θ leads to

$$\hat{\theta}_{\mathrm{ML}}(\chi) = \frac{1}{P}T(\chi) = \frac{1}{P}\sum_{i=1}^{P} x_i \,.$$

This estimator is therefore efficient.

7.10.4 Estimating the Variance of the Gaussian Distribution

The probability distribution is given by

$$P_X(x) = \frac{1}{\sqrt{2\pi\theta}} \exp\left[-\frac{(x-m)^2}{2\theta^2}\right] \,,$$

where θ is the parameter to be estimated and we assume that m is given. The log-likelihood is

$$\ell(\chi) = -\frac{1}{2\theta^2}T(\chi) - P\ln\theta - P\ln\sqrt{2\pi} \,,$$

where the sufficient statistic $T(\chi)$ is now $T(\chi) = \sum_{i=1}^{P}(x_i - m)^2$. We thus have

$$\left.\begin{array}{ll} a(\theta) = -1/(2\theta^2) \,, & a'(\theta) = 1/\theta^3 \\ b(\theta) = -\ln\theta \,, & b'(\theta) = -1/\theta \end{array}\right\} \implies -\frac{b'(\theta)}{a'(\theta)} = \theta^2 \,.$$

We see that the maximum likelihood estimator of θ^2 (and not θ) leads to

$$\hat{\theta}_{\mathrm{ML}}^2(\chi) = \frac{1}{P}T(\chi) = \frac{1}{P}\sum_{i=1}^{P}(x_i - m)^2 \,.$$

This estimator is unbiased and hence efficient. Indeed, we have

$$\langle\hat{\theta}_{\mathrm{ML}}^2(\chi)\rangle = \frac{1}{P}\sum_{i=1}^{P}\langle(x_i - m)^2\rangle \,.$$

Now, $\langle(x_i - m)^2\rangle = \theta^2$ and hence $\langle\hat{\theta}_{\mathrm{ML}}^2(\chi)\rangle = \theta^2$.

We thus observe that these probability distributions hold no surprises. The efficient estimators are precisely those we would expect to be efficient. The situation is not always so simple, however. For example, we will consider the case of the Weibull probability density function. This probability distribution is often used to describe the probability of breakdown in complex systems and it is therefore widely used when we need to study the reliability of components.

7.10.5 Estimating the Mean of the Weibull Distribution

The Weibull probability distribution is given by

$$P_X(x) = \frac{\alpha x^{\alpha-1}}{\theta^\alpha} \exp\left[-\left(\frac{x}{\theta}\right)^\alpha\right],$$

where θ is the parameter to be estimated and we assume that α is given. We can express the mean m as a function of θ by $m = \theta \Gamma\left[(\alpha+1)/\alpha\right]$. We consider once again the observation of a P-sample $\chi = \{x_1, x_2, \ldots, x_P\}$. Identifying θ by the moments method would lead to

$$\hat{\theta}_{\text{moment}}(\chi) = \frac{1}{P\Gamma\left(\dfrac{\alpha+1}{\alpha}\right)} \sum_{i=1}^{P} x_i.$$

The log-likelihood is

$$\ell(\chi) = -P\alpha \ln\theta - \frac{1}{\theta^\alpha} T(\chi) + (\alpha-1)\sum_{i=1}^{P} \ln x_i + P\ln\alpha,$$

where the sufficient statistic $T(\chi)$ is $T(\chi) = \sum_{i=1}^{P} x_i^\alpha$. We have

$$\left.\begin{array}{l} a(\theta) = -1/\theta^\alpha,\ a'(\theta) = \alpha/\theta^{\alpha+1} \\ b(\theta) = -\alpha\ln\theta,\ b'(\theta) = -\alpha/\theta \end{array}\right\} \implies -\frac{b'(\theta)}{a'(\theta)} = \theta^\alpha.$$

We thus find that the maximum likelihood estimator of θ leads to

$$\hat{\theta}_{\text{ML}}(\chi) = \left[\frac{1}{P} T(\chi)\right]^{1/\alpha} = \left[\frac{1}{P}\sum_{i=1}^{P} (x_i)^\alpha\right]^{1/\alpha}.$$

This estimator is efficient for θ^α (but not for θ). It should also be noted that the moment method and the maximum likelihood method do not lead to the same estimator. The Weibull distribution belongs to the exponential family so it is better to consider the maximum likelihood estimator.

To illustrate the differences that are effectively obtained with the moment and maximum likelihood methods, we have displayed the results of several numerical simulations in Table 7.2. Figure 7.8 shows histograms of the parameter θ estimated from 1000 independent samples of 5000 realizations each. The value of the parameter α is 0.25 and the true value of θ is 10.

The continuous curve shows the histogram of values obtained using the maximum likelihood method, whilst the dotted curve shows the same obtained using the moment method. The superiority of the maximum likelihood method is clear. We estimated θ from 100 independent samples of variable size P and Table 7.2 shows the means and variances of the values obtained. We thus

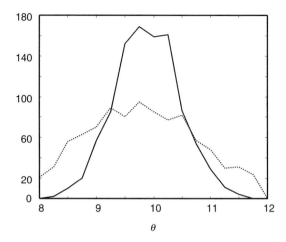

Fig. 7.8. Histograms of estimated values of the parameter in the Weibull distribution using the maximum likelihood method (*continuous line*) and the moment method (*dotted line*)

Table 7.2. Comparison of estimators for the Weibull distribution

Sample size	100	1000	10 000
Mean by the moment method	8.37	10.2	9.96
Mean by the maximum likelihood method	9.94	10.25	10.03
Variance by the moment method	28	8.16	0.65
Variance by the maximum likelihood method	15	1.77	0.15

observe that to obtain the same estimation variance with the method which involves identifying the first moment as with the maximum likelihood method, we would need a sample roughly four times larger.

We can find the Cramer–Rao bound which is attained by the statistic $T(\chi) = \sum_{i=1}^{P} x_i^\alpha$. If we consider this statistic, it constitutes an unbiased estimator of θ^α, but a biased estimator of θ. Let us show that it is indeed an unbiased estimator of θ^α. Setting $y = x^\alpha$, we thus have $\mathrm{d}y = \alpha x^{\alpha-1}\mathrm{d}x$. Now $P_Y(y)\mathrm{d}y = P_X(x)\mathrm{d}x$, which implies

$$P_Y(y)\alpha x^{\alpha-1} = \frac{\alpha x^{\alpha-1}}{\theta^\alpha} \exp\left(-\frac{x^\alpha}{\theta^\alpha}\right) .$$

Putting $\mu = \theta^\alpha$, we then see that $P_Y(y) = (1/\mu)\exp(-y/\mu)$, we deduce that $\langle y \rangle = \langle x^\alpha \rangle = \mu = \theta^\alpha$ and hence $\langle T(\chi)/P \rangle = \theta^\alpha$, which does indeed mean that $T(\chi)$ is an unbiased estimator of θ^α, as claimed. Now we must determine

$$
\mathrm{CRB} = \frac{\left| \dfrac{\partial}{\partial \mu} h(\mu) \right|^2}{I_\mathrm{F}} ,
$$

where $h(\mu) = \langle T(\chi) \rangle$ and

$$
I_\mathrm{F} = - \int \frac{\partial^2 \ln \left[L(\chi|\mu) \right]}{\partial \mu^2} L(\chi|\mu) \mathrm{d}\chi .
$$

It is easy to show that $\langle (x_i)^\alpha \rangle = \theta^\alpha = \mu$ and hence that $\partial h(\mu)/\partial \mu = P$. Moreover, a direct calculation shows that $I_\mathrm{F} = P/\mu$ and hence that $\mathrm{CRB} = P\mu^2$. The variance of the statistic $T_R(\chi) = T(\chi)/P$ is therefore $\sigma_{T_R}^2 = \theta^{2\alpha}/P$. Table 7.3 shows experimentally determined values and this theoretical value for samples of different sizes. Experimental conditions are as described above.

Table 7.3. Comparing experimental variances with the Cramer–Rao bound

Sample size	Cramer–Rao bound	Experimental variance of $T_R(\chi)$
100	3.16×10^{-2}	3.09×10^{-2}
1000	3.16×10^{-3}	3.1×10^{-3}
10 000	3.16×10^{-4}	3.4×10^{-4}

7.11 Robustness of Estimators

We should not end this chapter without discussing the robustness problems associated with estimation techniques using the maximum likelihood method. An estimator $\theta_{\mathrm{ML}}(\chi)$ is optimal in the sense of maximum likelihood for a given probability law (or probability density function) $P_\theta(x)$. In other words, we are concerned here with parameter estimation, since we assume that the observed data obey a law of a form that is known *a priori*. However, it may be that the observed data are distributed according to a law $\tilde{P}_\theta(x)$ which is slightly different from those in the family $P_\theta(x)$. An estimator is said to be robust if its variance changes only very slightly when it is evaluated for a sample arising from $\tilde{P}_\theta(x)$ rather than from $P_\theta(x)$.

Let us illustrate with an example. The maximum likelihood estimator for the mean of a Gaussian distribution has already been determined to be

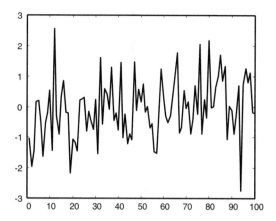

Fig. 7.9. Example of 100 realizations of Gaussian variables with mean 0 and variance 1

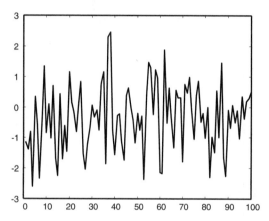

Fig. 7.10. Example of 100 realizations of variables distributed according to $\tilde{P}_\theta(x)$ with $\varepsilon = 10^{-2}$

$\hat{\theta}_{\mathrm{ML}}(\chi) = (\sum_{i=1}^{P} x_i)/P$. We have also seen that this estimator is efficient. Suppose now that the P-sample $\chi = \{x_1, x_2, \ldots, x_P\}$ arises from the probability law $\tilde{P}_\theta(x)$ rather than $P_\theta(x)$, where

$$\tilde{P}_\theta(x) = (1 - \varepsilon)N(x) + \varepsilon C(x) ,$$

$N(x)$ is the Gaussian distribution with mean 0 and variance 1, and $C(x)$ is the Cauchy distribution with probability density function $C(x) = 1/[\pi(1 + x^2)]$. Figures 7.9 and 7.10 show examples of 100 realizations of Gaussian variables of mean 0 and variance 1 and variables distributed according to $\tilde{P}_\theta(x) = (1 - \varepsilon)N(x) + \varepsilon C(x)$ with $\varepsilon = 10^{-2}$.

Although the data may appear to be very similar, we shall soon see that the performance of the estimator $\hat{\theta}_{\mathrm{ML}}(\chi) = (\sum_{i=1}^{P} x_i)/P$ is very different. To

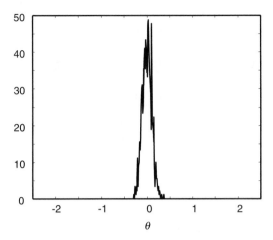

Fig. 7.11. Histogram of values obtained with $\hat{\theta}_{\mathrm{ML}}(\chi) = (1/P)\sum_{i=1}^{P} x_i$ when the samples are generated by pure Gaussian variables

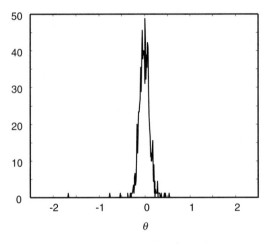

Fig. 7.12. Histogram of values obtained with $\hat{\theta}_{\mathrm{ML}}(\chi) = (1/P)\sum_{i=1}^{P} x_i$ when the samples are distributed according to $\tilde{P}_\theta(x) = (1-\varepsilon)N(x) + \varepsilon C(x)$

this end, we have estimated θ for 100 independent samples made up of 1000 realizations each. Figures 7.11 and 7.12 show the histograms of values obtained with $\hat{\theta}_{\mathrm{ML}}(\chi) = (\sum_{i=1}^{P} x_i)/P$ when the samples are generated by pure Gaussian variables with mean 0 and variance 1 or by variables distributed according to $\tilde{P}_\theta(x) = (1-\varepsilon)N(x) + \varepsilon C(x)$ with $\varepsilon = 10^{-2}$.

It should be observed that, although the realizations seem similar for pure Gaussian variables and variables distributed according to $\tilde{P}_\theta(x)$, there are spurious peaks in the second case for large values of $|\theta|$. The variances of the estimator are also very different, as can be seen from Table 7.4, where the values have been estimated for various configurations. [To be perfectly rigorous, the mean and the variance of a Cauchy random variable do not exist. This same is therefore true for our own problem as soon as $\varepsilon \neq 0$. The figures mentioned only have a meaning for the numerical experiments carried out.] It is quite clear then that the estimator $\hat{\theta}_{\mathrm{ML}}(\chi) = (\sum_{i=1}^{P} x_i)/P$ is not robust.

Table 7.4. Empirical variance in the presence of $\varepsilon\,\%$ Cauchy variables

Sample size	$\varepsilon = 0$	$\varepsilon = 10^{-2}$
100	9.4×10^{-3}	3.0×10^{-1}
1000	1.1×10^{-3}	1.2×10^{-1}

It is important to note that the perturbation we have considered is a Cauchy distribution. Indeed, its probability density function decreases very slowly and has no finite moments (see Section 2.3).

Another way of viewing the problem that we have just analyzed is to consider that the sample generated with $\tilde{P}_\theta(x) = (1-\varepsilon)N(x)+\varepsilon C(x)$ contains atypical data, known as outliers. Indeed, let χ be the P-sample generated with $\tilde{P}_\theta(x)$. It can be obtained from a P-sample χ_{N} itself generated with $P_\theta(x)$. We simply replace with a probability ε each sample x_i of χ_{N} by a Cauchy variable whose probability density is $C(x) = 1/[\pi(1+x^2)]$. We can then treat χ_{N} as a pure sample and say that χ contains outliers.

These atypical data have no mean but they do have a median value. Let us therefore analyze the results obtained if we estimate θ using the median value of the sample χ. To define the median value of χ, we put the set in increasing order. In other words, we carry out the permutation $\{x_1, x_2, \ldots, x_P\} \to \{x'_1, x'_2, \ldots, x'_P\}$ of the elements of χ in such a way that $x'_j \leq x'_{j+1}$, $\forall j$, and we consider $x'_{P/2}$ if p is even and $x'_{(P+1)/2}$ if p is odd. This new estimator will be written $\hat{\theta}_{\mathrm{median}}(\chi)$. Table 7.5 gives the values of the variances of the median estimator for the different configurations analyzed previously.

Table 7.5. Variances estimated with the median in the presence of $\varepsilon\,\%$ Cauchy variables

Sample size	$\varepsilon = 0$	$\varepsilon = 10^{-2}$
100	1.4×10^{-2}	1.5×10^{-2}
1000	1.7×10^{-3}	1.6×10^{-3}

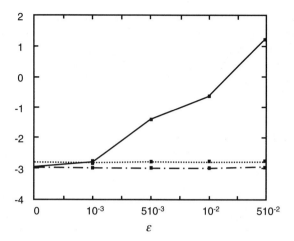

Fig. 7.13. Base 10 logarithm of the variances of several estimators for $P = 1000$ as a function of ε: mean (*continuous line*), median (*dotted line*), 4% truncated mean (*dot-dashed line*)

Although the variance of this estimator is slightly greater than that of $\hat{\theta}_{\mathrm{ML}}(\chi) = (\sum_{i=1}^{P} x_i)/P$ when $\varepsilon = 0$, we note that it is less sensitive to the presence of atypical values, i.e., drawn according to $\tilde{P}_\theta(x) = (1 - \varepsilon)N(x) + \varepsilon C(x)$ with $\varepsilon = 10^{-2}$.

We can define an intermediate method between the two previous estimators. To do so, starting with χ, we determine a new sample χ_a by eliminating the $(a/2)\%$ greatest values and the $(a/2)\%$ smallest values. We then simply calculate $\hat{\theta}_a(\chi) = \hat{\theta}_{\mathrm{ML}}(\chi_a) = (\sum_{i=1}^{P_a} x_i)/P_a$, where P_a is the number of elements in χ_a. We call this the $a\%$ truncated mean estimator. Figure 7.13 shows the base 10 logarithm of the variances, i.e., $\log_{10}(\sigma_\theta^2)$, of these estimators as a function of ε and for $a = 4\%$.

Note that the truncated mean performs extremely well. It is easy to generalize this method to the estimation of parameters other than the mean. We can say that the estimator has been robustified. This is an important point in applications as soon as there is any risk of atypical data. In particular, it is very important if the atypical data can exhibit large deviations, even if the probability of this happening is extremely low.

7.12 Appendix: Scalar Cramer–Rao Bound

Consider two statistics $T(\chi_\lambda)$ and $U(\chi_\lambda)$. Once again, and analogously to what is done in probability theory, we distinguish the notation $T(\chi_\lambda)$ and $U(\chi_\lambda)$, where we consider the statistics for a random sample, and the functions $T(x_1, x_2, \ldots, x_P)$ and $U(x_1, x_2, \ldots, x_P)$, which are simply functions of the variables $x_1, x_2, \ldots, x_P$.

Let us begin by showing something that will be particularly important in the following:

$$[\langle T(\chi_\lambda)U(\chi_\lambda)\rangle_\theta]^2 \leqslant \langle T(\chi_\lambda)^2\rangle_\theta \langle U(\chi_\lambda)^2\rangle_\theta .$$

This result is obtained by considering the quadratic form $\left[\alpha T(\chi_\lambda) - U(\chi_\lambda)\right]^2$ in α. As this form is positive, its expectation value must also be positive, i.e.,

$$\langle \left[\alpha T(\chi_\lambda) - U(\chi_\lambda)\right]^2\rangle_\theta \geqslant 0 .$$

Expanding this out, we obtain

$$\alpha^2 \langle T(\chi_\lambda)^2\rangle_\theta - 2\alpha\langle T(\chi_\lambda)U(\chi_\lambda)\rangle_\theta + \langle U(\chi_\lambda)^2\rangle_\theta \geqslant 0 .$$

The discriminant of this quadratic form in α must be negative since it has no root. This implies

$$[\langle T(\chi_\lambda)U(\chi_\lambda)\rangle_\theta]^2 - \langle T(\chi_\lambda)^2\rangle_\theta \langle U(\chi_\lambda)^2\rangle_\theta \leqslant 0 ,$$

thus proving the above claim.

The result can also be expressed in terms of the standard deviations of the statistic $T(\chi_\lambda)$. To do so, write $\delta T(\chi_\lambda) = T(\chi_\lambda) - \langle T(\chi_\lambda)\rangle$ and $\sigma_T^2(\theta) = \langle [\delta T(\chi_\lambda)]^2\rangle_\theta$, whereupon

$$[\langle \delta T(\chi_\lambda)U(\chi_\lambda)\rangle_\theta]^2 \leqslant \sigma_T^2(\theta)\langle [U(\chi_\lambda)]^2\rangle_\theta .$$

We now use this property, which is actually a Cauchy–Schwartz inequality, to find a lower bound for the variance we can hope to attain with a statistic $T(\chi_\lambda)$. For this purpose, we assume that the probability distribution of the sample has support independent of the parameter θ. In other words, the possible regions of variation of the random variables of χ_λ are assumed to be the same for all the laws $L(\chi|\theta)$, whatever the value of θ. To find the Cramer–Rao bound, we consider the inequality

$$[\langle \delta T(\chi_\lambda)U(\chi_\lambda)\rangle_\theta]^2 \leqslant \sigma_T^2(\theta)\langle [U(\chi_\lambda)]^2\rangle_\theta ,$$

with $U(\chi_\lambda) = \partial V(\chi_\lambda|\theta)/\partial\theta$ and $V(\chi_\lambda|\theta) = \ln [L(\chi_\lambda|\theta)]$. We note that $V(\chi_\lambda|\theta)$ has mean independent of θ. Indeed, we have

$$\left\langle \frac{\partial V(\chi_\lambda|\theta)}{\partial\theta}\right\rangle_\theta = \int \frac{\partial \ln [L(\chi|\theta)]}{\partial\theta} L(\chi|\theta)\mathrm{d}\chi .$$

Now $\partial \ln [L(\chi|\theta)]/\partial\theta = [\partial L(\chi|\theta)/\partial\theta]/L(\chi|\theta)$, so that

$$\left\langle \frac{\partial V(\chi_\lambda|\theta)}{\partial\theta}\right\rangle_\theta = \int \frac{\partial L(\chi|\theta)}{\partial\theta}\mathrm{d}\chi .$$

Exchanging the integration and differentiation, it follows that

$$\left\langle \frac{\partial \ln\left[L(\chi_\lambda|\theta)\right]}{\partial\theta} \right\rangle_\theta = \frac{\partial\left[\int L(\chi|\theta)\mathrm{d}\chi\right]}{\partial\theta} .$$

However, since $L(\chi|\theta)$ is a probability law, we have $\int L(\chi|\theta)\mathrm{d}\chi = 1$, $\forall\theta$, and therefore

$$\left\langle \frac{\partial V(\chi_\lambda|\theta)}{\partial\theta} \right\rangle_\theta = \frac{\partial\langle\ln\left[L(\chi_\lambda|\theta)\right]\rangle_\theta}{\partial\theta} = 0 ,$$

or in other words, $\langle U(\chi_\lambda)\rangle_\theta = 0$.

We thus deduce that $\langle\delta T(\chi_\lambda)U(\chi_\lambda)\rangle_\theta = \langle T(\chi_\lambda)U(\chi_\lambda)\rangle_\theta$. Indeed,

$$\langle\delta T(\chi_\lambda)U(\chi_\lambda)\rangle_\theta = \left\langle\left[T(\chi_\lambda) - \langle T(\chi_\lambda)\rangle_\theta\right]U(\chi_\lambda)\right\rangle_\theta ,$$

or

$$\langle\delta T(\chi_\lambda)U(\chi_\lambda)\rangle_\theta = \langle T(\chi_\lambda)U(\chi_\lambda)\rangle_\theta - \langle T(\chi_\lambda)\rangle_\theta\langle U(\chi_\lambda)\rangle_\theta ,$$

and hence,

$$\langle\delta T(\chi_\lambda)U(\chi_\lambda)\rangle_\theta = \langle T(\chi_\lambda)U(\chi_\lambda)\rangle_\theta .$$

This can be written explicitly as

$$\langle\delta T(\chi_\lambda)U(\chi_\lambda)\rangle_\theta = \int T(\chi_\lambda)\frac{\partial \ln\left[L(\chi|\theta)\right]}{\partial\theta}L(\chi|\theta)\mathrm{d}\chi .$$

Using the same properties as before, we obtain

$$\langle\delta T(\chi_\lambda)U(\chi_\lambda)\rangle_\theta = \int T(\chi)\frac{\partial L(\chi|\theta)}{\partial\theta}\mathrm{d}\chi .$$

Exchanging the integration and differentiation once again,

$$\langle\delta T(\chi_\lambda)U(\chi_\lambda)\rangle_\theta = \frac{\partial}{\partial\theta}\int T(\chi)L(\chi|\theta)\mathrm{d}\chi .$$

Previously we set $h(\theta) = \int T(\chi)L(\chi|\theta)\mathrm{d}\chi$. Using the inequality,

$$\left[\langle\delta T(\chi_\lambda)U(\chi_\lambda)\rangle_\theta\right]^2 \leqslant \sigma_T^2(\theta)\langle[U(\chi_\lambda)]^2\rangle_\theta ,$$

we finally obtain

$$\sigma_T^2(\theta) \geqslant \frac{|\gamma(\theta)|^2}{I_\mathrm{F}} ,$$

where

$$\gamma(\theta) = \frac{\partial h(\theta)}{\partial\theta} \quad \text{and} \quad I_\mathrm{F} = \int \left[\frac{\partial \ln\left[L(\chi|\theta)\right]}{\partial\theta}\right]^2 L(\chi|\theta)\mathrm{d}\chi .$$

I_F is known classically as the Fisher information.

If the estimator is unbiased, we must have $\langle T(\chi_\lambda)\rangle = \theta$ and thus $\gamma(\theta) = 1$. The Cramer–Rao bound is then

$$\sigma_T^2(\theta) \geqslant \frac{1}{I_{\mathrm{F}}} \ .$$

We can obtain a new expression for the Fisher information once again using the fact that the likelihood defines a probability density function on χ. Indeed, we have $\int L(\chi|\theta)\mathrm{d}\chi = 1$. We have already seen that

$$\int \frac{\partial \ln\left[L(\chi|\theta)\right]}{\partial\theta} L(\chi|\theta)\mathrm{d}\chi = 0 \ .$$

If we differentiate a second time with respect to θ, we obtain

$$\int \frac{\partial^2 \ln\left[L(\chi|\theta)\right]}{\partial\theta^2} L(\chi|\theta)\mathrm{d}\chi + \int \frac{\partial \ln\left[L(\chi|\theta)\right]}{\partial\theta}\frac{\partial L(\chi|\theta)}{\partial\theta}\mathrm{d}\chi = 0 \ .$$

Using the same calculation as before, we can write

$$\int \frac{\partial \ln\left[L(\chi|\theta)\right]}{\partial\theta}\frac{\partial L(\chi|\theta)}{\partial\theta}\mathrm{d}\chi = \int \left[\frac{\partial \ln\left[L(\chi|\theta)\right]}{\partial\theta}\right]^2 L(\chi|\theta)\mathrm{d}\chi \ ,$$

and hence,

$$\int \left[\frac{\partial \ln\left[L(\chi|\theta)\right]}{\partial\theta}\right]^2 L(\chi|\theta)\mathrm{d}\chi = -\int \frac{\partial^2 \ln\left[L(\chi|\theta)\right]}{\partial\theta^2} L(\chi|\theta)\mathrm{d}\chi \ .$$

We thereby obtain a new expression for the Fisher information:

$$I_{\mathrm{F}} = -\int \frac{\partial^2 \ln\left[L(\chi|\theta)\right]}{\partial\theta^2} L(\chi|\theta)\mathrm{d}\chi \ .$$

7.13 Appendix: Efficient Statistics

Let us again refer to the demonstration in Section 7.12. The Cramer–Rao inequality is an equality if

$$[\langle\delta T(\chi_\lambda)U(\chi_\lambda)\rangle]^2 - \langle\delta T(\chi_\lambda)^2\rangle_\theta\langle U(\chi_\lambda)^2\rangle_\theta = 0 \ ,$$

which implies that there exists α_0 such that

$$\langle[\alpha_0\delta T(\chi_\lambda) - U(\chi_\lambda)]^2\rangle_\theta = 0 \ .$$

The random variables $\delta T(\chi_\lambda)$ and $U(\chi_\lambda)$ in χ_λ are therefore equal in quadratic mean. Since α_0 can depend on θ, we write $\alpha_0 = \alpha_0(\theta)$ and hence $U(\chi_\lambda) = \alpha_0(\theta)\delta T(\chi_\lambda)$, or

$$\frac{\partial \ln\big[L(\chi_\lambda|\theta)\big]}{\partial\theta} = \alpha_0(\theta)\delta T(\chi_\lambda)\,.$$

We put $\beta(\theta) = -\langle\alpha_0(\theta)T(\chi_\lambda)\rangle_\theta$, and consider a given sample χ. We have $\partial\ln\big[L(\chi|\theta)\big]/\partial\theta = \alpha_0(\theta)T(\chi) + \beta(\theta)$ and if we integrate this expression with respect to θ, we obtain $\ln\big[L(\chi|\theta)\big] = a(\theta)T(\chi) + F(\chi) + Pb(\theta)$, or

$$L(\chi|\theta) = \frac{\exp\big[a(\theta)T(\chi) + F(\chi)\big]}{Z_P(\theta)}\,,$$

where $Z_P(\theta) = \exp\big[-Pb(\theta)\big]$.

If χ corresponds to an independent realization of random variables, we have $T(\chi) = \sum_{n=1}^P t(x_n)$ and $F(\chi) = \sum_{n=1}^P f(x_n)$ and the probability density function of X_λ must therefore be

$$P_\theta(x) = \frac{\exp\big[a(\theta)t(x) + f(x)\big]}{Z(\theta)}\,,$$

with $Z(\theta) = \exp\big[-b(\theta)\big] = \int \exp\big[a(\theta)t(x) + f(x)\big]\mathrm{d}x$. We now determine the variance of the estimator. Since the Cramer–Rao bound is attained, we have

$$\sigma_T^2(\theta) = \frac{|\partial h(\theta)/\partial\theta|^2}{\displaystyle\int \left[\frac{\partial}{\partial\theta}\ln\big[L(\chi|\theta)\big]\right]^2 L(\chi|\theta)\mathrm{d}\chi}\,,$$

where $h(\theta) = \int T(\chi)L(\chi|\theta)\mathrm{d}\chi$. We have $\partial\ln\big[L(\chi|\theta)\big]/\partial\theta = \alpha_0(\theta)\delta T(\chi_\lambda)$ and hence

$$\int \left[\frac{\partial}{\partial\theta}\ln\big[L(\chi|\theta)\big]\right]^2 L(\chi|\theta)\mathrm{d}\chi = \alpha_0^2(\theta)\sigma_T^2(\theta)\,,$$

which shows that

$$\sigma_T^2(\theta) = \left|\frac{\partial h(\theta)/\partial\theta}{\alpha_0(\theta)}\right|\,.$$

7.14 Appendix: Vectorial Cramer–Rao Bound

We use the notation $\boldsymbol{\theta} = (\theta_1, \theta_2, \ldots, \theta_n)^\dagger$, $\delta\boldsymbol{T}(\chi) = \boldsymbol{T}(\chi) - \langle\boldsymbol{T}(\chi)\rangle_\theta$, where the vector statistic $\boldsymbol{T}(\chi_\lambda)$ has the same dimension as $\boldsymbol{\theta}$. We will also assume that $\boldsymbol{T}(\chi)$ is an unbiased estimator of $\boldsymbol{\theta}$. The covariance matrix $\overline{\overline{\Gamma}}$ of $\boldsymbol{T}(\chi_\lambda)$ is

$$\Gamma_{ij} = \langle\delta T_i(\chi)\delta T_j(\chi)\rangle_\theta\,,$$

and the Fisher information matrix $\overline{\overline{J}}$ is

$$J_{ij} = -\left\langle\frac{\partial^2\ell(\chi|\boldsymbol{\theta})}{\partial\theta_i\partial\theta_j}\right\rangle_\theta\,,$$

where $\ell(\chi|\boldsymbol{\theta}) = \ln\left[L(\chi|\boldsymbol{\theta})\right]$. We also write $\boldsymbol{U}(\chi_\lambda)$ for the vector with components $U_i(\chi) = \partial\ell(\chi|\boldsymbol{\theta})/\partial\theta_i$. Then,

$$\langle \boldsymbol{U}(\chi_\lambda)\,[\boldsymbol{\delta T}(\chi_\lambda)]^\dagger\rangle_\theta = \overline{\overline{\mathrm{Id}_n}}\ ,$$

where $[\boldsymbol{\delta T}(\chi_\lambda)]^\dagger$ is the vector transpose of $\boldsymbol{\delta T}(\chi_\lambda)$ and $\overline{\overline{\mathrm{Id}_n}}$ the n-dimensional identity matrix. Indeed, we have

$$\left[\langle \boldsymbol{U}(\chi_\lambda)\,[\boldsymbol{\delta T}(\chi_\lambda)]^\dagger\rangle_\theta\right]_{ij} = \int \delta T_j(\chi)\frac{\partial\ell(\chi|\boldsymbol{\theta})}{\partial\theta_i}L(\chi|\boldsymbol{\theta})\mathrm{d}\chi\ ,$$

or

$$\left[\langle \boldsymbol{U}(\chi_\lambda)\,[\boldsymbol{\delta T}(\chi_\lambda)]^\dagger\rangle_\theta\right]_{ij} = \int T_j(\chi)\frac{\partial}{\partial\theta_i}\ell(\chi|\boldsymbol{\theta})L(\chi|\boldsymbol{\theta})\mathrm{d}\chi\ .$$

This follows because

$$\int \theta_j\frac{\partial\ell(\chi|\boldsymbol{\theta})}{\partial\theta_i}L(\chi|\boldsymbol{\theta})\mathrm{d}\chi = \theta_j\int \frac{\partial\ell(\chi|\boldsymbol{\theta})}{\partial\theta_i}L(\chi|\boldsymbol{\theta})\mathrm{d}\chi\ ,$$

or

$$\theta_j\int \frac{\partial\ell(\chi|\boldsymbol{\theta})}{\partial\theta_i}L(\chi|\boldsymbol{\theta})\mathrm{d}\chi = \theta_j\int \frac{\partial}{\partial\theta_i}L(\chi|\boldsymbol{\theta})\mathrm{d}\chi = \theta_j\frac{\partial}{\partial\theta_i}\int L(\chi|\boldsymbol{\theta})\mathrm{d}\chi\ .$$

Now,

$$\int L(\chi|\boldsymbol{\theta})\mathrm{d}\chi = 1\ ,\quad \text{and hence,}\quad \theta_j\int \frac{\partial\ell(\chi|\boldsymbol{\theta})}{\partial\theta_i}L(\chi|\boldsymbol{\theta})\mathrm{d}\chi = 0\ ,$$

as required.

An analogous calculation to the one carried out previously shows that

$$\left[\langle \boldsymbol{U}(\chi_\lambda)\,[\boldsymbol{\delta T}(\chi_\lambda)]^\dagger\rangle_\theta\right]_{ij} = \frac{\partial}{\partial\theta_i}\int T_j(\chi)L(\chi|\boldsymbol{\theta})\mathrm{d}\chi\ .$$

Now as the estimator $\boldsymbol{T}(\chi_\lambda)$ was assumed to be unbiased, we have

$$\frac{\partial}{\partial\theta_i}\int T_j(\chi)L(\chi|\boldsymbol{\theta})\mathrm{d}\chi = \delta_{i-j}\ ,$$

where δ_{i-j} is the Kronecker symbol. Indeed, we have $\int T_j(\chi)L(\chi|\boldsymbol{\theta})\mathrm{d}\chi = \theta_j$ and

$$\frac{\partial}{\partial\theta_i}\int T_j(\chi)L(\chi|\boldsymbol{\theta})\mathrm{d}\chi = \begin{cases} 1 & \text{if } i = j\ , \\ 0 & \text{if } i \neq j\ . \end{cases}$$

This equation can be written in matrix form, viz., $\langle \boldsymbol{U}(\chi_\lambda)\,[\boldsymbol{\delta T}(\chi_\lambda)]^\dagger\rangle_\theta = \overline{\overline{\mathrm{Id}_n}}$ and we deduce that

$$\overline{\overline{J}}^{-1}\langle \boldsymbol{U}(\chi_\lambda)\,[\boldsymbol{\delta T}(\chi_\lambda)]^\dagger\rangle_\theta = \overline{\overline{J}}^{-1}\ .$$

It follows that, $\forall u \in \mathbb{C}^n \setminus \{0\}$,

$$u^\dagger \left[\overline{\overline{J}}^{-1} \langle U(\chi_\lambda) [\delta T(\chi_\lambda)]^\dagger \rangle_\theta \right] u = u^\dagger \overline{\overline{J}}^{-1} u \,,$$

where $u^\dagger$ is the conjugated transpose of u. Note further that

$$u^\dagger \overline{\overline{J}}^{-1} u > 0 \,.$$

This is shown as follows. We have $\int L(\chi|\boldsymbol{\theta}) \mathrm{d}\chi = 1$ and hence,

$$\frac{\partial}{\partial \theta_i} \int L(\chi|\boldsymbol{\theta}) \mathrm{d}\chi = 0 \,,$$

or alternatively,

$$\int \frac{\partial}{\partial \theta_i} \ell(\chi|\boldsymbol{\theta}) L(\chi|\boldsymbol{\theta}) \mathrm{d}\chi = 0 \,.$$

Differentiating a second time,

$$\frac{\partial}{\partial \theta_j} \int \frac{\partial \ell(\chi|\boldsymbol{\theta})}{\partial \theta_i} L(\chi|\boldsymbol{\theta}) \mathrm{d}\chi = 0 \,,$$

or

$$\int \frac{\partial^2 \ell(\chi|\boldsymbol{\theta})}{\partial \theta_j \partial \theta_i} L(\chi|\boldsymbol{\theta}) \mathrm{d}\chi + \int \frac{\partial \ell(\chi|\boldsymbol{\theta})}{\partial \theta_i} \frac{\partial L(\chi|\boldsymbol{\theta})}{\partial \theta_j} \mathrm{d}\chi = 0 \,,$$

and hence,

$$\int \frac{\partial^2 \ell(\chi|\boldsymbol{\theta})}{\partial \theta_j \partial \theta_i} L(\chi|\boldsymbol{\theta}) \mathrm{d}\chi + \int \frac{\partial \ell(\chi|\boldsymbol{\theta})}{\partial \theta_j} \frac{\partial \ell(\chi|\boldsymbol{\theta})}{\partial \theta_i} L(\chi|\boldsymbol{\theta}) \mathrm{d}\chi = 0 \,.$$

This result can also be written

$$\left\langle \frac{\partial^2 \ell(\chi|\boldsymbol{\theta})}{\partial \theta_j \partial \theta_i} \right\rangle_\theta = - \left\langle \frac{\partial \ell(\chi|\boldsymbol{\theta})}{\partial \theta_j} \frac{\partial \ell(\chi|\boldsymbol{\theta})}{\partial \theta_i} \right\rangle_\theta \,.$$

As we saw above,

$$J_{ij} = - \left\langle \frac{\partial^2 \ell(\chi|\boldsymbol{\theta})}{\partial \theta_j \partial \theta_i} \right\rangle_\theta = \left\langle \frac{\partial \ell(\chi|\boldsymbol{\theta})}{\partial \theta_j} \frac{\partial \ell(\chi|\boldsymbol{\theta})}{\partial \theta_i} \right\rangle_\theta \,,$$

so that

$$u^\dagger \overline{\overline{J}} u = \sum_{j=1}^n \sum_{i=1}^n u_j \left\langle \frac{\partial \ell(\chi|\boldsymbol{\theta})}{\partial \theta_j} \frac{\partial \ell(\chi|\boldsymbol{\theta})}{\partial \theta_i} \right\rangle_\theta u_i \,,$$

and hence,

$$u^\dagger \overline{\overline{J}} u = \left\langle \left[\sum_{i=1}^n \frac{\partial \ell(\chi|\boldsymbol{\theta})}{\partial \theta_i} u_i \right]^2 \right\rangle_\theta \,.$$

It now follows that
$$\boldsymbol{u}^\dagger \overline{\overline{J}} \boldsymbol{u} > 0 \,,$$

assuming, of course, that $\overline{\overline{J}}$ is non-singular. Now $\boldsymbol{u}^\dagger \overline{\overline{J}} \boldsymbol{u} > 0$ implies that $\boldsymbol{u}^\dagger \overline{\overline{J}}^{-1} \boldsymbol{u} > 0$, as claimed.

We can thus write
$$\left\langle \boldsymbol{u}^\dagger \overline{\overline{J}}^{-1} \boldsymbol{U}(\chi_\lambda) \left[\delta\boldsymbol{T}(\chi_\lambda) \right]^\dagger \boldsymbol{u} \right\rangle_\theta = \boldsymbol{u}^\dagger \left\{ \overline{\overline{J}}^{-1} \left\langle \boldsymbol{U}(\chi_\lambda) \left[\delta\boldsymbol{T}(\chi_\lambda) \right]^\dagger \right\rangle_\theta \right\} \boldsymbol{u} \,,$$

which implies
$$\left\langle \boldsymbol{u}^\dagger \overline{\overline{J}}^{-1} \boldsymbol{U}(\chi_\lambda) \left[\delta\boldsymbol{T}(\chi_\lambda) \right]^\dagger \boldsymbol{u} \right\rangle_\theta = \boldsymbol{u}^\dagger \overline{\overline{J}}^{-1} \boldsymbol{u} \,,$$

and finally,
$$\left| \left\langle \boldsymbol{u}^\dagger \overline{\overline{J}}^{-1} \boldsymbol{U}(\chi_\lambda) \left[\delta\boldsymbol{T}(\chi_\lambda) \right]^\dagger \boldsymbol{u} \right\rangle_\theta \right|^2 = \left| \boldsymbol{u}^\dagger \overline{\overline{J}}^{-1} \boldsymbol{u} \right|^2 \,.$$

As in Section 7.12, we apply the inequality
$$\left| \langle F(\chi_\lambda) G(\chi_\lambda) \rangle_\theta \right|^2 \leqslant \langle |F(\chi_\lambda)|^2 \rangle_\theta \langle |G(\chi_\lambda)|^2 \rangle_\theta \,,$$

with $F(\chi_\lambda) = \boldsymbol{u}^\dagger \overline{\overline{J}}^{-1} \boldsymbol{U}(\chi_\lambda)$ and $G(\chi_\lambda) = \left[\delta\boldsymbol{T}(\chi_\lambda) \right]^\dagger \boldsymbol{u}$. This implies that
$$\left| \boldsymbol{u}^\dagger \overline{\overline{J}}^{-1} \boldsymbol{u} \right|^2 \leqslant \langle |F(\chi_\lambda)|^2 \rangle_\theta \langle |G(\chi_\lambda)|^2 \rangle_\theta \,,$$

where
$$\langle |F(\chi_\lambda)|^2 \rangle_\theta = \left\langle \boldsymbol{u}^\dagger \overline{\overline{J}}^{-1} \boldsymbol{U}(\chi_\lambda) \left(\boldsymbol{U}(\chi_\lambda) \right)^\dagger \overline{\overline{J}}^{-1} \boldsymbol{u} \right\rangle_\theta \,,$$

and
$$\langle |G(\chi_\lambda)|^2 \rangle_\theta = \left\langle \boldsymbol{u}^\dagger \delta\boldsymbol{T}(\chi_\lambda) \left[\delta\boldsymbol{T}(\chi_\lambda) \right]^\dagger \boldsymbol{u} \right\rangle_\theta \,.$$

We can analyze each term on the right-hand side. We have for the first term
$$\langle |F(\chi_\lambda)|^2 \rangle_\theta = \boldsymbol{u}^\dagger \overline{\overline{J}}^{-1} \left\langle \boldsymbol{U}(\chi_\lambda) \left[\boldsymbol{U}(\chi_\lambda) \right]^\dagger \right\rangle_\theta \overline{\overline{J}}^{-1} \boldsymbol{u} \,.$$

Now we have already seen that
$$\left\langle \boldsymbol{U}(\chi_\lambda) \left[\boldsymbol{U}(\chi_\lambda) \right]^\dagger \right\rangle_\theta = \overline{\overline{J}} \,,$$

and therefore,
$$\boldsymbol{u}^\dagger \overline{\overline{J}}^{-1} \left\langle \boldsymbol{U}(\chi_\lambda) \left[\boldsymbol{U}(\chi_\lambda) \right]^\dagger \right\rangle_\theta \overline{\overline{J}}^{-1} \boldsymbol{u} = \boldsymbol{u}^\dagger \overline{\overline{J}}^{-1} \boldsymbol{u} \,.$$

We now analyze the second term on the right-hand side of the above equation:

$$\langle |G(\chi_\lambda)|^2\rangle_\theta = \boldsymbol{u}^\dagger \left\langle \delta\boldsymbol{T}(\chi_\lambda)\,[\delta\boldsymbol{T}(\chi_\lambda)]^\dagger \right\rangle_\theta \boldsymbol{u}\;.$$

Now

$$\left\langle \delta\boldsymbol{T}(\chi_\lambda)\,[\delta\boldsymbol{T}(\chi_\lambda)]^\dagger \right\rangle_\theta = \overline{\overline{T}}\;,$$

so that

$$\left\langle \boldsymbol{u}^\dagger\delta\boldsymbol{T}(\chi_\lambda)\,[\delta\boldsymbol{T}(\chi_\lambda)]^\dagger\,\boldsymbol{u} \right\rangle_\theta = \boldsymbol{u}^\dagger\overline{\overline{T}}\boldsymbol{u}\;.$$

The inequality

$$\left|\boldsymbol{u}^\dagger\overline{\overline{J}}^{-1}\boldsymbol{u}\right|^2 \leqslant \langle |F(\chi_\lambda)|^2\rangle_\theta\langle |G(\chi_\lambda)|^2\rangle_\theta$$

thus becomes

$$\boldsymbol{u}^\dagger\overline{\overline{J}}^{-1}\boldsymbol{u} \leqslant \boldsymbol{u}^\dagger\overline{\overline{T}}\boldsymbol{u}\;.$$

This inequality gives equality if $F(\chi_\lambda) = \alpha(\theta)G(\chi_\lambda)$ with

$$F(\chi_\lambda) = \boldsymbol{u}^\dagger\overline{\overline{J}}^{-1}\boldsymbol{U}(\chi_\lambda)\;,$$

and $G(\chi_\lambda) = [\delta\boldsymbol{T}(\chi_\lambda)]^\dagger\,\boldsymbol{u}$. This leads to

$$\boldsymbol{u}^\dagger\overline{\overline{J}}^{-1}\boldsymbol{U}(\chi_\lambda) = \alpha(\boldsymbol{\theta})\boldsymbol{u}^\dagger\delta\boldsymbol{T}(\chi_\lambda)\;,$$

which we shall write

$$\boldsymbol{u}^\dagger\overline{\overline{J}}^{-1}\boldsymbol{U}(\chi_\lambda) = \alpha(\boldsymbol{\theta})\boldsymbol{u}^\dagger\boldsymbol{T}(\chi_\lambda) + \boldsymbol{u}^\dagger\boldsymbol{c}(\boldsymbol{\theta})\;.$$

This equality is true $\forall\boldsymbol{u} \in \mathbb{C}^n$ and therefore $\overline{\overline{J}}^{-1}\boldsymbol{U}(\chi_\lambda) = \alpha(\boldsymbol{\theta})\boldsymbol{T}(\chi_\lambda)+\boldsymbol{c}(\boldsymbol{\theta})$ or alternatively, $\boldsymbol{U}(\chi_\lambda) = \overline{\overline{A}}(\boldsymbol{\theta})\boldsymbol{T}(\chi_\lambda) + \boldsymbol{\beta}(\boldsymbol{\theta})$, where $\overline{\overline{A}}(\boldsymbol{\theta}) = \overline{\overline{J}}\alpha(\boldsymbol{\theta})$ and $\boldsymbol{\beta}(\boldsymbol{\theta}) = \overline{\overline{J}}\boldsymbol{c}(\boldsymbol{\theta})$. Expanding out this equation, we obtain

$$U_i(\chi) = \frac{\partial}{\partial\theta_i}\ell(\chi|\theta) = \sum_{j=1}^{n} A_{i,j}(\boldsymbol{\theta})T_j(\chi) + \beta_i(\boldsymbol{\theta})\;.$$

This is only possible if $\ell(\chi|\theta)$ can be written

$$\ell(\chi|\theta) = \sum_{j=1}^{n} a_j(\boldsymbol{\theta})T_j(\chi) + Pb(\boldsymbol{\theta}) + g(\chi)\;.$$

Note that in the case of observations corresponding to P independent realizations, we must have

$$L(\chi|\boldsymbol{\theta}) = \prod_{i=1}^{P} P(\boldsymbol{x}_i|\boldsymbol{\theta})\;.$$

[If we have $T_j(\chi) = (1/P) \sum_{i=1}^{P} t_j(x_i)$ rather than $T_j(\chi) = \sum_{i=1}^{P} t_j(x_i)$, this is irrelevant here, because the parameters in the probability law are only defined up to a multiplicative constant.] We thus have

$$T_j(\chi) = \sum_{i=1}^{P} t_j(\boldsymbol{x}_i) ,$$

and

$$g(\chi) = \sum_{i=1}^{P} f(\boldsymbol{x}_i) .$$

The probability or density of the law thus has the form

$$P(\boldsymbol{x}|\boldsymbol{\theta}) = \exp \left[\sum_{j=1}^{n} a_j(\boldsymbol{\theta}) t_j(\boldsymbol{x}) + b(\boldsymbol{\theta}) + f(\boldsymbol{x}) \right] ,$$

which defines the exponential family in the vectorial case.

Exercises

Exercise 7.1. Cramer–Rao Bound

By analyzing the general expression for the Cramer–Rao bound in the case where the estimator may be biased, explain qualitatively why the variance of this estimator might actually be less than the Cramer–Rao bound of an unbiased estimator for the same parameter.

Exercise 7.2. Parameter Estimation

Consider a random variable X that can take the three values -1, 0, 1. The probabilities of each of these values are $P(1) = P(-1) = a$ and $P(0) = b$.

(1) Express a as a function of b.
(2) Show that
$$P_X(x) = \exp \left[f(b) + x^2 g \left(\tfrac{1-b}{2b} \right) \right] .$$
(3) Find an unbiased estimator for b with minimal variance.

Exercise 7.3. Parameter Estimation

Consider a real-valued random variable X and the following two situations:

$$P_A(x) = \frac{1}{2a} \exp \left(-\frac{|x|}{a} \right) ,$$

where $a > 0$, and

$$P_B(x) = \frac{1}{2} \exp \left(-|x - a| \right) .$$

(1) Do these probability density functions belong to the exponential family?
(2) In what situation would it be easy to find an unbiased estimator for a with minimal variance? Is it efficient?

Exercise 7.4. Beta Distributions of Type I and II

Consider a random variable X_λ taking real values in the interval $[0, 1]$ with a beta probability law of type I:

$$P_X(x) = \frac{1}{B(n, p)} x^{n-1}(1 - x)^{p-1} \, ,$$

where $0 \leq x \leq 1$, $B(n, p) = \Gamma(n)\Gamma(p)/\Gamma(n + p)$ and $\Gamma(n)$ is the Gamma function.

(1) Does this probability density function belong to the exponential family with regard to its parameters n and p?
(2) Determine the maximum likelihood estimators of n and p, but without seeking an explicit form for n and p.
(3) Consider now the random variable $Y_\lambda = X_\lambda/(1 - X_\lambda)$, and determine the probability density function of Y_λ. This is a type II beta distribution.
(4) Find the maximum likelihood estimators of n and p for the type II beta distribution, but without seeking an explicit form for n and p.

Exercise 7.5. Uniform Distribution

Consider a real random variable X_λ with uniform probability distribution over the interval $[0, \theta]$.

(1) Write down the probability density function of X_λ.
(2) Find the estimator of the first order moment of θ.
(3) Find the maximum likelihood estimator for θ.
(4) Can it be asserted that this estimator has minimal variance?

Consider now a real random variable X_λ with uniform probability distribution over the interval $[-\theta, \theta]$.

(5) Write down the probability density function of X_λ.
(6) Suggest an estimator for θ in the sense of moments.
(7) Find the maximum likelihood estimator for θ.

Exercise 7.6. Cramer–Rao Bound for Additive Noise

Consider a measurement made in the presence of additive noise which is not exactly Gaussian:

$$X_\lambda = \theta + B_\lambda \, .$$

The real random variable B_λ is assumed to have a probability density function of the form

$$P_B(b) = A(\sigma_0, c) \exp\left[-\frac{1}{2\sigma_0^2}b^2 - cb^4\right] ,$$

where $c \geq 0$.

(1) Calculate the Cramer–Rao bound for the estimator of the empirical mean.
(2) Compare the Cramer–Rao bounds when $c = 0$ and when $c > 0$.

8

Examples of Estimation in Physics

In this chapter we provide examples of the estimation techniques described in Chapter 7, showing how to apply them to simple and typical problems encountered in physics.

8.1 Measurement of Optical Flux

We now analyze the consequences of the mathematical results in Chapter 7 for the simple situation in which we wish to measure an optical flux. We will consider successively the cases where the flux is measured in the presence of Poisson noise, Gamma noise, or additive Gaussian noise. Recall first that Poisson noise describes particle noise and is generally present at low fluxes. Gamma noise provides a simple model to describe speckle phenomena, whilst additive Gaussian noise is the model often adopted when the domninant noise is electronic (see Chapter 4).

We propose to begin with flux constant in time, and hence stationary. We then analyze the situation in which the relaxation parameter of a decreasing flux is measured.

We will use the variance as the quality criterion for the estimate. As we saw in Chapter 7, the best way to estimate the flux when it is constant in time is to determine its average. Indeed, for Gaussian or Poisson noise, we have $\hat{\theta}_{\mathrm{ML}}(\chi) = (\sum_{i=1}^{P} x_i)/P$, and for Gamma noise $\hat{\theta}_{\mathrm{ML}}(\chi) = (\sum_{i=1}^{P} x_i)/(\alpha P)$. This is just what our intuition would have suggested, i.e., to use the method in which we identify the first moment. We have also seen that these estimators have minimal variance. Indeed, in the exponential family, the maximum likelihood estimators have minimal variance when they have no bias. Let us determine their Cramer–Rao bounds. We know that, when we observe a P-sample $\chi = \{x_1, x_2, \ldots, x_P\}$, we have $\sigma^2(\theta) \geqslant 1/I_{\mathrm{F}}(\chi)$, where $I_{\mathrm{F}} = -\langle \partial^2 \ell(\chi)/\partial \theta^2 \rangle$ and where $\sigma^2(\theta)$ is the variance of the sufficient statistic $\hat{\theta}_{\mathrm{ML}}(\chi) = (\sum_{i=1}^{P} x_i)/P$ for the Gauss and Poisson distributions and

$\hat{\theta}_{\mathrm{ML}}(\chi) = (\sum_{i=1}^{P} x_i)/(\alpha P)$ is the sufficient statistic for the Gamma distribution. Since the estimators we are considering here are efficient, the last inequality is actually an equality and hence $\sigma^2(\theta) = 1/I_{\mathrm{F}}(\chi)$.

The log-likelihood of the Poisson distribution is

$$\ell(\chi) = -P\theta + T(\chi) \ln \theta - \sum_{i=1}^{P} \ln(x_i!) \; ,$$

where $T(\chi) = \sum_{i=1}^{P} x_i$. (In contrast to the last chapter, we denote discrete and continuous random variables in the same way here, for reasons of simplicity.) We have $\langle T(\chi) \rangle = P\theta$ and, as the Fisher information is given by $I_{\mathrm{F}} = \langle T(\chi) \rangle / \theta^2$, we deduce that $I_{\mathrm{F}} = P/\theta$. We thus obtain

$$\sigma_T^2(\theta) = \frac{\theta}{P} \; .$$

The log-likelihood of the Gamma distribution is

$$\ell(\chi) = -P\alpha \ln \theta - \frac{1}{\theta} T(\chi) + (\alpha - 1) \sum_{i=1}^{P} \ln x_i - P \ln \Gamma(\alpha) \; ,$$

where $T(\chi) = \sum_{i=1}^{P} x_i$. The Fisher information is thus

$$I_{\mathrm{F}} = -P \frac{\alpha}{\theta^2} + \frac{2}{\theta^3} \langle T(\chi) \rangle \; .$$

Now $\langle T(\chi) \rangle = \alpha P \theta$, and hence, $I_{\mathrm{F}} = P\alpha/\theta^2$. Since the estimator $\hat{\theta}_{\mathrm{ML}}(\chi) = (\sum_{i=1}^{P} x_i)/(\alpha P)$ is unbiased, we deduce that

$$\sigma_T^2(\theta) = \frac{\theta^2}{P\alpha} \; .$$

The log-likelihood of the Gaussian distribution is

$$\ell(\chi) = -\frac{1}{2\sigma^2} \left[\sum_{i=1}^{P} x_i^2 - 2\theta T(\chi) + P\theta^2 \right] - P \ln \sigma - P \ln \sqrt{2\pi} \; ,$$

where $T(\chi) = \sum_{i=1}^{P} x_i$. The Fisher information is therefore $I_{\mathrm{F}} = P/\sigma^2$ and we find

$$\sigma_T^2(\theta) = \frac{\sigma^2}{P} \; .$$

We thus see that the variance of the estimate will be different in each case. Although it is, of course, dependent on the noise power σ^2 in the case of additive Gaussian noise, it only depends on the mean in the case of Poisson noise and on the mean and the parameter α in the case of Gamma noise. We have $\sigma_T^2(\theta) = \theta/P$ and $\sigma_T^2(\theta) = \theta^2/(P\alpha)$, respectively. The signal-to-noise ratio, which can be defined as the square of the mean over the variance, is then equal to $\rho = P\theta$ for Poisson noise and $\rho = P\alpha$ for noise associated with a Gamma distribution. The signal-to-noise ratio is therefore independent of the mean for Gamma noise, whereas it increases linearly with the mean in the case of Poisson noise. For additive Gaussian noise, we would have $\rho = P\theta^2/\sigma^2$, which shows that the signal-to-noise ratio would increase in this case as the square of the signal mean.

Let us now analyze what happens if we make an estimate of the relaxation parameter of a time-varying flux using the maximum likelihood method. To simplify the analysis, we assume that the signal is measured at discrete time intervals $t = 1, 2, \ldots, P$ and denote the P-sample as usual by $\chi = \{x_1, x_2, \ldots, x_P\}$. When there is no noise, the signal would be equal to $\{s_i^\theta\}_{i=1,\ldots,P}$, and we will assume that it is equal to the mean value of the measured signal: $s_i^\theta = \langle x_i \rangle$. $\{s_i^\theta\}_{i=1,\ldots,P}$ then constitutes a parametric model of the signal. For example, for an exponential relaxation of the flux, we would have $s_i^\theta = s_0 \exp(-i/\theta)$ and for a flux varying sinusoidally with time, $s_i^\theta = s_0 \sin(\omega i + \theta)$, where we have assumed that the phase of the signal is an unknown parameter. We will assume that the measurement noise is uncorrelated in such a way that we can write the log-likelihood in the form

$$\ell(\chi) = \sum_{i=1}^{P} \ln\left[P_{s_i^\theta}(x_i) \right] .$$

First of all, the relation between the parameter θ and the measurements $\chi = \{x_1, x_2, \ldots, x_P\}$ is no longer as simple as in the last case and we cannot use the previous arguments to guarantee that the maximum likelihood technique will lead to the estimator with minimal variance, even if it is unbiased. Let us consider the three cases of Poisson, Gamma and additive Gaussian noise in turn.

For Poisson noise, the log-likelihood is

$$\ell(\chi) = -\sum_{i=1}^{P} s_i^\theta + \sum_{i=1}^{P} x_i \ln s_i^\theta - \sum_{i=1}^{P} \ln(x_i!) .$$

The maximum likelihood estimate of θ is thus obtained by minimizing

$$E(\chi, \theta) = \sum_{i=1}^{P} \left(s_i^\theta - x_i \ln s_i^\theta \right) .$$

For Gamma noise, the log-likelihood is

$$\ell(\chi) = (\alpha - 1) \sum_{i=1}^{P} \ln x_i - \alpha \sum_{i=1}^{P} \ln s_i^\theta - \sum_{i=1}^{P} \frac{x_i}{s_i^\theta} - \sum_{i=1}^{P} \ln \Gamma(\alpha) .$$

The maximum likelihood estimate of θ is thus obtained by minimizing

$$E(\chi, \theta) = \sum_{i=1}^{P} \left(\frac{x_i}{s_i^\theta} + \alpha \ln s_i^\theta \right).$$

For additive Gaussian noise, the log-likelihood is

$$\ell(\chi) = -\frac{1}{\sigma^2} \sum_{i=1}^{P} (x_i - s_i^\theta)^2 - \ln(\sigma \sqrt{2\pi}).$$

The maximum likelihood estimate of θ is thus obtained by minimizing

$$E(\chi, \theta) = \sum_{i=1}^{P} (x_i - s_i^\theta)^2.$$

We see that it is only in the case of additive Gaussian noise that the least squares criterion is optimal as far as likelihood is concerned. Table 8.1 gives the various quantities we must minimize to achieve the maximum likelihood estimate in the case of an exponential relaxation $s_i^\theta = s_0 \exp(-i/\theta)$.

Table 8.1. Fitting an exponential law by the maximum likelihood criterion

Distribution	Quantity to be minimized
Poisson	$\sum_{i=1}^{P} \left[s_0 \exp\left(-\frac{i}{\theta} \right) + \frac{i x_i}{\theta} \right]$
Gamma	$\sum_{i=1}^{P} \left(\frac{x_i}{s_0} \exp \frac{i}{\theta} - \frac{\alpha i}{\theta} \right)$
Gauss	$\sum_{i=1}^{P} \left[x_i - s_0 \exp\left(-\frac{i}{\theta} \right) \right]^2$

8.2 Measurement Accuracy
in the Presence of Gaussian Noise

A fundamental question for the physicist concerns the estimation of experimental error. He or she will use such estimates to draw important conclusions, for example, concerning the adequacy of a theoretical model in the face of experimental results, whether two results are really different, or whether some measured characteristics satisfy manufacturing requirements. The situation is relatively simple when we are concerned with measurements made in the presence of Gaussian noise. Indeed, this corresponds to a fairly general situation, as explained in Chapter 4. We shall therefore discuss this case in detail.

Consider first measurements made in the presence of additive Gaussian noise. Suppose we wish to estimate the true value a of the physical signal when we have P measurements $x_1, x_2, \ldots, x_P$. We thus set $\chi = \{x_1, x_2, \ldots, x_P\}$. Since we assume that the measurements are marred by additive Gaussian noise, we can consider the following model:

$$x_i = a + n_i, \quad \forall i = 1, \ldots, P,$$

where n_i is a Gaussian variable with zero mean and unknown variance σ_0^2. x_i is thus a Gaussian variable with mean a and unknown variance. We saw in Section 7.10 that the maximum likelihood estimator of the mean of a Gaussian distribution is unbiased and efficient:

$$\hat{a}_{\mathrm{ML}} = \frac{1}{P} \sum_{i=1}^{P} x_i .$$

Let us now calculate the variance of this estimator, viz., $\left\langle [\hat{a}_{\mathrm{ML}} - \langle \hat{a}_{\mathrm{ML}} \rangle]^2 \right\rangle$, where, as in previous chapters, $\langle \ \rangle$ represents the expectation value. If a_0 denotes the true value of a, since the estimator is unbiased, we have $\langle \hat{a}_{\mathrm{ML}} \rangle = a_0$. We obtain

$$\langle |\delta \hat{a}_{\mathrm{ML}}|^2 \rangle = \langle [\hat{a}_{\mathrm{ML}} - \langle \hat{a}_{\mathrm{ML}} \rangle]^2 \rangle = \left\langle \left(\frac{1}{P} \sum_{i=1}^{P} x_i - a_0 \right)^2 \right\rangle ,$$

where $\delta \hat{a}_{\mathrm{ML}} = \hat{a}_{\mathrm{ML}} - a_0$, and hence,

$$\langle |\delta \hat{a}_{\mathrm{ML}}|^2 \rangle = \frac{1}{P^2} \sum_{i=1}^{P} \sum_{j=1}^{P} \langle (x_j - a_0)(x_i - a_0) \rangle .$$

Now $\langle (x_j - a_0)(x_i - a_0) \rangle = \sigma_0^2 \delta_{i-j}$, where δ_n is the Kronecker delta, and therefore,

$$\langle |\delta \hat{a}_{\mathrm{ML}}|^2 \rangle = \frac{\sigma_0^2}{P} .$$

We may therefore say that the standard deviation σ_a of the estimator $\hat{a}_{\mathrm{ML}}$ is

$$\sigma_a = \frac{\sigma_0}{\sqrt{P}} .$$

Figure 8.1 shows the variances of the random variable and the estimator of the mean.

To estimate the accuracy of the estimate of a, we need to know σ_0. Let us therefore turn to the problem of estimating the variance of the x_i. This will allow us to deduce the standard deviation σ_a and hence the accuracy of the estimate of a. We will then be able to plot an error bar of plus or minus σ_a on either side of $\hat{a}_{\mathrm{ML}}$. We do this in two stages. First of all, we consider the case

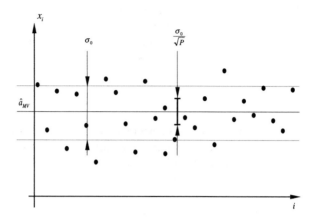

Fig. 8.1. Comparing the variances of the random variable and the estimator of the mean

where the mean is assumed to be known, and then we turn to the situation when it too is unknown. In the first case, only the variance is unknown and the likelihood is

$$L(\chi|\sigma) = \prod_{i=1}^{P} \left\{ \frac{1}{\sqrt{2\pi}\sigma} \exp\left[-\frac{1}{2\sigma^2}(x_i - a_0)^2 \right] \right\}.$$

This leads to a log-likelihood

$$\ell(\chi|\sigma) = -\frac{1}{2\sigma^2} \sum_{i=1}^{P} (x_i - a_0)^2 - P\ln(\sqrt{2\pi}\sigma).$$

The maximum is reached when $\partial\ell(\chi|\sigma)/\partial\sigma = 0$, which yields

$$\frac{1}{\sigma^3} \sum_{i=1}^{P} (x_i - a_0)^2 - P\frac{1}{\sigma} = 0,$$

and hence,

$$\hat{\sigma}_{\mathrm{ML}}^2 = \frac{1}{P} \sum_{i=1}^{P} (x_i - a_0)^2.$$

We have already seen in Section 7.10 that this maximum likelihood estimator of the variance is unbiased and efficient for σ^2.

Needless to say, this simple scenario does not correspond to the one usually encountered in reality. Indeed, the mean is generally unknown, precisely because it is the very thing we seek to estimate. In this case, we must write

$$\ell(\chi|a,\sigma) = -\frac{1}{2\sigma^2}\sum_{i=1}^{P}(x_i - a)^2 - P\ln(\sqrt{2\pi}\sigma) \ .$$

The maximum is attained when $\partial\ell(\chi|a,\sigma)/\partial a = 0$ and $\partial\ell(\chi|a,\sigma)/\partial\sigma = 0$, so that

$$\hat{a}_{\mathrm{ML}} = \frac{1}{P}\sum_{i=1}^{P}x_i \ ,$$

and

$$\hat{\sigma}_{\mathrm{ML}}^2 = \frac{1}{P}\sum_{i=1}^{P}(x_i - \hat{a}_{\mathrm{ML}})^2 \ ,$$

or alternatively,

$$\hat{\sigma}_{\mathrm{ML}}^2 = \frac{1}{P}\sum_{i=1}^{P}\left[x_i - \left(\frac{1}{P}\sum_{i=1}^{P}x_i\right)\right]^2 \ .$$

This estimator of the variance is no longer an unbiased estimator. Indeed, whatever the probability law of the x_i, provided it has a finite second moment, we have

$$\langle\hat{\sigma}_{\mathrm{ML}}^2\rangle = \frac{1}{P}\sum_{i=1}^{P}\left\langle\left[x_i - \left(\frac{1}{P}\sum_{j=1}^{P}x_j\right)\right]^2\right\rangle \ .$$

This equation can be written in the form

$$\langle\hat{\sigma}_{\mathrm{ML}}^2\rangle = \frac{1}{P}\sum_{i=1}^{P}\left\langle\left[(x_i - a) - \left(\frac{1}{P}\sum_{j=1}^{P}(x_j - a)\right)\right]^2\right\rangle \ .$$

Setting $\delta x_i = x_i - a$ and expanding, it is easy to show that

$$\langle\hat{\sigma}_{\mathrm{ML}}^2\rangle = \frac{1}{P}\sum_{i=1}^{P}\langle\delta x_i^2\rangle - \frac{1}{P^2}\sum_{i=1}^{P}\sum_{j=1}^{P}\langle\delta x_i\delta x_j\rangle \ .$$

Now $\langle\delta x_i^2\rangle = \sigma_0^2$ and $\langle\delta x_i\delta x_j\rangle = \sigma_0^2\delta_{i-j}$, where δ_n is the Kronecker delta, so that

$$\langle\hat{\sigma}_{\mathrm{ML}}^2\rangle = \frac{P-1}{P}\sigma_0^2 \ .$$

This shows clearly that $\hat{\sigma}_{\mathrm{ML}}^2$ is a biased estimator of σ_0^2. In contrast,

$$\hat{\sigma}_{\mathrm{cor}}^2 = \frac{1}{P-1}\sum_{i=1}^{P}\left[x_i - \left(\frac{1}{P}\sum_{i=1}^{P}x_i\right)\right]^2$$

is an unbiased estimator. We thus see that we can estimate the error bar on a by

$$\langle |\delta \hat{a}_{\mathrm{ML}}|^2 \rangle \simeq \frac{1}{P(P-1)} \sum_{i=1}^{P} \left[x_i - \left(\frac{1}{P} \sum_{i=1}^{P} x_i \right) \right]^2 ,$$

and hence,

$$\sigma_a \simeq \frac{1}{\sqrt{P(P-1)}} \sqrt{ \sum_{i=1}^{P} \left[x_i - \left(\frac{1}{P} \sum_{i=1}^{P} x_i \right) \right]^2 } .$$

We can calculate the Fisher information matrix in this case where we make a joint estimate of the mean a and the variance of the distribution. We have seen that the log-likelihood is

$$\ell(\chi|a,\sigma) = -\frac{1}{2\sigma^2} \sum_{i=1}^{P} (x_i - a)^2 - P \ln(\sqrt{2\pi}\sigma) .$$

Setting $T_1(\chi) = \sum_{i=1}^{P}(x_i - a)$ and $T_2(\chi) = \sum_{i=1}^{P}(x_i - a)^2$, we deduce that

$$\frac{\partial \ell}{\partial a}(\chi|a,\sigma) = \frac{1}{\sigma^2} T_1(\chi) ,$$

$$\frac{\partial \ell}{\partial \alpha}(\chi|a,\sigma) = \frac{1}{2\sigma^4} T_2(\chi) - \frac{P}{2\sigma^2} ,$$

where we have put $\alpha = \sigma^2$. We thus obtain

$$\left\langle \frac{\partial^2 \ell}{\partial a^2}(\chi|a,\sigma) \right\rangle = -\frac{P}{\sigma^2} ,$$

$$\left\langle \frac{\partial^2 \ell}{\partial \alpha^2}(\chi|a,\sigma) \right\rangle = -\frac{1}{\sigma^6} \langle T_2(\chi) \rangle + \frac{P}{2\sigma^4} ,$$

$$\left\langle \frac{\partial^2 \ell}{\partial a \partial \alpha}(\chi|a,\sigma) \right\rangle = -\frac{1}{\sigma^4} \langle T_1(\chi) \rangle .$$

When $a = a_0$ and $\sigma = \sigma_0$, we have $\langle T_1(\chi) \rangle = 0$ and $\langle T_2(\chi) \rangle = P\sigma_0^2$. The Fisher information matrix is thus

$$\overline{\overline{J}} = \begin{pmatrix} \dfrac{P}{\sigma_0^2} & 0 \\ 0 & \dfrac{P}{2\sigma_0^4} \end{pmatrix} ,$$

and therefore,

$$\overline{\overline{J}}^{-1} = \begin{pmatrix} \dfrac{\sigma_0^2}{P} & 0 \\ 0 & \dfrac{2\sigma_0^4}{P} \end{pmatrix} .$$

Consider an unbiased estimator $\hat{a}(\chi)$ of a_0 and an unbiased estimator $\hat{\sigma}^2(\chi)$ of σ_0^2. Put $\delta\hat{a}(\chi) = \hat{a}(\chi) - a_0$ and $\delta\hat{\sigma}^2(\chi) = \hat{\sigma}^2(\chi) - \sigma_0^2$. The Cramer–Rao bounds in the vector case then give

$$\langle[\delta\hat{a}(\chi)]^2\rangle \geq \frac{\sigma_0^2}{P} \; ,$$

$$\langle[\delta\hat{\sigma}^2(\chi)]^2\rangle \geq \frac{2\sigma_0^4}{P} \; .$$

We thus find that $\hat{a}_{\mathrm{ML}}$ is an efficient estimator of a_0. We also see that there is no reason why the fluctuations in the estimations $\hat{a}(\chi)$ and $\hat{\sigma}^2(\chi)$ of a_0 and σ_0^2 should be correlated. Moreover, the Cramer–Rao bound of an estimator of σ_0^2 is simply $2\sigma_0^4/P$.

8.3 Estimating a Detection Efficiency

We now consider the type of experiment in which we seek to estimate a success rate. For concreteness, suppose that we wish to estimate the probability τ_{D} of detection in a particle detector. Let x_i be the binary variable which is equal to 1 if the system has detected the particle in measurement number i and 0 otherwise (see Fig. 8.2).

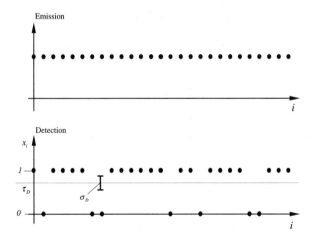

Fig. 8.2. Estimating a detection rate using a Bernoulli process

The statistical sample is then $\chi = \{x_1, x_2, \ldots, x_P\}$. This problem corresponds to estimating the parameter of the Bernoulli probability law

$$x = \begin{cases} 1 & \text{with probability } \tau_{\mathrm{D}} \; , \\ 0 & \text{with probability } 1 - \tau_{\mathrm{D}} \; . \end{cases}$$

We write the Bernoulli law in a form which shows that it does belong to the exponential family. For this purpose, we use the fact that x is binary-valued:

$$P(x) = \exp\left\{x\left[\ln \tau_\mathrm{D} - \ln(1 - \tau_\mathrm{D})\right] + \ln(1 - \tau_\mathrm{D})\right\} .$$

The log-likelihood is then

$$\ell(\chi|\tau_\mathrm{D}) = \sum_{i=1}^{P} \left[x_i \ln \tau_\mathrm{D} + (1 - x_i) \ln(1 - \tau_\mathrm{D})\right] .$$

The maximum likelihood estimator of τ_D is found by writing

$$\frac{\partial \ell(\chi|\tau_\mathrm{D})}{\partial \tau_\mathrm{D}} = 0 ,$$

so that

$$\frac{1}{\tau_\mathrm{D}} \sum_{i=1}^{P} x_i - \frac{1}{1 - \tau_\mathrm{D}} \left(P - \sum_{i=1}^{P} x_i\right) = 0 ,$$

and hence,

$$[\hat{\tau}_\mathrm{D}]_\mathrm{ML} = \frac{1}{P} \sum_{i=1}^{P} x_i ,$$

which finally turns out to be very simple. It is easy to check that this estimator is unbiased. Let us now consider its variance. To this end we set

$$\delta[\hat{\tau}_\mathrm{D}]_\mathrm{ML} = [\hat{\tau}_\mathrm{D}]_\mathrm{ML} - \langle[\hat{\tau}_\mathrm{D}]_\mathrm{ML}\rangle \quad \text{and} \quad \sigma_\mathrm{D}^2 = \langle|\delta[\hat{\tau}_\mathrm{D}]_\mathrm{ML}|^2\rangle .$$

We then have

$$\sigma_\mathrm{D}^2 = \left\langle \left(\frac{1}{P} \sum_{i=1}^{P} x_i - \tau_0\right)^2 \right\rangle ,$$

where τ_0 is the true but unknown value of τ_D. Expanding out, we obtain

$$\sigma_\mathrm{D}^2 = \left\langle \frac{1}{P^2} \sum_{i=1}^{P} \sum_{j=1}^{P} x_i x_j - \tau_0^2 \right\rangle .$$

Now $\langle x_i x_j \rangle = \tau_0^2(1 - \delta_{i-j}) + \tau_0 \delta_{i-j}$, which implies

$$\sigma_\mathrm{D}^2 = \tau_0^2 - \frac{1}{P}\tau_0^2 + \frac{1}{P}\tau_0 - \tau_0^2 ,$$

and so finally,

$$\sigma_\mathrm{D}^2 = \frac{\tau_0(1 - \tau_0)}{P} .$$

In other words, the standard deviation σ_D of the estimator of τ_D is

$$\sigma_D = \sqrt{\frac{\tau_0(1 - \tau_0)}{P}} \, .$$

This immediately raises the problem that we do not know τ_0 and that if we replace this value directly by its estimate $[\hat{\tau}_D]_{ML}$, we may be led to underestimate the error bar when $[\hat{\tau}_D]_{ML}$ is large. Indeed, if we find $[\hat{\tau}_D]_{ML} = 1$, we will conclude that $\sigma_D = 0$ and attribute a zero error bar to this estimate of τ_D. A very pessimiztic view would lead us to choose an upper bound for this error bar, and such a thing can be obtained by setting $\sigma_D^2 \simeq 0.25/P$, or $\sigma_D \simeq 1/(2\sqrt{P})$.

A less radical solution can be implemented when $\tau_0 \approx 1$. If we find $[\hat{\tau}_D]_{ML} = 1$, we can consider that this value is situated at the maximum of the error bar. In other words, we will assume that, for the estimator $\hat{\tau}_D$ of the parameter τ_D for determining D, it is reasonable to choose the value such that $1 = \hat{\tau}_D + \sigma_G$, where $\sigma_G^2 = \hat{\tau}_D(1 - \hat{\tau}_D)/P$. We then find $\sigma_G \approx 1/(P + 1)$.

Note that, although these two approaches are somewhat arbitrary, they are nevertheless better than an over-optimiztic attitude as far as the conclusions that can be drawn from our experiments are concerned.

We now consider a numerical example. Assume that we have made 100 measurements and find $[\hat{\tau}_D]_{ML} = 0.8$. Taking $\sigma_{MG}^2 = [\hat{\tau}_D]_{ML}(1 - [\hat{\tau}_D]_{ML})/P$, we obtain $\sigma_{MG} = 0.04$, corresponding to an accuracy of 4%. Note that if we had estimated σ_G with $\sigma_D \simeq 1/2\sqrt{P}$, we would have obtained $\sigma_D \simeq 5\%$. On the other hand, if we find $[\hat{\tau}_D]_{ML} = 0.99$, where $\sigma_{MG}^2 = [\hat{\tau}_D]_{ML}(1 - [\hat{\tau}_D]_{ML})/P$, we obtain $\sigma_{MG} = 0.01$, corresponding to an accuracy of 1%, which is indeed of the order of $1/(P + 1)$. We have just seen that the most cautious attitude leads to $\sigma_D \simeq 5\%$. We find that τ_D could be of the order of 0.94. This would then lead to $\sigma_{MG} \approx 0.02$, or 2%, which is more reasonable.

8.4 Estimating the Covariance Matrix

In this section, we shall estimate the covariance matrix $\overline{\overline{\Gamma}}$ of a zero mean Gaussian stochastic vector. Let $\boldsymbol{X}_\lambda$ be a stochastic vector with values in $\mathbb{R}^n$. Its probability density is

$$P(\boldsymbol{x}) = \frac{1}{(\sqrt{2\pi})^n \sqrt{\left|\overline{\overline{\Gamma}}\right|}} \exp\left(-\frac{1}{2}\boldsymbol{x}^\dagger \overline{\overline{\Gamma}}^{-1} \boldsymbol{x}\right) ,$$

where $\boldsymbol{x}^\dagger$ is the transpose of $\boldsymbol{x}$, $\left|\overline{\overline{\Gamma}}\right|$ is the determinant of $\overline{\overline{\Gamma}}$, and $\overline{\overline{\Gamma}}^{-1}$ is the inverse of the matrix $\overline{\overline{\Gamma}}$. Suppose we have a sample χ comprising P measurements, i.e., $\chi = \{\boldsymbol{x}_1, \boldsymbol{x}_2, \dots, \boldsymbol{x}_P\}$. The log-likelihood is then

$$\ell\left(\chi|\overline{\overline{\Gamma}}\right) = -\frac{1}{2}\sum_{i=1}^{P}\left(\boldsymbol{x}_i^\dagger \overline{\overline{\Gamma}}^{-1} \boldsymbol{x}_i\right) - \frac{P}{2}\ln\left|\overline{\overline{\Gamma}}\right| - Pn\ln\sqrt{2\pi} \, .$$

We know that any covariance matrix is positive and diagonalizable. We shall assume further that it is non-singular and denote its eigenvalues by μ_j, corresponding to eigenvectors $\boldsymbol{u}_j$ chosen here with norm equal to 1. We will thus have

$$\overline{\overline{\Gamma}} = \sum_{j=1}^{n} \mu_j \left(\boldsymbol{u}_j \boldsymbol{u}_j^\dagger \right) ,$$

and hence,

$$\sum_{i=1}^{P} \boldsymbol{x}_i^\dagger \overline{\overline{\Gamma}}^{-1} \boldsymbol{x}_i = \sum_{i=1}^{P} \sum_{j=1}^{n} \frac{\boldsymbol{x}_i^\dagger \boldsymbol{u}_j \boldsymbol{u}_j^\dagger \boldsymbol{x}_i}{\mu_j} .$$

The likelihood then becomes

$$\ell\left(\chi|\overline{\overline{\Gamma}}\right) = -\frac{1}{2} \sum_{i=1}^{P} \sum_{j=1}^{n} \frac{\boldsymbol{x}_i^\dagger \boldsymbol{u}_j \boldsymbol{u}_j^\dagger \boldsymbol{x}_i}{\mu_j} - \frac{P}{2} \sum_{j=1}^{n} \ln \mu_j - Pn \ln \sqrt{2\pi} ,$$

where $\boldsymbol{u}_j^\dagger \boldsymbol{u}_j = 1$. Let us first estimate the eigenvalues. For this purpose, we put

$$\frac{\partial \ell\left(\chi|\overline{\overline{\Gamma}}\right)}{\partial \mu_j} = 0 ,$$

which implies that

$$\frac{1}{2} \sum_{i=1}^{P} \frac{\boldsymbol{x}_i^\dagger \boldsymbol{u}_j \boldsymbol{u}_j^\dagger \boldsymbol{x}_i}{\mu_j^2} - \frac{P}{2} \frac{1}{\mu_j} = 0 ,$$

or

$$\mu_j = \frac{1}{P} \sum_{i=1}^{P} \boldsymbol{x}_i^\dagger \boldsymbol{u}_j \boldsymbol{u}_j^\dagger \boldsymbol{x}_i . \tag{8.1}$$

To determine the eigenvectors $\boldsymbol{u}_j$, we must maximize $\ell\left(\chi|\overline{\overline{\Gamma}}\right)$ with the constraints $\| \boldsymbol{u}_j \| = 1$. To do so, we introduce the Lagrange function

$$\mathcal{L} = \ell\left(\chi|\overline{\overline{\Gamma}}\right) + \sum_{j=1}^{n} \alpha_j \boldsymbol{u}_j^\dagger \boldsymbol{u}_j .$$

Writing $\partial \mathcal{L}/\partial [\boldsymbol{u}_j]_k = 0$, where $[\boldsymbol{u}_j]_k$ is the kth coordinate of $\boldsymbol{u}_j$, we obtain

$$\frac{1}{2} \sum_{i=1}^{P} \sum_{m=1}^{n} \left\{ [\boldsymbol{x}_i]_k [\boldsymbol{x}_i]_m [\boldsymbol{u}_j]_m + [\boldsymbol{u}_j]_m [\boldsymbol{x}_i]_m [\boldsymbol{x}_i]_k \right\} = \mu_j \alpha_j [\boldsymbol{u}_j]_k . \tag{8.2}$$

In order to analyze (8.1) and (8.2), we introduce the covariance matrix of the measurements:

$$\overline{\overline{C}}_{m,k} = \frac{1}{P} \sum_{i=1}^{P} [\boldsymbol{x}_i]_m [\boldsymbol{x}_i]_k ,$$

or

$$\overline{\overline{C}} = \frac{1}{P}\sum_{i=1}^{P} x_i x_i^{\dagger} .$$

Equations (8.1) and (8.2) then become

$$\mu_j = u_j^{\dagger}\overline{\overline{C}}u_j ,$$

and

$$\overline{\overline{C}}u_j = \mu_j \alpha_j u_j / P .$$

We can deduce from these two last equations that u_j is the j th eigenvector of $\overline{\overline{C}}$, corresponding to the eigenvalue μ_j. In other words we have

$$\overline{\overline{C}} = \sum_{j=1}^{n} \mu_j (u_j u_j^{\dagger}) .$$

We thus observe that the maximum likelihood estimator of $\overline{\overline{\Gamma}}$ is simply $\overline{\overline{C}}$, or

$$\overline{\overline{\Gamma}}_{\mathrm{ML}} = \frac{1}{P}\sum_{i=1}^{P} x_i x_i^{\dagger} .$$

This is not surprising in itself, but it is worth noting that it is obtained under the hypothesis that the stochastic vectors are distributed according to a Gaussian law.

8.5 Application to Coherency Matrices

The results of Section 8.4 generalize to complex-valued Gaussian stochastic vectors. In this case we have

$$\overline{\overline{\Gamma}}_{\mathrm{ML}} = \frac{1}{P}\sum_{i=1}^{P} x_i x_i^{\dagger} ,$$

where $x_i^{\dagger}$ is the complex conjugate transpose of x_i.

Let E be the electric field of a plane electromagnetic wave propagating in the direction parallel to some vector k. (In order to simplify the equations, we will not indicate the dependence on the random variable λ. We will simply distinguish random variables from deterministic variables by denoting the former with upper case letters and the latter with lower case.) We project this field onto two mutually orthonormal vectors which are also orthogonal to k, viz.,

$$E = (U_X i + U_Y j)e^{-i2\pi\nu_0 t} .$$

The coherency matrix is (see Section 3.14)

$$\overline{\overline{\Gamma}} = \begin{pmatrix} \langle U_X^* U_X \rangle & \langle U_Y^* U_X \rangle \\ \langle U_X^* U_Y \rangle & \langle U_Y^* U_Y \rangle \end{pmatrix} \, ,$$

which we shall also write

$$\overline{\overline{\Gamma}} = \begin{pmatrix} I_X & \rho \\ \rho^* & I_Y \end{pmatrix} \, .$$

We now assume that the field $\boldsymbol{E}$ is Gaussian, i.e., that U_X and U_Y are complex Gaussian variables. We then have

$$P(\boldsymbol{u}) = \frac{1}{\pi^2 |\overline{\overline{\Gamma}}|} \exp\left(-\boldsymbol{u}^\dagger \overline{\overline{\Gamma}}^{-1} \boldsymbol{u} \right) \, ,$$

where $|\overline{\overline{\Gamma}}|$ is the determinant of $\overline{\overline{\Gamma}}$ and $\boldsymbol{u} = u_X \boldsymbol{i} + u_Y \boldsymbol{j}$.

Suppose now that we have a sample $\chi = \{\boldsymbol{u}_1, \boldsymbol{u}_2, \ldots, \boldsymbol{u}_P\}$ comprising P measurements. The log-likelihood is thus

$$\ell\left(\chi | \overline{\overline{\Gamma}} \right) = -\sum_{n=1}^{P} (\boldsymbol{u}_n^\dagger \overline{\overline{\Gamma}}^{-1} \boldsymbol{u}_n) - P \ln |\overline{\overline{\Gamma}}| - 2P \ln \pi \, ,$$

and the maximum likelihood estimate of $\overline{\overline{\Gamma}}$ is

$$\overline{\overline{\Gamma}}_{\mathrm{ML}} = \frac{1}{P} \sum_{n=1}^{P} \boldsymbol{u}_n \boldsymbol{u}_n^\dagger \, .$$

We set $\boldsymbol{u}_n = [u_X]_n \boldsymbol{i} + [u_Y]_n \boldsymbol{j}$ and

$$\left[u_X^R \right]_n = \mathrm{Re}\left([u_X]_n \right) \, , \quad \left[u_X^I \right]_n = \mathrm{Im}\left([u_X]_n \right) \, ,$$
$$\left[u_Y^R \right]_n = \mathrm{Re}\left([u_Y]_n \right) \, , \quad \left[u_Y^I \right]_n = \mathrm{Im} = \left([u_Y]_n \right) \, ,$$

where Re( ) and Im( ) denote extraction of the real and imaginary parts of the argument, respectively. Expanding out the expression for the maximum likelihood estimate of $\overline{\overline{\Gamma}}$,

$$[I_X]_{\mathrm{ML}} = \frac{1}{P} \sum_{n=1}^{P} \left[\left(\left[u_X^R \right]_n \right)^2 + \left(\left[u_X^I \right]_n \right)^2 \right] \, ,$$

$$[I_Y]_{\mathrm{ML}} = \frac{1}{P} \sum_{n=1}^{P} \left[\left(\left[u_Y^R \right]_n \right)^2 + \left(\left[u_Y^I \right]_n \right)^2 \right] \, ,$$

$$\mathrm{Re}\left([\rho]_{\mathrm{ML}} \right) = \frac{1}{P} \sum_{n=1}^{P} \left\{ \left[u_X^R \right]_n \left[u_Y^R \right]_n + \left[u_X^I \right]_n \left[u_Y^I \right]_n \right\} \, ,$$

$$\mathrm{Im}\left([\rho]_{\mathrm{ML}} \right) = \frac{1}{P} \sum_{n=1}^{P} \left\{ \left[u_X^I \right]_n \left[u_Y^R \right]_n - \left[u_X^R \right]_n \left[u_Y^I \right]_n \right\} \, .$$

In optics, electric fields are not measured directly. However, it is possible to measure the instantaneous Stokes parameters,

$$S_0^{(n)} = [I_X]_n + [I_Y]_n \ , \qquad S_1^{(n)} = [I_X]_n - [I_Y]_n \ ,$$

$$S_2^{(n)} = 2\left\{ \left[u_X^R\right]_n \left[u_Y^R\right]_n + \left[u_X^I\right]_n \left[u_Y^I\right]_n \right\} \ ,$$

$$S_3^{(n)} = 2\left\{ \left[u_X^R\right]_n \left[u_Y^I\right]_n - \left[u_X^I\right]_n \left[u_Y^R\right]_n \right\} \ ,$$

where

$$[I_X]_n = \left|[u_X]_n\right|^2 = \left[u_X^R\right]_n^2 + \left[u_X^I\right]_n^2 \ ,$$

$$[I_Y]_n = \left|[u_Y]_n\right|^2 = \left[u_Y^R\right]_n^2 + \left[u_Y^I\right]_n^2 \ .$$

Indeed, the first two components are easily measured for they are the sums and differences of intensities measured along the linear polarization directions i and j. For the other two components, this comes out more easily if we observe that

$$S_2^{(n)} = \left[I_{\pi/4}\right]_n - \left[I_{-\pi/4}\right]_n \ ,$$

$$S_3^{(n)} = \left[I_{(+)}\right]_n - \left[I_{(-)}\right]_n \ ,$$

where

$$\left[I_{\pi/4}\right]_n = \left|\left[U_{\pi/4}\right]_n\right|^2 \qquad \text{with} \quad \left[U_{\pi/4}\right]_n = \frac{1}{\sqrt{2}} \left([U_X]_n + [U_Y]_n\right) \ ,$$

$$\left[I_{-\pi/4}\right]_n = \left|\left[U_{-\pi/4}\right]_n\right|^2 \qquad \text{with} \quad \left[U_{-\pi/4}\right]_n = \frac{1}{\sqrt{2}} \left([U_X]_n - [U_Y]_n\right) \ ,$$

$$\left[I_{(+)}\right]_n = \left|\left[U_{(+)}\right]_n\right|^2 \qquad \text{with} \quad \left[U_{(+)}\right]_n = \frac{1}{\sqrt{2}} \left([U_X]_n - i\,[U_Y]_n\right) \ ,$$

$$\left[I_{(-)}\right]_n = \left|\left[U_{(-)}\right]_n\right|^2 \qquad \text{with} \quad \left[U_{(-)}\right]_n = \frac{1}{\sqrt{2}} \left([U_X]_n + i\,[U_Y]_n\right) \ .$$

We can measure $\left[I_{\pi/4}\right]_n$ and $\left[I_{-\pi/4}\right]_n$ since these are the intensities in the linear polarization directions $(i + j)/\sqrt{2}$ and $(i - j)/\sqrt{2}$. Regarding $\left[I_{(+)}\right]_n$ and $\left[I_{(-)}\right]_n$, we associate half-wave plates [12] to introduce phase differences $+i$ and $-i$ between the components U_X and U_Y.

The Stokes parameters are the expectation values of the instantaneous Stokes parameters, i.e.,

$$S_0 = \left\langle \left(U_X^R\right)^2 + \left(U_X^I\right)^2 + \left(U_Y^R\right)^2 + \left(U_Y^I\right)^2 \right\rangle \ ,$$

$$S_1 = \left\langle \left(U_X^R\right)^2 + \left(U_X^I\right)^2 - \left(U_Y^R\right)^2 - \left(U_Y^I\right)^2 \right\rangle \ ,$$

$$S_2 = 2 \left\langle U_X^R U_Y^R + U_X^I U_Y^I \right\rangle ,$$

$$S_3 = 2 \left\langle U_X^R U_Y^I - U_X^I U_Y^R \right\rangle ,$$

using upper case letters because we must consider the field components as random variables whose expectation values we seek to determine. It is easy to express the coherency matrix in terms of these parameters and conversely. Here, we simply note that the maximum likelihood estimates of the Stokes parameters are

$$[S_k]_{\text{ML}} = \frac{1}{P} \sum_{n=1}^{P} S_k^{(n)} , \quad k = 0, 1, 2, 3 .$$

8.6 Making Estimates in the Presence of Speckle

We discuss here the simultaneous measurement of the average intensity and the order of the Gamma distribution which describes the fluctuations observed when measurements are made in the presence of speckle. We have

$$P_{\alpha,\beta}(x) = \frac{\beta^\alpha x^{\alpha-1}}{\Gamma(\alpha)} \exp(-\beta x) ,$$

where the function $\Gamma(\alpha)$ is defined for positive α by

$$\Gamma(\alpha) = \int_0^\infty x^{\alpha-1} e^{-x} dx .$$

Setting $\boldsymbol{\theta} = (\alpha, \beta)^\dagger$, the log-likelihood is

$$\ell(\chi|\boldsymbol{\theta}) = -\beta T_1(\chi) + (\alpha - 1)T_2(\chi) + P\alpha \ln \beta - P \ln \Gamma(\alpha) ,$$

where $T_1(\chi) = \sum_{i=1}^{P} x_i$ and $T_2(\chi) = \sum_{i=1}^{P} \ln x_i$. The Fisher matrix is obtained from

$$\frac{\partial^2}{\partial \alpha^2} \ell(\chi|\boldsymbol{\theta}) = -P \frac{\partial^2}{\partial \alpha^2} \ln \Gamma(\alpha) , \quad \frac{\partial^2}{\partial \beta^2} \ell(\chi|\boldsymbol{\theta}) = -P \frac{\alpha}{\beta^2} ,$$

and

$$\frac{\partial^2}{\partial \alpha \partial \beta} \ell(\chi|\boldsymbol{\theta}) = \frac{P}{\beta} .$$

The Fisher matrix is therefore

$$\overline{\overline{J}} = P \begin{pmatrix} \partial^2 \ln \Gamma(\alpha)/\partial \alpha^2 & -1/\beta \\ -1/\beta & \alpha/\beta^2 \end{pmatrix} .$$

We thus obtain

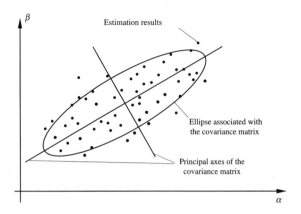

Fig. 8.3. Intuitive meaning of the covariance matrix for joint estimation of α and β

$$\overline{\overline{J^{-1}}} = \frac{1}{P(\alpha A_\alpha - 1)} \begin{pmatrix} \alpha & \beta \\ \beta & \beta^2 A_\alpha \end{pmatrix} ,$$

where we have set $A_\alpha = \partial^2 \ln \Gamma(\alpha)/\partial \alpha^2$. The Cramer–Rao bound is then

$$u_1^2 \Gamma_{11} + u_2^2 \Gamma_{22} + 2u_1 u_2 \Gamma_{12} \geqslant \frac{1}{P(\alpha A_\alpha - 1)} \left(u_1^2 \alpha + u_2^2 \beta^2 A_\alpha + 2u_1 u_2 \beta \right) .$$

In particular, we have $\Gamma_{11} \geqslant \alpha/P(\alpha A_\alpha - 1)$ and $\Gamma_{22} \geqslant \beta^2 A_\alpha/P(\alpha A_\alpha - 1)$. Let us compare this bound with the one obtained when α is known and we only need to estimate β. In this case, the Fisher information is

$$I_F = -\frac{\partial^2 \langle \ell(\chi|\theta) \rangle_\theta}{\partial \theta^2} = P \frac{\alpha}{\beta^2} ,$$

and hence $\sigma_T^2 \geqslant \beta^2/(P\alpha)$. Since $(\alpha A_\alpha - 1)/\beta^2$ is the determinant of the Fisher matrix which is positive, as has been shown in Section 7.14, we have $\alpha A_\alpha > 1$ and we can deduce that $\alpha A_\alpha/(\alpha A_\alpha - 1) > 1$. We thus see that the bound is greater when α is unknown than when α is known.

This is an important result because it shows that the introduction of extra parameters can lead to an increase in the variance of their estimator. In other words, there is a price to pay for having complex models in which there are a large number of parameters to be estimated, namely, the difficulty in accurately estimating those parameters.

8.7 Fluctuation–Dissipation and Estimation

The exponential family plays an important role because, as we have seen, the probability laws belonging to it possess simple optimality properties. In this section, we shall study the analogies between probability laws in the exponential family and the Gibbs distributions discussed in Chapter 6. For this purpose, we consider the canonical form of the laws in the exponential family, using the notation $X = \{x_1, x_2, \ldots, x_P\}$, $M(X) = \sum_{i=1}^{P} t(x_i)$, $h = -\theta$, and $H_0(X) = \sum_{i=1}^{P} f(x_i)$ and setting $\beta = -1$. It is then quite clear that the laws in the exponential family can be written in the form of the Gibbs distributions:

$$P_\theta(X) = \frac{\exp\left\{-\beta\left[H_0(X) - hM(X)\right]\right\}}{Z_\beta} ,$$

with $Z_\beta = \left[Z(\theta)\right]^P$.

As we saw in Section 6.6, the total fluctuation theorem stipulates that

$$\left\langle\left[M(x)\right]^2\right\rangle - \left[\langle M(X)\rangle\right]^2 = \frac{\chi_\beta}{\beta} ,$$

where χ_β is the susceptibility defined by $\chi_\beta = \partial M_\beta/\partial h$ with $M_\beta = \langle M(x)\rangle$. Let us examine what this says for probability laws in the exponential family. To this end, we return to the canonical notation, whereupon

$$\left\langle\left[M(x)\right]^2\right\rangle - \left[\langle M(X)\rangle\right]^2 = \left\langle\left[\sum_{i=1}^{P} t(x_i)\right]^2\right\rangle - \left[\left\langle\sum_{i=1}^{P} t(x_i)\right\rangle\right]^2$$

$$= P^2\left\{\langle[t(x)]^2\rangle - [\langle t(x)\rangle]^2\right\} .$$

Furthermore,

$$\frac{\partial\langle M(x)\rangle}{\partial h} = -\frac{\partial}{\partial\theta}\left\langle\sum_{i=1}^{P} t(x_i)\right\rangle_\theta = -P\frac{\partial}{\partial\theta}\langle t(x)\rangle_\theta ,$$

and $\beta = -1$, so that the fluctuation–dissipation theorem can be given in the form

$$\langle[t(x)]^2\rangle_\theta - [\langle t(x)\rangle_\theta]^2 = \frac{1}{P}\frac{\partial}{\partial\theta}\langle t(x)\rangle_\theta .$$

We recognize the result obtained in Section 7.6 concerning the Cramer–Rao bound of the efficient estimators in the exponential family.

This result simply shows that the total fluctuation theorem corresponds to the Cramer–Rao bound in the exponential family. The main difference is that, in statistical physics, we know the applied field h and we seek to determine the properties of $M(x)$, whereas in statistics, the problem is the opposite one, since we know $\sum_{i=1}^{P} t(x_i)$ and we seek to estimate θ. It is in this sense that statistical physics is more probabilistic than statistical.

Exercises

Exercise 8.1. Speckle at Low Fluxes

An intensity measurement is made in the presence of speckle and at low flux. We saw in Section 4.10 that the measured intensity is proportional to a random variable N taking positive integer values according to a probability law

$$P(n) = Aa^n .$$

(1) Express A in terms of a and specify the allowed values of a.
(2) Suggest a function of a that can be estimated efficiently and without bias.

Exercise 8.2. Estimating the Mean of Multilook SAR Images I

Suppose we have M synthetic aperture radar (SAR) images of the same region with the same reflectivities and independent speckle realizations. The gray levels are therefore described by random variables obeying a Gamma probability distribution. The aim is to estimate the expectation value of the gray levels of a region comprising P pixels. Show that it suffices to obtain the image corresponding to the mean of the M SAR images for this estimation.

Exercise 8.3. Estimating the Mean of Multilook SAR Images II

Consider an analogous situation to the one in the last exercise, with the only difference being that the gray levels are now assumed to be described by independent random variables obeying the Weibull probability density function with parameter α.

Exercise 8.4. Random Attenuation

The intensity of an optical wave is measured. The wave has undergone a great many attenuations with random coefficients. As we saw in Exercise 4.9, the measured intensity can be described by a real positive random variable X whose probability density is log-normal, i.e.,

$$P_X(x) = \frac{1}{x\sigma\sqrt{2\pi}} \exp\left[-\frac{(\ln x - m)^2}{2\sigma^2}\right] .$$

Find an unbiased estimator for m with minimal variance.

Exercise 8.5. Amplitude and Phase

The aim here is to find a lower bound for the estimation accuracy of the amplitude and phase of a sinusoidal signal of known frequency. We have N independent measurements at times iT/N, where T is the period of the signal and $i = 1, 2, \ldots, N$. Determine the Cramer–Rao bound (CRB) of the unbiased estimators when the measurements are perturbed by Gaussian additive white noise of variance σ^2.

Exercise 8.6. Degree of Polarization in Coherent Illumination

Consider a partially polarized coherent light source. Assume that the intensities along the horizontal and vertical axes are described by independent random variables with exponential probability density functions, so that the coherency matrix is diagonal. P intensity measurements X_i and Y_i $(i = 1, \ldots, P)$ have been made along the horizontal and vertical axes, respectively. These measurements correspond to independent speckle realizations. The degree of polarization of the light for measurement number i is

$$\rho_i = \frac{X_i - Y_i}{X_i + Y_i} .$$

Write $I_X = \langle X_i \rangle$ and $I_Y = \langle Y_i \rangle$.

(1) Calculate the probability density of ρ_i. Does it belong to the exponential family when we consider that the unknown parameter is $u = (I_X - I_Y)/(I_X + I_Y)$?
(2) Defining $\beta_i = \ln(X_i/Y_i)$, calculate the probability density of β_i. Does it belong to the exponential family when we consider that the unknown parameter is $\gamma = \ln(I_X/I_Y)$?
(3) Calculate the estimator of the first order moment of γ. Is it biased?
(4) Calculate the Cramer–Rao bound (CRB) of the unbiased estimators of γ as a function of

$$I = \int_{-\infty}^{+\infty} \frac{1}{[\exp(x/2) + \exp(-x/2)]^4} \, dx .$$

Exercise 8.7. Accuracy of Maximum Likelihood Fitting

In this exercise, we shall use the maximum likelihood method to estimate the variation of a flux that is assumed to vary linearly with time. We assume that the signal is measured at discrete time intervals $t = 1, 2, \ldots, P$ and denote the P-sample by $\chi = \{x_1, x_2, \ldots, x_P\}$ as usual.

The signal without noise is assumed to evolve according to the model $s_i^\theta = i\theta$. The parameter θ is estimated in the presence of independent noise for each measurement. In other words, the measurement noise is uncorrelated in such a way that we can write the log-likelihood in the form

$$\ell(\chi) = \sum_{i=1}^{P} \ln \left[P_{s_i^\theta}(x_i) \right] .$$

Calculate the Cramer–Rao bound for estimating θ in the following cases:

(1) the noise is additive Gaussian noise,
(2) the noise is Poisson noise,
(3) the noise is Gamma noise.

(4) Are the maximum likelihood estimators efficient?

(5) Apply the least squares estimator, i.e., the one for the Gaussian case, to the measurements perturbed by Poisson noise and compare the variances of the estimators.

(6) Modify the least squares estimator so that it becomes unbiased when applied to measurements perturbed by Gamma noise and compare the variances of the estimators.

Solutions to Exercises

9.1 Chapter Two. Random Variables

Solution to Exercise 2.1

Let $P_X(x)$ be the probability density function of Y_λ. For $-a/2 < X_\lambda < a/2$, we have $P_Y(y) = P_X(x)$. The probability of having $-a \le X_\lambda \le -a/2$ is $1/4$, as is the probability that $a/2 \le X_\lambda \le a$. We thus see that, for Y_λ, we must consider a joint probability density and discrete probability law. This can be simply achieved using the Dirac distribution $\delta(y)$. We can then write

$$P_Y(y) = \frac{1}{4}\delta(y + a/2) + \frac{1}{2a}\text{Rect}_{-a/2,a/2}(y) + \frac{1}{4}\delta(y - a/2)\,,$$

where

$$\text{Rect}_{-a/2,a/2}(y) = \begin{cases} 1 & \text{if } -a/2 < X_\lambda < a/2\,, \\ 0 & \text{otherwise}\,. \end{cases}$$

We thus observe that simple transformations can lead to mixtures of discrete and continuous probability distributions. The Dirac distribution then provides an extremely useful tool.

Solution to Exercise 2.2

This is a variable change and the transformation is

$$y = g(x) = \int_{-\infty}^{x} P_X(\eta)\mathrm{d}\eta\,.$$

Clearly, the range of variation of Y_λ is $[0, 1]$. The function g is increasing and we can therefore apply the relation $P_Y(y)\mathrm{d}y = P_X(x)\mathrm{d}x$. Now it is immediately clear that $\mathrm{d}y = P_X(x)\mathrm{d}x$, and we deduce that $P_Y(y) = 1$, corresponding to a uniform probability density between 0 and 1. The above transformation $y = g(x)$ is often used in data processing to obtain a good distribution of the values of the random variables over a given region.

Solution to Exercise 2.3

We have

$$P_X(x) = \frac{1}{\sqrt{2\pi}\sigma} \exp\left[-\frac{(x-m)^2}{2\sigma^2}\right] \,,$$

and hence,

$$\langle (x-m)^n \rangle = \int_{-\infty}^{\infty} (x-m)^n \frac{1}{\sqrt{2\pi}\sigma} \exp\left[-\frac{(x-m)^2}{2\sigma^2}\right] dx \,.$$

Note first that, by symmetry, we must have $\langle (x-m)^n \rangle = 0$ when n is odd. Setting $u = (x-m)/\sigma$, we find

$$\frac{\langle (x-m)^n \rangle}{\sigma^n} = \int_{-\infty}^{\infty} u^n \frac{1}{\sqrt{2\pi}} \exp\left(-\frac{u^2}{2}\right) du \,.$$

If we put

$$I(\alpha) = \int_{-\infty}^{\infty} \exp\left(-\alpha\frac{u^2}{2}\right) du = \sqrt{2\pi}\alpha^{-1/2} \,,$$

we then have

$$\frac{\langle (x-m)^{2n} \rangle}{\sigma^{2n}} = \frac{2^n}{\sqrt{2\pi}}(-1)^n \frac{d^n}{d\alpha^n} I(\alpha)\Big|_{\alpha=1} \,.$$

We deduce that

$$n = 1: \qquad \frac{\langle (x-m)^2 \rangle}{\sigma^2} = 2\left(\frac{1}{2}\right)\alpha^{-3/2}\Big|_{\alpha=1} \,,$$

$$n = 2: \qquad \frac{\langle (x-m)^4 \rangle}{\sigma^4} = 2^2\left(\frac{1}{2}\frac{3}{2}\right)\alpha^{-5/2}\Big|_{\alpha=1} \,,$$

$$n = 3: \qquad \frac{\langle (x-m)^6 \rangle}{\sigma^6} = 2^3\left(\frac{1}{2}\frac{3}{2}\frac{5}{2}\right)\alpha^{-7/2}\Big|_{\alpha=1} \,,$$

$$n: \qquad \frac{\langle (x-m)^{2n} \rangle}{\sigma^{2n}} = 2^n\left(\frac{1}{2}\frac{3}{2}\cdots\frac{2n-1}{2}\right)\alpha^{-(2n+1)/2}\Big|_{\alpha=1} \,.$$

Finally,

$$\langle (x-m)^{2n} \rangle = 1\cdot 3\cdot 5\cdots(2n-1)\sigma^{2n} \,.$$

Solution to Exercise 2.4

Using the covariance matrix $\overline{\overline{\Gamma}}$, the probability density function can be written

$$P_{X_1,X_2}(x_1,x_2) = \frac{1}{2\pi\sqrt{|\overline{\overline{\Gamma}}|}} \exp\left(-\frac{1}{2}\boldsymbol{x}^{\mathrm{T}}\overline{\overline{\Gamma}}^{-1}\boldsymbol{x}\right) \,,$$

where $|\overline{\overline{\Gamma}}|$ is the determinant of $\overline{\overline{\Gamma}}$ and

$$x = \begin{pmatrix} x_1 \\ x_2 \end{pmatrix} .$$

We set $\sigma_1^2 = \Gamma_{11}$, $\sigma_2^2 = \Gamma_{22}$ and $\rho = \Gamma_{12}/(\sigma_1\sigma_2)$. Then,

$$\overline{\overline{\Gamma}} = \begin{pmatrix} \sigma_1^2 & \rho\sigma_1\sigma_2 \\ \rho\sigma_1\sigma_2 & \sigma_2^2 \end{pmatrix} ,$$

and so $|\overline{\overline{\Gamma}}| = \sigma_1^2\sigma_2^2(1 - \rho^2)$. We thus obtain

$$\overline{\overline{\Gamma}}^{-1} = \begin{pmatrix} \left[\sigma_1^2(1 - \rho^2)\right]^{-1} & -\rho/\left[\sigma_1\sigma_2(1 - \rho^2)\right] \\ -\rho/\left[\sigma_1\sigma_2(1 - \rho^2)\right] & \left[\sigma_2^2(1 - \rho^2)\right]^{-1} \end{pmatrix} ,$$

whence,

$$P_{X_1,X_2}(x_1, x_2) = \frac{1}{2\pi\sigma_1\sigma_2\sqrt{(1 - \rho^2)}} \exp\left[-\frac{x_1^2/\sigma_1^2 + x_2^2/\sigma_2^2 - 2x_1x_2\rho/\sigma_1\sigma_2}{2(1 - \rho^2)} \right] .$$

Solution to Exercise 2.5

We have $G(x, y) = \int_x^y P_X(\eta)\mathrm{d}\eta$ and hence, $\partial G(x, y)/\partial y = P_X(y)$, which can also be written $P_X(x) = -\partial G(x, y)/\partial x$.

Solution to Exercise 2.6

The probability of observing an atom of species A_1 is c_1, whilst for species A_2, it is c_2. Applying Bayes' rule, we obtain $p = c_1p_1 + c_2p_2$. When there are N species, we have $p = \sum_{i=1}^{N} c_ip_i$.

Solution to Exercise 2.7

We have $P(x, y) = P_X(x)P_Y(y)$ and hence,

$$P(x, y) = \frac{1}{2\pi\sigma^2} \exp\left[-\frac{(x - m_X)^2}{2\sigma^2} - \frac{(y - m_Y)^2}{2\sigma^2} \right] .$$

We set $M = m_X + im_Y$, so that

$$|z - M|^2 = (z - M)^*(z - M) = (x - m_X)^2 + (y - m_Y)^2 ,$$

where z^* is the complex conjugate of z. We can thus write

$$P_Z(z) = \frac{1}{2\pi\sigma^2} \exp\left(-\frac{1}{2\sigma^2}|z - M|^2 \right) .$$

Putting $\Sigma = 2\sigma^2$, we can then write

$$P_Z(z) = \frac{1}{\pi\Sigma} \exp\left(-\frac{|z - M|^2}{\Sigma} \right) .$$

Solution to Exercise 2.8

The probability density function of the Gamma probability law, defined for $x \geq 0$, is

$$P_X(x) = \frac{x^{\alpha-1}}{m^\alpha \Gamma(\alpha)} \exp\left(-\frac{x}{m}\right) .$$

The function $y = x^\beta$ is increasing and we can therefore apply the relation $P_Y(y)\mathrm{d}y = P_X(x)\mathrm{d}x$. We deduce that $\mathrm{d}y = \beta x^{(\beta-1)}\mathrm{d}x$ and $x = y^{1/\beta}$. We then obtain

$$P_Y(y) = \frac{y^{\alpha/\beta-1}}{\beta m^\alpha \Gamma(\alpha)} \exp\left(-\frac{y^{1/\beta}}{m}\right) .$$

When $\alpha = 1$, we obtain

$$P_Y(y) = \frac{\gamma y^{\gamma-1}}{\mu^\gamma} \exp\left[-\left(\frac{y}{\mu}\right)^\gamma\right] ,$$

where $\gamma = 1/\beta$ and $\mu = m^\beta$. $P_Y(y)$ is the Weibull probability density function.

Solution to Exercise 2.9

(1) We have

$$P_B(b) = \frac{1}{\sqrt{2\pi}\sigma} \exp\left(-\frac{b^2}{2\sigma^2}\right) ,$$

Now $\langle Y \rangle = g$ and $\langle (Y - \langle Y \rangle)^2 \rangle = \sigma^2/N$, so that

$$P_Y(y) = \frac{\sqrt{N}}{\sqrt{2\pi}\sigma} \exp\left[-\frac{N}{2\sigma^2}(y-g)^2\right] .$$

(2) As the variance of Y decreases with N, the accuracy in the determination of g increases.

Solution to Exercise 2.10

The distribution function of Z_{λ_T} can be determined from

$$F_Z(z) = \int_{-\infty}^{+\infty} \int_{-\infty}^{+\infty} \theta\left(z - \frac{x}{y}\right) P_{X,Y}(x,y)\mathrm{d}x\mathrm{d}y ,$$

where $\theta(u)$ is the Heaviside step function

$$\theta(u) = \begin{cases} 1 & \text{if } u \geq 0 , \\ 0 & \text{otherwise} , \end{cases}$$

and $P_{X,Y}(x,y)$ is the joint probability density function

$$P_{X,Y}(x,y) = \frac{1}{2\pi\sigma^2} \exp\left(-\frac{x^2+y^2}{2\sigma^2}\right).$$

Since $dF_Z(z)/dz = P_Z(z)$ and $d\theta(u)/du = \delta(u)$, where $\delta(u)$ is the Dirac distribution, we have

$$P_Z(z) = \int_{-\infty}^{+\infty}\int_{-\infty}^{+\infty} \delta\left(z - \frac{x}{y}\right) P_{X,Y}(x,y)\mathrm{d}x\mathrm{d}y$$

$$= 2\int_0^{+\infty}\left[\int_{-\infty}^{+\infty}\delta\left(z - \frac{x}{y}\right)P_{X,Y}(x,y)\mathrm{d}x\right]\mathrm{d}y.$$

If we put $\nu = x/y$, we can then write

$$P_Z(z) = 2\int_0^{+\infty} y\left[\int_{-\infty}^{+\infty}\delta(z-\nu)P_{X,Y}(y\nu,y)\mathrm{d}\nu\right]\mathrm{d}y$$

$$= 2\int_0^{+\infty} yP_{X,Y}(yz,y)\mathrm{d}y$$

$$= 2\int_0^{+\infty} y\frac{1}{2\pi\sigma^2}\exp\left(-\frac{y^2\,z^2+y^2}{2\sigma^2}\right)\mathrm{d}y.$$

A direct calculation then leads to

$$P_Z(z) = \frac{1}{\pi(1+z^2)},$$

which corresponds to a Cauchy variable.

9.2 Chapter Three. Fluctuations and Covariance

Solution to Exercise 3.1

Since $\langle(X_\lambda^c - Y_\lambda^c)^2\rangle = \langle(X_\lambda^c)^2\rangle + \langle(Y_\lambda^c)^2\rangle - 2\langle X_\lambda^c Y_\lambda^c\rangle \geq 0$, it follows that $\langle(X_\lambda^c)^2\rangle + \langle(Y_\lambda^c)^2\rangle \geq 2\Gamma_{XY}$.

Solution to Exercise 3.2

To begin with, we do not indicate the dependence on time t. We have $Y_\lambda = xB_\lambda$, and hence $P_Y(y)\mathrm{d}y = P_B(b)\mathrm{d}b$. Since $\mathrm{d}y = x\mathrm{d}b$, we deduce that $P_Y(y) = (1/x)P_B(y/x)$, or

$$P_Y(y) = \frac{y^{r-1}}{(ax)^r\Gamma(r)}\exp\left(-\frac{y}{ax}\right).$$

The second transformation to consider is $Z_\lambda = \ln Y_\lambda$. We can thus write $P_{Z,t}(z)\mathrm{d}z = P_Y(y)\mathrm{d}y$, with $\mathrm{d}z = \mathrm{d}y/y$. Hence,

$$P_{Z,t}(z) = \frac{e^{rz}}{(ax)^r\Gamma(r)}\exp\left(-\frac{e^z}{ax}\right).$$

Solution to Exercise 3.3

We have

$$P_{X,t}(x) = \frac{1}{\sqrt{2\pi}\sigma} \exp\left(-\frac{x^2}{2\sigma^2}\right) .$$

Furthermore, $P_{Y,t}(y)\mathrm{d}y = P_{X,t}(x)\mathrm{d}x$ and $Y_\lambda(t) = g(t)X_\lambda(t)$. As $g(t)$ is strictly positive, the transformation $y = g(t)x$ is bijective and $\mathrm{d}y = g(t)\mathrm{d}x$. We thus have

$$P_{Y,t}(y) = \frac{1}{\sqrt{2\pi}g(t)\sigma} \exp\left[-\frac{y^2}{2g(t)^2\sigma^2}\right] .$$

Solution to Exercise 3.4

(1) We find that

$$\langle f_T(t - \tau_\lambda)\rangle = \int_0^T f_T(t - \tau)P_\tau(\tau)\mathrm{d}\tau = \frac{1}{T}\int_0^T f_T(t - \tau)\mathrm{d}\tau ,$$

and so

$$\langle f_T(t - \tau_\lambda)\rangle = \frac{1}{T}\int_0^T f_T(\xi)\mathrm{d}\xi ,$$

which leads to a result independent of t. Likewise,

$$\langle f_T(t - \tau_\lambda)f_T(t + \mu - \tau_\lambda)\rangle = \int_0^T f_T(t - \tau)f_T(t + \mu - \tau)P_\tau(\tau)\mathrm{d}\tau$$

$$= \frac{1}{T}\int_0^T f_T(t - \tau)f_T(t + \mu - \tau)\mathrm{d}\tau ,$$

and hence,

$$\langle f_T(t - \tau_\lambda)f_T(t + \mu - \tau_\lambda)\rangle = \frac{1}{T}\int_0^T f_T(\xi)f_T(\xi + \mu)\mathrm{d}\xi ,$$

which also leads to a result independent of t. $f_T(t - \tau_\lambda)$ is therefore weakly stationary.

(2) We have

$$\overline{f_T(t - \tau_\lambda)} = \lim_{\substack{T_1 \to -\infty \\ T_2 \to \infty}} \frac{1}{T_2 - T_1}\int_{T_1}^{T_2} f_T(t - \tau_\lambda)\mathrm{d}t ,$$

and as the function is periodic,

$$\overline{f_T(t - \tau_\lambda)} = \lim_{\substack{T_1 \to -\infty \\ T_2 \to \infty}} \frac{1}{T_2 - T_1}\int_{T_1}^{T_2} f_T(\xi)\mathrm{d}\xi ,$$

which leads to a result independent of λ. Likewise,

$$
\overline{f_T(t - \tau_\lambda)f_T(t + \mu - \tau_\lambda)}
$$

$$
= \lim_{\substack{T_1 \to -\infty \\ T_2 \to \infty}} \frac{1}{T_2 - T_1} \int_{T_1}^{T_2} f_T(t - \tau_\lambda)f_T(t + \mu - \tau_\lambda)dt \ ,
$$

and hence,

$$
\overline{f_T(t - \tau_\lambda)} = \lim_{\substack{T_1 \to -\infty \\ T_2 \to \infty}} \frac{1}{T_2 - T_1} \int_{T_1}^{T_2} f_T(\xi)f_T(\xi + \mu)d\xi \ ,
$$

which also leads to a result independent of λ. $f_T(t - \tau_\lambda)$ is therefore weakly ergodic.

Solution to Exercise 3.5

We have $\langle f_\lambda(t) \rangle = F$, where F is independent of t. Since $\langle h_\lambda(t) \rangle = \langle g(t)f_\lambda(t) \rangle$, we deduce that $\langle h_\lambda(t) \rangle = g(t)\langle f_\lambda(t) \rangle$ and so $\langle h_\lambda(t) \rangle = g(t)F$. Therefore $g(t)$ must be independent of t if $h_\lambda(t)$ is to be stationary to order 1. In this case, put $g(t) = g_0$, so that $\langle h_\lambda(t)h_\lambda(t + \tau) \rangle = \langle g_0^2 f_\lambda(t)f_\lambda(t + \tau) \rangle$, or $\langle h_\lambda(t)h_\lambda(t + \tau) \rangle = g_0^2\langle f_\lambda(t)f_\lambda(t + \tau) \rangle$. Since $f_\lambda(t)$ is assumed to be weakly stationary, $\langle f_\lambda(t)f_\lambda(t + \tau) \rangle$ is independent of t and we deduce that the same is true for $\langle h_\lambda(t)h_\lambda(t + \tau) \rangle$.

Solution to Exercise 3.6

It is enough for $X_\lambda(t)$ to be stationary and ergodic up to second order moments. The proof is immediate.

Solution to Exercise 3.7

We begin by analyzing the stationarity up to second order moments. We have $\langle Y_\lambda(t) \rangle = a_1\langle X_\lambda(t) \rangle + a_2\langle [X_\lambda(t)]^2 \rangle$ and

$$
\langle Y_\lambda(t)Y_\lambda(t + \tau) \rangle
$$

$$
= \left\langle \left\{ a_1 X_\lambda(t) + a_2 [X_\lambda(t)]^2 \right\} \left\{ a_1 X_\lambda(t + \tau) + a_2 [X_\lambda(t + \tau)]^2 \right\} \right\rangle \ ,
$$

or

$$
\langle Y_\lambda(t)Y_\lambda(t + \tau) \rangle = \left\langle \left\{ a_1^2 X_\lambda(t)X_\lambda(t + \tau) + a_2 a_1 [X_\lambda(t)]^2 X_\lambda(t + \tau) \right. \right.
$$

$$
\left. \left. + a_1 a_2 X_\lambda(t) [X_\lambda(t + \tau)]^2 + a_2^2 [X_\lambda(t)]^2 [X_\lambda(t + \tau)]^2 \right\} \right\rangle \ ,
$$

and hence,

$$\langle Y_\lambda(t)Y_\lambda(t+\tau)\rangle = \left\langle \left\{ a_1^2 \langle X_\lambda(t)X_\lambda(t+\tau)\rangle + a_2 a_1 \langle [X_\lambda(t)]^2 X_\lambda(t+\tau)\rangle \right.\right.$$
$$\left.\left. + a_1 a_2 \langle X_\lambda(t)[X_\lambda(t+\tau)]^2\rangle + a_2^2 \langle [X_\lambda(t)]^2 [X_\lambda(t+\tau)]^2\rangle \right\}\right\rangle .$$

We thus see that weak stationarity is not enough. On the other hand, if $X_\lambda(t)$ is stationary up to fourth order moments, the quantities

$$\langle X_\lambda(t)\rangle ,$$
$$\langle X_\lambda(t)X_\lambda(t+\tau_1)\rangle ,$$
$$\langle X_\lambda(t)X_\lambda(t+\tau_1)X_\lambda(t+\tau_2)\rangle ,$$
$$\langle X_\lambda(t)X_\lambda(t+\tau_1)X_\lambda(t+\tau_2)X_\lambda(t+\tau_3)\rangle$$

are independent of t, and in this case $Y_\lambda(t)$ is stationary up to second order moments.

Let us now address the question of weak ergodicity. We have $\overline{Y_\lambda(t)} = a_1 \overline{X_\lambda(t)\rangle} + a_2 \overline{[X_\lambda(t)]^2}$ and

$$\overline{Y_\lambda(t)Y_\lambda(t+\tau)} = \left\{ a_1^2 \overline{X_\lambda(t)X_\lambda(t+\tau)} + a_2 a_1 \overline{[X_\lambda(t)]^2 X_\lambda(t+\tau)} \right.$$
$$\left. + a_1 a_2 \overline{X_\lambda(t)[X_\lambda(t+\tau)]^2} + a_2^2 \overline{[X_\lambda(t)]^2 [X_\lambda(t+\tau)]^2} \right\} .$$

We thus see that weak ergodicity is not sufficient. However, if $X_\lambda(t)$ is ergodic up to fourth order moments, the quantities

$$\overline{X_\lambda(t)} ,$$
$$\overline{X_\lambda(t)X_\lambda(t+\tau_1)} ,$$
$$\overline{X_\lambda(t)X_\lambda(t+\tau_1)X_\lambda(t+\tau_2)} ,$$
$$\overline{X_\lambda(t)X_\lambda(t+\tau_1)X_\lambda(t+\tau_2)X_\lambda(t+\tau_3)}$$

are independent of λ, and in this case $Y_\lambda(t)$ is ergodic up to second order moments.

Solution to Exercise 3.8

(1) We have

$$\hat{f}_\lambda(\nu) = a\delta\left(\nu - \frac{1}{T}\right)\exp\left(-\mathrm{i}\phi_\lambda\right) ,$$

where $\delta(x)$ is the Dirac distribution. The phase of $\hat{f}_\lambda(\nu)$ is thus $-\phi_\lambda$.

(2) We have $\langle f_\lambda(t)\rangle = a\exp(2\mathrm{i}\pi t/T)\langle \exp(-\mathrm{i}\phi_\lambda)\rangle$. $f_\lambda(t)$ is not therefore weakly stationary in general. However, if $\langle \exp(-\mathrm{i}\phi_\lambda)\rangle = 0$, we have $\langle f_\lambda(t)\rangle = 0$ and it is then worth taking the analysis to second order.

In this case, $\langle f_\lambda^*(t)f_\lambda(t+\tau)\rangle = a^2 \exp(2\mathrm{i}\pi\tau/T)$ and we thus observe that $f_\lambda(t)$ is stationary up to second order moments. The condition $\langle \exp(-\mathrm{i}\phi_\lambda)\rangle = 0$ is equivalent to

$$\int_0^{2\pi} \exp(-i\phi) P_\Phi(\phi) d\phi = 0 \; .$$

This condition is fulfilled, for example, if ϕ_λ is a random variable uniformly distributed over the interval $[0, 2\pi]$, i.e., $P_\Phi(\phi) = 1/2\pi$ in the interval $[0, 2\pi]$.

(3) $\langle \hat{f}_\lambda(\nu) \rangle = a\delta(\nu - 1/T)\langle \exp(-i\phi_\lambda) \rangle$. When $f_\lambda(t)$ is weakly stationary, we thus have $\langle \hat{f}_\lambda(\nu) \rangle = 0$. $\hat{f}_\lambda(\nu)$ is therefore a complex random variable with zero mean. In particular, if ϕ_λ is a random variable distributed uniformly over the interval $[0, 2\pi]$, $\hat{f}_\lambda(\nu)$ is an isotropic complex random variable, i.e., the probability density functions of $\hat{f}_\lambda(\nu)$ and $\exp(-i\varphi)\hat{f}_\lambda(\nu)$ are equal, $\forall \varphi \in [0, 2\pi]$.

(4) We have

$$\hat{f}_\lambda(\nu) = \sum_{n=-\infty}^{\infty} a_n \delta \left(\nu - \frac{n}{T} \right) \exp \left(-i\phi_{n,\lambda} \right) \; ,$$

and we can therefore generalize the last result for each frequency $\nu = n/T$.

(5) We have

$$\hat{f}_\lambda^*(\nu_1) \hat{f}_\lambda(\nu_2)$$
$$= \sum_{n=-\infty}^{\infty} \sum_{m=-\infty}^{\infty} a_n^* a_m \delta \left(\nu_1 - \frac{n}{T} \right) \delta \left(\nu_2 - \frac{m}{T} \right) \exp(i\phi_{n,\lambda} - i\phi_{m,\lambda}) \; ,$$

and hence,

$$\langle \hat{f}_\lambda^*(\nu_1) \hat{f}_\lambda(\nu_2) \rangle$$
$$= \sum_{n=-\infty}^{\infty} \sum_{m=-\infty}^{\infty} a_n^* a_m \delta \left(\nu_1 - \frac{n}{T} \right) \delta \left(\nu_2 - \frac{m}{T} \right) \langle \exp(i\phi_{n,\lambda} - i\phi_{m,\lambda}) \rangle \; .$$

We first analyze the case when $\nu_1 \neq \nu_2$. Since the $\phi_{n,\lambda}$ are independent random variables distributed uniformly over the interval $[0, 2\pi]$, when $n \neq m$, we have

$$\langle \exp(i\phi_{n,\lambda} - i\phi_{m,\lambda}) \rangle = \langle \exp(i\phi_{n,\lambda}) \rangle \langle \exp(-i\phi_{m,\lambda}) \rangle = 0 \; .$$

We thus have

$$\langle \hat{f}_\lambda^*(\nu_1) \hat{f}_\lambda(\nu_2) \rangle = \sum_{n=-\infty}^{\infty} a_n^* a_n \delta \left(\nu_1 - \frac{n}{T} \right) \delta \left(\nu_2 - \frac{n}{T} \right) = 0 \; ,$$

since $\nu_1 \neq \nu_2$, by hypothesis. $\hat{f}_\lambda(\nu_1)$ and $\hat{f}_\lambda(\nu_2)$ are then uncorrelated. If $\nu_1 = \nu_2 = \nu$, $\hat{f}_\lambda^*(\nu) \hat{f}_\lambda(\nu)$ is not defined because $\delta(\nu - n/T)\delta(\nu - n/T)$ is not a distribution. However, the coefficient in front of this term will be $|a_n|^2$.

Solution to Exercise 3.9

(1) Let $f(t) \otimes g(t)$ be the convolution of $f(t)$ with $g(t)$ defined by

$$f(t) \otimes g(t) = \int_{-\infty}^{\infty} f(t - \xi)g(\xi)\mathrm{d}\xi .$$

We have $s(t) = (1 - a)r(t) + ar(t) \otimes \delta(t - \tau)$, or

$$s(t) = r(t) \otimes [(1 - a)\delta(t) + a\delta(t - \tau)] .$$

We shall write $h(t) = (1 - a)\delta(t) + a\delta(t - \tau)$.

(2) Let $B_\lambda(t)$ be the emitted signal, which is filtered white noise, and let σ_B^2 be its power. The power spectral density of $B_\lambda(t)$ is then

$$\hat{S}_{BB}(\nu) = \begin{cases} \sigma_B^2/2\nu_B & \text{if } \nu \in [-\nu_B, \nu_B] , \\ 0 & \text{otherwise} . \end{cases}$$

The transfer function of the filter is $\hat{h}(\nu) = (1 - a) + a\exp(-\mathrm{i}2\pi\nu\tau)$ so that $|\hat{h}(\nu)|^2 = [(1 - a) + a\cos(2\pi\nu\tau)]^2 + [a\sin(2\pi\nu\tau)]^2$. The power spectral density of the measured signal is given by

$$\hat{S}_{SS}(\nu) = |\hat{h}(\nu)|^2 \hat{S}_{BB}(\nu) ,$$

or

$$\hat{S}_{SS}(\nu) = \frac{\sigma_B^2}{2\nu_B} \left[(1 - a) + a\cos(2\pi\nu\tau) \right]^2 + \left[a\sin(2\pi\nu\tau) \right]^2 ,$$

if $\nu \in [-\nu_B, \nu_B]$ and 0 otherwise.

Solution to Exercise 3.10

(1) We have $Y_\lambda(t) = \int_{-\infty}^{\infty} X_\lambda(t - \xi)h(\xi)\mathrm{d}\xi$ and hence,

$$\langle Y_\lambda^*(t)Y_\lambda(t + \tau)\rangle$$
$$= \int_{-\infty}^{\infty} \int_{-\infty}^{\infty} \langle X_\lambda^*(t - \xi_1)X_\lambda(t + \tau - \xi_2)\rangle h^*(\xi_1)h(\xi_2)\mathrm{d}\xi_1\mathrm{d}\xi_2$$
$$= \int_{-\infty}^{\infty} \int_{-\infty}^{\infty} \Gamma_{XX}(\xi_1 + \tau - \xi_2)h^*(\xi_1)h(\xi_2)\mathrm{d}\xi_1\mathrm{d}\xi_2 .$$

Now $\Gamma_{XX}(\xi_1 + \tau - \xi_2)$ is the Fourier transform of $\hat{S}_{XX}(\nu)$ so that, when $B \to +\infty$,

$$\Gamma_{XX}(\xi_1 + \tau - \xi_2) \to \sigma_B^2\delta(\xi_1 + \tau - \xi_2) ,$$

where $\delta(t)$ is the Dirac distribution. We then have

$$\langle Y_\lambda^*(t)Y_\lambda(t+\tau)\rangle = \sigma_B^2 \int_{-\infty}^{\infty} \int_{-\infty}^{\infty} \delta(\xi_1 + \tau - \xi_2)h^*(\xi_1)h(\xi_2)\mathrm{d}\xi_1\mathrm{d}\xi_2$$

$$= \sigma_B^2 \int_{-\infty}^{\infty} h^*(\xi)h(\xi+\tau)\mathrm{d}\xi .$$

Now

$$\int_{-\infty}^{\infty} h^*(\xi)h(\xi+\tau)\mathrm{d}\xi = \int_{0}^{\infty} a^2 \exp(-2a\xi - a\tau)\mathrm{d}\xi ,$$

if $\tau > 0$. We thus obtain

$$\int_{-\infty}^{\infty} h^*(\xi)h(\xi+\tau)\mathrm{d}\xi = \frac{a}{2}\exp(-a\tau) .$$

If τ is negative, we have

$$\int_{-\infty}^{\infty} h^*(\xi)h(\xi+\tau)\mathrm{d}\xi = \frac{a}{2}\exp(-a|\tau|) ,$$

and hence,

$$\langle Y_\lambda^*(t)Y_\lambda(t+\tau)\rangle = \frac{a\sigma_B^2}{2}\exp(-a|\tau|) .$$

(2) The total power of the fluctuations after filtering is thus

$$P_Y = \langle Y_\lambda^*(t)Y_\lambda(t)\rangle = \frac{a\sigma_B^2}{2} .$$

(3) We observe that $P_Y \to +\infty$ if $a \to +\infty$. This is understandable since the noise power $X_\lambda(t)$ diverges when $B \to +\infty$, and $h(t) \to \delta(t)$ when $a \to +\infty$. This means that no frequency is attenuated by the filter $h(t)$.

Solution to Exercise 3.11

(1) Let $f(t) \otimes g(t)$ be the convolution of $f(t)$ with $g(t)$. We first note that

$$Y_\lambda(t) = X_\lambda(t) \otimes \mathrm{Rect}_{0,T}(t) = \int_{-\infty}^{\infty} X_\lambda(\xi)\mathrm{Rect}_{0,T}(t-\xi)\mathrm{d}\xi ,$$

where

$$\mathrm{Rect}_{0,T}(t) = \begin{cases} 1 & \text{if } 0 < t < T , \\ 0 & \text{otherwise} . \end{cases}$$

We then have

$$\hat{S}_{YY}(\nu) = |\widehat{\mathrm{Rect}_{0,T}}(\nu)|^2 \hat{S}_{XX}(\nu) ,$$

where $\widehat{\mathrm{Rect}_{0,T}}(\nu)$ is the Fourier transform of $\mathrm{Rect}_{0,T}(t)$, i.e.,

$$\widehat{\mathrm{Rect}_{0,T}}(\nu) = \int_{0}^{T} \exp(-\mathrm{i}2\pi\nu t)\mathrm{d}t = \exp(-\mathrm{i}\pi\nu T)\left[\frac{\sin(\pi\nu T)}{\pi\nu}\right] ,$$

and hence,

$$\hat{S}_{YY}(\nu) = \left[\frac{\sin(\pi\nu T)}{\pi\nu}\right]^2 \hat{S}_{XX}(\nu) .$$

(2) If $\hat{S}_{XX}(\nu) = \sigma^2 \delta(\nu - n/T)$, we have

$$\hat{S}_{YY}(\nu) = \sigma^2 \left[\frac{\sin(\pi\nu T)}{\pi\nu}\right]^2 \delta\left(\nu - \frac{n}{T}\right) = 0 ,$$

because

$$\frac{T\sin(\pi n)}{\pi n} = 0 .$$

We can interpret this result by observing that $X_\lambda(t)$ is then a sinusoidal signal with period T, and that $Y_\lambda(t)$ is the integral of $X_\lambda(t)$ over a period T.

(3) If $\langle X_\lambda(t_1)X_\lambda(t_2)\rangle = \delta(t_1 - t_2)$, we then have

$$\hat{S}_{XX}(\nu) = 1 ,$$

and hence,

$$\hat{S}_{YY}(\nu) = \left[\frac{\sin(\pi\nu T)}{\pi\nu}\right]^2 .$$

(4) The inverse transform of $\left[(1/\pi\nu)\sin(\pi\nu T)\right]^2$ is the autocorrelation function of $\mathrm{Rect}_{0,T}(t)$. The value of this autocorrelation function at 0 is therefore T. We deduce that $\int_{-\infty}^{\infty} \hat{S}_{YY}(\nu)\mathrm{d}\nu = T$. The power of $Y_\lambda(t)$ is then proportional to T. This result should be compared with the one which says that the variance of the sum of N independent and identically distributed random variables is proportional to N.

Solution to Exercise 3.12

(1) We have

$$Y_\lambda(t) = \int_{-\infty}^{\infty} X_\lambda(t - \xi)h(\xi)\mathrm{d}\xi ,$$

whereupon

$$\langle X_\lambda(t)Y_\lambda(t + \tau)\rangle = \int_{-\infty}^{\infty} \langle X_\lambda(t)X_\lambda(t + \tau - \xi)\rangle h(\xi)\mathrm{d}\xi ,$$

or

$$\Gamma_{XY}(\tau) = \int_{-\infty}^{\infty} \Gamma_{XX}(\tau - \xi)h(\xi)\mathrm{d}\xi .$$

(2) The last equation is a convolution relation, so Fourier transforming yields

$$\hat{S}_{XY}(\nu) = \hat{S}_{XX}(\nu)\hat{h}(\nu) ,$$

where we have assumed that the Fourier transforms $\hat{S}_{XY}(\nu)$ and $\hat{S}_{XX}(\nu)$ of $\Gamma_{XY}(\tau)$ and $\Gamma_{XX}(\tau)$ exist.

(3) If we know $\hat{h}(\nu)$, we automatically know $h(t)$, at least in principle. Now the last equation allows us to find $\hat{h}(\nu)$ from $\hat{S}_{XY}(\nu)$ if $\hat{S}_{XX}(\nu) \neq 0$.

(4) If $X_\lambda(t)$ is white noise in the frequency band between $-B$ and B, its power spectral density (denoted σ^2) is constant in this band. We can then immediately determine $\hat{h}(\nu)$ for frequencies between $-B$ and B:

$$\hat{h}(\nu) = \hat{S}_{XY}(\nu)/\sigma^2 \ .$$

9.3 Chapter Four. Limit Theorems and Fluctuations

To simplify the notation, the dependence of random variables on the random events λ is not indicated in the solutions to the exercises here.

Solution to Exercise 4.1

The sum of two Gaussian variables is a Gaussian variable. The mean of $x = (x_1 + x_2)/2$ is equal to the mean of x_1 and the variance of $x = (x_1 + x_2)/2$ is equal to half that of x_1.

Solution to Exercise 4.2

(1) $\langle S_i \rangle = 0$ and $\langle S_i^2 \rangle = \sigma_B^2 = (1/2\alpha) \int_{-\alpha}^{\alpha} s^2 \mathrm{d}s$, so that $\sigma_B^2 = \alpha^2/3$.

(2) We have

$$\langle S_i S_j \rangle = \frac{1}{N} \sum_{n=1}^{N} \sum_{m=1}^{N} \langle B_{i+n} B_{j+m} \rangle = \frac{1}{N} \sum_{n=1}^{N} \sum_{m=1}^{N} \sigma_B^2 \delta_{i+n-j-m} \ .$$

Now,

$$\sum_{m=1}^{N} \delta_{i+n-j-m} = \begin{cases} 1 & \text{if } 0 < i+n-j \leq N \ , \\ 0 & \text{otherwise} \ , \end{cases}$$

and hence,

$$\langle S_i S_j \rangle = \begin{cases} \sigma_B^2 \left(1 - \dfrac{|i-j|}{N} \right) & \text{if } |i-j| \leq N \ , \\ 0 & \text{otherwise} \ . \end{cases}$$

(3) The probability density function of the sum of two independent random variables is obtained by convolution of the probability density functions of each random variable. The convolution of the two indicator functions $\mathrm{Rect}_{-\alpha,\alpha}(b)$ defined by

$$\mathrm{Rect}_{-\alpha,\alpha}(b) = \begin{cases} 1 & \text{if } |b| \leq \alpha \ , \\ 0 & \text{otherwise} \ , \end{cases}$$

leads to the probability density function

$$P_S(s) = \begin{cases} \dfrac{1}{2\alpha}\left(1 - \dfrac{|s|}{2\alpha}\right) & \text{if } |s| \leq 2\alpha , \\ 0 & \text{otherwise} . \end{cases}$$

(4) If $N \to +\infty$, the central limit theorem is applicable and we deduce that the probability density of S_i will tend toward the normal probability density function. It has zero mean and variance $\langle S_i^2 \rangle = \sigma_B^2$.

(5) The central limit theorem is still applicable and we deduce that the probability density function of S_i will tend toward the normal distribution. The mean is still zero and its variance can be found as follows. We have

$$\langle S_i^2 \rangle = \frac{1}{N}\sum_{j=1}^{N}\sum_{k=1}^{N} a_j a_k \langle B_{i+j} B_{i+k}\rangle = \frac{1}{N}\sum_{j=1}^{N}\sum_{k=1}^{N} a_j a_k \sigma_B^2 \delta_{k-j} .$$

Hence,

$$\langle S_i^2 \rangle = \frac{\sigma_B^2}{N}\sum_{j=1}^{N} a_j^2 ,$$

and finally, $\langle S_i^2 \rangle = \sigma_B^2$.

(6) We have already encountered this type of variable change in Section 2.7. We thus have

$$P_Y(y) = \frac{1}{\sqrt{2\pi y}\sigma_B} \exp\left(-\frac{y}{2\sigma_B^2}\right) .$$

Solution to Exercise 4.3

(1) The result of an experiment can be represented by a sequence of N binary digits in such a way that the jth digit is equal to 0 if no electrons passed during the jth time interval of length $\delta\tau$, and equal to 1 if one electron went through. The probability of observing 1 a total of m times is $p^m q^{N-m}$ in a given sequence. All such sequences containing m 1s contribute to the probability that m electrons pass, and the total number of such sequences is $N!/[(N-m)!m!]$. It follows that the probability $p(m)$ that exactly m electrons pass during the time interval τ is equal to

$$\frac{N!}{(N-m)!m!}p^m q^{N-m} .$$

(2) Consider the last model and let X_j be the binary random variable which is equal to 1 if an electron has passed during the jth time interval of length $\delta\tau$ and 0 otherwise. We thus have $m = \sum_{i=1}^{N} X_i$ and therefore $\langle m \rangle = \sum_{i=1}^{N}\langle X_i \rangle = Np$ and

$$\langle m^2 \rangle = \sum_{i=1}^{N}\sum_{j=1}^{N}\langle X_i X_j \rangle = Np + N(N-1)p^2 .$$

We can write the last result in the form $\langle m^2 \rangle = Np\big[1 + (N-1)p\big]$.

(3) Put $\mu = \lim_{N \to \infty} Np$. We have $\langle m \rangle \to \mu$ when $N \to \infty$. Likewise, $\langle m^2 \rangle \to \mu(1+\mu)$ when $N \to \infty$.

(4) When $N \to \infty$, we see that $\langle I \rangle = \mu e/\tau$. Moreover, $\langle m^2 \rangle - \langle m \rangle^2 = \mu$ and so $\langle I^2 \rangle - \langle I \rangle^2 = \langle I \rangle e/\tau$.

Solution to Exercise 4.4

(1) We have $\langle I_{\text{hor}} \rangle = \langle I_{\text{ver}} \rangle = a$.

(2) The probability density function of the sum of two independent random variables is obtained by convoluting the probability density functions of each random variable. We have

$$P_X(x) = \int_{-\infty}^{\infty} \theta(x - \xi)\theta(\xi)\frac{1}{a}\exp\left(-\frac{x-\xi}{a}\right)\frac{1}{a}\exp\left(-\frac{\xi}{a}\right)\,\mathrm{d}\xi\,,$$

where

$$\theta(u) = \begin{cases} 1 & \text{if } u > 0\,, \\ 0 & \text{otherwise}\,. \end{cases}$$

We thus obtain

$$P_X(x) = \theta(x)\frac{1}{a^2}\exp\left(-\frac{x}{a}\right)\int_{-\infty}^{\infty}\theta(x-\xi)\theta(\xi)\,\mathrm{d}\xi\,,$$

or

$$P_X(x) = \theta(x)\frac{x}{a^2}\exp\left(-\frac{x}{a}\right)\,.$$

The probability density function of the difference of two independent random variables is obtained by correlation of the probability density functions for each random variable. We have

$$P_Y(y) = \int_{-\infty}^{\infty}\theta(y+\xi)\theta(\xi)\frac{1}{a}\exp\left(-\frac{y+\xi}{a}\right)\frac{1}{a}\exp\left(-\frac{\xi}{a}\right)\,\mathrm{d}\xi\,,$$

or

$$P_Y(y) = \frac{1}{a^2}\exp\left(-\frac{y}{a}\right)\int_{-\infty}^{\infty}\theta(y+\xi)\theta(\xi)\exp\left(-2\frac{\xi}{a}\right)\,\mathrm{d}\xi\,,$$

Let us consider the two cases $y \geq 0$ and $y \leq 0$ separately. When $y \geq 0$,

$$P_Y(y) = \frac{1}{a^2}\exp\left(-\frac{y}{a}\right)\int_{-\infty}^{\infty}\theta(\xi)\exp\left(-2\frac{\xi}{a}\right)\,\mathrm{d}\xi$$

$$= \frac{1}{a^2}\exp\left(-\frac{y}{a}\right)\int_{0}^{\infty}\exp\left(-2\frac{\xi}{a}\right)\,\mathrm{d}\xi$$

$$= \frac{1}{2a}\exp\left(-\frac{y}{a}\right)\,.$$

When $y \leq 0$,

$$P_Y(y) = \frac{1}{a^2} \exp\left(-\frac{y}{a}\right) \int_{-\infty}^{\infty} \theta(y+\xi) \exp\left(-2\frac{\xi}{a}\right) d\xi$$

$$= \frac{1}{a^2} \exp\left(-\frac{y}{a}\right) \int_{-y}^{\infty} \exp\left(-2\frac{\xi}{a}\right) d\xi$$

$$= \frac{a}{2a^2} \exp\left(-\frac{y}{a}\right) \exp\left(2\frac{y}{a}\right)$$

$$= \frac{1}{2a} \exp\left(\frac{y}{a}\right) .$$

This result can be written in the form

$$P_Y(y) = \frac{1}{2a} \exp\left(-\frac{|y|}{a}\right) .$$

for any value of y.

Solution to Exercise 4.5

(1) Put $P_r(r) = p\delta(r-1) + s\delta(r) + q\delta(r+1)$, where $\delta(x)$ is the Dirac distribution. The characteristic function is

$$\hat{P}_r(\nu) = \int_{-\infty}^{\infty} P_r(r) \exp(i\nu r) dr ,$$

and we deduce that

$$\hat{P}_r(\nu) = p\exp(i\nu) + s + q\exp(-i\nu) .$$

Since

$$\hat{P}_{R_n}(\nu) = \left[\hat{P}_r(\nu)\right]^n ,$$

we have

$$\hat{P}_{R_n}(\nu) = \left[p\exp(i\nu) + s + q\exp(-i\nu)\right]^n .$$

(2) First method: direct calculation. We have $R_n = \sum_{i=1}^{n} r_i$ and hence, $\langle R_n \rangle = \sum_{i=1}^{n} \langle r_i \rangle$. Now $\langle r_i \rangle = p-q$, whereupon $\langle R_n \rangle = n(p-q)$. Moreover, $\langle R_n^2 \rangle = \sum_{i=1}^{n} \sum_{j=1}^{n} \langle r_i r_j \rangle$. Now

$$\langle r_i r_j \rangle = (p+q)\delta_{i-j} + (p-q)^2(1 - \delta_{i-j}) ,$$

and thus

$$\langle R_n^2 \rangle = \sum_{i=1}^{n} \sum_{j=1}^{n} \left\{ [p+q - (p-q)^2]\delta_{i-j} + (p-q)^2 \right\} .$$

We then obtain

$$\langle R_n^2 \rangle = n\left\{ [p+q - (p-q)^2] + n(p-q)^2 \right\} ,$$

which gives the variance $\langle R_n^2 \rangle - \langle R_n \rangle^2 = n[p + q - (p - q)^2]$.

Second method: calculation using the characteristic function. We have

$$\hat{P}_{R_n}(\nu) = \left[p\exp(i\nu) + s + q\exp(-i\nu) \right]^n ,$$

$$\frac{\partial}{\partial \nu}\hat{P}_{R_n}(\nu) = n\left[ip\exp(i\nu) - iq\exp(-i\nu) \right]\left[p\exp(i\nu) + s + q\exp(-i\nu) \right]^{n-1} ,$$

and therefore

$$\frac{\partial}{\partial \nu}\hat{P}_{R_n}(0) = in(p - q) ,$$

from which we retrieve $\langle R_n \rangle = n(p - q)$. We also have

$$\frac{\partial^2}{\partial \nu^2}\hat{P}_{R_n}(\nu)$$

$$= n\left[-p\exp(i\nu) - q\exp(-i\nu) \right]\left[p\exp(i\nu) + s + q\exp(-i\nu) \right]^{n-1}$$
$$+ n(n-1)\left[ip\exp(i\nu) - iq\exp(-i\nu) \right]^2\left[p\exp(i\nu) + s + q\exp(-i\nu) \right]^{n-2},$$

and hence,

$$\frac{\partial^2}{\partial \nu^2}\hat{P}_{R_n}(0) = -n(p + q) - n(n-1)(p - q)^2 .$$

We do indeed retrieve $\langle R_n^2 \rangle = n(p + q) + n(n-1)(p - q)^2$.

Solution to Exercise 4.6

(1) $P(r)$ has finite first and second moments. We deduce that $R_n/\sqrt{n}$ will
be normally distributed. From the symmetry of $P(r)$, we have $\langle r_i \rangle = 0$
and so $\langle R_n \rangle = 0$. Further, $\langle r_i^2 \rangle = 2\int_0^\infty r^2 P(r)dr$, or $\langle r_i^2 \rangle = 2$, and hence,
$\langle R_n^2 \rangle = 2n$.
(2) $P(r)$ does not have finite first and second moments. The characteristic
function of $P(r)$ is (see Section 4.2)

$$\hat{P}(\nu) = \exp(-|\nu|) .$$

Since $R_n = \sum_{i=1}^n r_i$, we have

$$\hat{P}_{R_n}(\nu) = \exp(-n|\nu|) .$$

If we put $M_n = R_n/n$, we see that

$$\hat{P}_{M_n}(\nu) = \exp(-|\nu|) .$$

M_n therefore has the same probability density function as r. In part (1), we
had $\langle R_n^2 \rangle = 2n$ and hence, $\langle M_n^2 \rangle = 2/n$, which means that M_n converges
in quadratic mean toward a deterministic variable equal to zero.

Solution to Exercise 4.7

(1) We have

$$P(x,t) = \sum_{n=-\infty}^{+\infty} \frac{1}{\sqrt{2\pi t\sigma}} \exp\left[-\frac{(x-na)^2}{2\sigma^2 t}\right] .$$

(2) The restriction of the last solution to the interval $[0,1]$ is a solution of the partial differential equation which describes diffusion, with boundary condition that the derivatives of the concentration should be equal at 0 and 1 (for the problem is invariant under translation by whole numbers n). Consider a circle of unit circumference and let x be curvilinear coordinates defined on $[-1/2, 1/2]$. From the symmetry of the problem, the concentrations are equal at $-1/2$ and $1/2$ and the derivative is continuous. We thus have two problems governed by the same partial differential equation with the same boundary conditions and the same initial conditions. The solutions must therefore be the same. We deduce that, for a circle of radius R,

$$P(x,t) = \sum_{n=-\infty}^{+\infty} \frac{1}{\sqrt{2\pi t\sigma}} \exp\left[-\frac{(x-2n\pi R)^2}{2\sigma^2 t}\right] .$$

Solution to Exercise 4.8

(1) The characteristic functions of X_i and $Y_{i,\ell}$ are

$$\hat{P}_X(\nu) = \exp(-a|\nu|) ,$$

and

$$\hat{P}_Y(\nu) = \exp\left(-\frac{1}{2}\sigma^2\nu^2\right) .$$

Since $\hat{P}_Z(\nu) = \hat{P}_X(\nu)\left[\hat{P}_Y(\nu)\right]^L$, we obtain

$$\hat{P}_Z(\nu) = \exp\left(-a|\nu| - \frac{L}{2}\sigma^2\nu^2\right) ,$$

(2) Set $S_n = \sum_{i=1}^n Z_i$ and $R_n = S_n/n$. We deduce immediately that

$$\hat{P}_{S_n}(\nu) = \exp\left(-an|\nu| - \frac{Ln}{2}\sigma^2\nu^2\right) .$$

Now $\hat{P}_{R_n}(\nu) = \hat{P}_{S_n}(\nu/n)$ and hence,

$$\hat{P}_{R_n}(\nu) = \exp\left(-a|\nu| - \frac{L}{2n}\sigma^2\nu^2\right) .$$

(3) We thus see that when $n \to +\infty$ we obtain

$$\hat{P}_{R_n}(\nu) \to \exp(-a|\nu|) .$$

This result shows that the asymptotic behavior of the random walk is totally conditioned by the Cauchy distribution, i.e., by the large deviations corresponding here to the flea's jumps.

Solution to Exercise 4.9

(1) As the X_n are strictly positive, we can set $Z_n = \ln Y_n$. We then find that

$$Z_n = \sum_{i=1}^{n} \ln X_i .$$

If $m_{\log} = \int_{-\infty}^{\infty} P_X(x) \ln x \, dx$ and $\sigma_{\log}^2 = \int_{-\infty}^{\infty} P_X(x) \ln x^2 dx$ exist, we can apply the central limit theorem. For large n, the probability density function of Z_n is approximately

$$P_Z(z) = \frac{1}{\sqrt{2\pi}\sigma_{\log}} \exp\left[-\frac{(z - m_{\log})^2}{2\sigma_{\log}^2}\right] .$$

We have $Y_n = \exp Z_n$ and hence $P_Y(y)dy = P_Z(z)dz$, with $dy = ydz$. We thus obtain

$$P_Y(y) = \frac{1}{\sqrt{2\pi}y\sigma_{\log}} \exp\left[-\frac{(\ln y - m_{\log})^2}{2\sigma_{\log}^2}\right] ,$$

which corresponds to the log-normal distribution.

(2) Set $X_n = \epsilon_n U_n$, where U_n is the absolute value of X_n and ϵ_n is its sign. Since the probability of the sign ϵ_n is assumed to be independent of the probability density of the modulus U_n, Y_n can thus be written

$$Y_n = \prod_{i=1}^{n} X_i = \prod_{i=1}^{n} \epsilon_i U_i = \left(\prod_{i=1}^{n} \epsilon_i\right)\left(\prod_{i=1}^{n} U_i\right) .$$

Let p be the probability that X_i is positive and $q = 1 - p$ the probability that it is negative. The probability that $\epsilon_i = 1$ is thus p and the probability that $\epsilon_i = -1$ is $1 - p$. If $\Upsilon_n = \prod_{i=1}^{n} \epsilon_i$, when $n \to \infty$, the probability a that $\Upsilon_n = 1$ must be equal to the probability that $\Upsilon_{n-1} = 1$. Likewise, it follows that $P(\Upsilon_n = -1) = P(\Upsilon_{n-1} = -1)$. We deduce that a must satisfy the equations

$$ap + (1 - a)(1 - p) = a ,$$
$$a(1 - p) + (1 - a)p = (1 - a) ,$$

which can also be written

$$a(p-1) + (1-a)(1-p) = 0 \Longrightarrow (1-2a)(1-p) = 0 \ ,$$
$$a(1-p) + (1-a)(p-1) = 0 \Longrightarrow (1-2a)(p-1) = 0 \ .$$

Therefore $a = 1/2$. The probability that X_n is positive is then equal to the probability that it is negative (if $p \neq 0$ and $p \neq 1$). The asymptotic distribution of $\prod_{i=1}^{n} U_i$ was determined in (1). We deduce that

$$P_Y(y) = \frac{1}{2\sqrt{2\pi}|y|\sigma_{\log}} \exp\left[-\frac{(\ln|y| - m_{\log})^2}{2\sigma_{\log}^2} \right] \ .$$

9.4 Chapter Five. Information and Fluctuations

Solution to Exercise 5.1

(1) Since the probability must be a positive quantity, we know that

$$1 \geq \frac{1}{N} + (N-1)\alpha \geq 0 \quad \text{and} \quad 1 \geq \frac{1}{N} - \alpha \geq 0 \ .$$

We thus deduce that

$$-\frac{1}{N(N-1)} \leq \alpha \leq \frac{1}{N} \ .$$

(2) We have

$$S = -\left[\frac{1}{N} + (N-1)\alpha \right] \ln \left[\frac{1}{N} + (N-1)\alpha \right]$$
$$- \left[1 - \frac{1}{N} - (N-1)\alpha \right] \ln \left(\frac{1}{N} - \alpha \right) \ .$$

(3) We know that the entropy is maximal when there is equiprobability, i.e., when $\alpha = 0$. This can be checked by setting $\partial S / \partial \alpha = 0$.

Solution to Exercise 5.2

(1) We have
$$p_E(e) = \frac{1}{\pi^2 \det \overline{\overline{\Gamma}}} \exp\left(-e^\dagger \overline{\overline{\Gamma}}^{-1} e \right) \ ,$$

where $\det \overline{\overline{\Gamma}}$ is the determinant of $\overline{\overline{\Gamma}}$.

(2) The entropy of the system is $S = -\int p_E(e) \ln p_E(e) \mathrm{d}e$, and hence,

$$S = \ln \left(\pi^2 \det \overline{\overline{\Gamma}} \right) + \int \left(e^\dagger \overline{\overline{\Gamma}}^{-1} e \right) p_E(e) \mathrm{d}e \ .$$

Now,

$$\int \left(e^{\dagger}\overline{\overline{\Gamma}}^{-1}e\right) p_E(e)\mathrm{d}e = 2 \,,$$

and therefore,

$$S = \ln\left(\pi^2 \det \overline{\overline{\Gamma}}\right) + 2 = \ln\left(\pi^2 e^2 \det \overline{\overline{\Gamma}}\right) \,.$$

(3) The degree of polarization is defined by

$$\mathcal{P}_2 = \sqrt{1 - 4\frac{\det \overline{\overline{\Gamma}}}{\mathrm{tr}\overline{\overline{\Gamma}}^2}} \,,$$

where $\mathrm{tr}\overline{\overline{\Gamma}}$ is the trace of $\overline{\overline{\Gamma}}$. We thus have $1 - \mathcal{P}_2^2 = 4\det \overline{\overline{\Gamma}}/\mathrm{tr}\overline{\overline{\Gamma}}^2$. Now the total intensity I_0 is equal to $\mathrm{tr}\overline{\overline{\Gamma}}$ and so

$$S = \ln\left[\pi^2 e^2 I_0^2 (1 - \mathcal{P}_2^2)/4\right] \,.$$

(4) In d dimensions, the probability density is

$$p_E(e) = \frac{1}{\pi^d \det \overline{\overline{\Gamma}}} \exp\left(-e^{\dagger}\overline{\overline{\Gamma}}^{-1}e\right) \,.$$

The entropy will therefore be

$$S_d = \ln\left(\pi^d e^d \det \overline{\overline{\Gamma}}\right) \,.$$

We now introduce the dimensionless quantity $\alpha = \det \overline{\overline{\Gamma}}/\mathrm{tr}\overline{\overline{\Gamma}}^d$, whereupon we obtain

$$S_d = \ln\left(\pi^d e^d \alpha I_0^d\right) \,.$$

This expression can be expanded to bring out a term more directly related to the disorder, since it turns out to be the same for all situations which differ only by a multiplicative factor:

$$S_d = d + d\ln(\pi I_0) + \ln \alpha \,.$$

There are several ways to define a degree of polarization. In order to push the analysis a little further, it is interesting to express the last term δS_d of the entropy in terms of the eigenvalues $\lambda_1, \lambda_2, \ldots, \lambda_d$ of the covariance matrix:

$$\delta S_d = \ln \alpha = \ln\left[\frac{\lambda_1 \lambda_2 \cdots \lambda_d}{(\lambda_1 + \lambda_2 + \cdots + \lambda_d)^d}\right] \,.$$

In 3 dimensions, we obtain

$$S_3 = 3 + 3\ln(\pi I_0) + \ln\left[\frac{\lambda_1\lambda_2\lambda_3}{(\lambda_1 + \lambda_2 + \lambda_3)^3}\right] \ .$$

By analogy with the 2-dimensional case, we seek a definition of the degree of polarization in the form

$$1 - \mathcal{P}_3^n = u\frac{\lambda_1\lambda_2\lambda_3}{(\lambda_1 + \lambda_2 + \lambda_3)^3} \ .$$

If we are to have $\mathcal{P}_3 = 0$ when $\lambda_1 = \lambda_2 = \lambda_3$, it is clear that we must choose $u = 27$. The entropy is then

$$S_3 = 3 + 3\ln(\pi I_0) + \ln\left[(1 - \mathcal{P}_3^n)/27\right] \ .$$

Let us analyze the case where the polarization vector varies in a plane. In this case, there will be one zero eigenvalue, say λ_3. We observe that the entropy tends to $-\infty$. This divergence arises because we are considering a probability density, i.e., continuous random variables. In reality there will always be a lower bound for λ_3, which we denote by ϵ. We then have

$$S_3 = 3 + 3\ln(\pi I_0) + \ln\left[\frac{\epsilon}{(\lambda_1 + \lambda_2 + \epsilon)}\right] + \ln\left[\frac{\lambda_1\lambda_2}{(\lambda_1 + \lambda_2 + \epsilon)^2}\right] \ .$$

This implies that

$$S_3 \simeq 3 + 3\ln\pi + 2\ln I_0 + \ln\epsilon + \ln\frac{1 - \mathcal{P}_2^2}{4} \ ,$$

where $\mathcal{P}_2$ is the degree of polarization in the plane. The entropy of the polarization in two dimensions is

$$S_2 = 2 + 2\ln\pi + 2\ln I_0 + \ln\frac{1 - \mathcal{P}_2^2}{4} \ ,$$

and hence,

$$S_3 \simeq 1 + \ln\pi + \ln\epsilon + S_2 \ .$$

Solution to Exercise 5.3

In the case of discrete probability laws, we have

$$K_{\mathrm{u}}(P_a\|P_b) = \sum_{n=-\infty}^{+\infty} P_a(n)\ln\left[\frac{P_a(n)}{P_b(n)}\right] \ .$$

Consider a continuous variable X_λ and two probability density functions $P_a^{(c)}(x)$ et $P_b^{(c)}(x)$ which are themselves continuous. We now quantize the range of variation of this random variable with a unit δ, which amounts to applying the transformation

$$X_\lambda \mapsto Y_\lambda = N_\lambda \delta ,$$

where N_λ is an integer-valued random variable defined in terms of X_λ by $X_\lambda \in [N_\lambda \delta - \delta/2, N_\lambda \delta + \delta/2]$. Y_λ is a discrete random variable isomorphic to N_λ. If the probability density function of X_λ is $P_a^{(c)}(x)$, the probability law of N_λ is

$$P_a(n) = \int_{n\delta-\delta/2}^{n\delta+\delta/2} P_a^{(c)}(x)\mathrm{d}x .$$

Likewise, if the probability density function of X_λ is $P_b^{(c)}(x)$, the probability law of N_λ is

$$P_b(n) = \int_{n\delta-\delta/2}^{n\delta+\delta/2} P_b^{(c)}(x)\mathrm{d}x .$$

The Kullback measure of the two probability laws for N_λ is

$$K_{\mathrm{u}}(P_a\|P_b) = \sum_{n=-\infty}^{+\infty} P_a(n) \ln\left[\frac{P_a(n)}{P_b(n)}\right] .$$

If δ is small enough, we can write $P_a(n) \approx P_a^{(c)}(n\delta)\delta$ and $P_b(n) \approx P_b^{(c)}(n\delta)\delta$. We then have

$$K_{\mathrm{u}}(P_a\|P_b) = \sum_{n=-\infty}^{+\infty} P_a^{(c)}(n\delta)\delta \ln\left[\frac{P_a^{(c)}(n\delta)\delta}{P_b^{(c)}(n\delta)\delta}\right]$$

$$= \sum_{n=-\infty}^{+\infty} P_a^{(c)}(n\delta)\delta \ln\left[\frac{P_a^{(c)}(n\delta)}{P_b^{(c)}(n\delta)}\right] .$$

Using the Riemann approximation to the integral, we then have

$$K_{\mathrm{u}}(P_a\|P_b) = \int_{-\infty}^{+\infty} P_a^{(c)}(x) \ln\left[\frac{P_a^{(c)}(x)}{P_b^{(c)}(x)}\right] \mathrm{d}x .$$

The Kullback measure is essentially a comparison beween two probability laws and behaves as the difference between two entropies. There is therefore no divergence problem such as arises in the definition of the entropy of a probability density for a continuous random variable.

Solution to Exercise 5.4

(1) **Scalar Gaussian Distributions.** The probability laws $P_a(x)$ and $P_b(x)$ are Gaussian. We thus have

$$P_a(x) = \frac{1}{\sqrt{2\pi}\sigma_a} \exp\left[-\frac{(x-m_a)^2}{2\sigma_a^2}\right] \ , \quad P_b(x) = \frac{1}{\sqrt{2\pi}\sigma_b} \exp\left[-\frac{(x-m_b)^2}{2\sigma_b^2}\right] \ ,$$

and we obtain

$$K_{\mathrm{u}}(P_a\|P_b) = \langle \ln P_a(x)\rangle_a - \langle \ln P_b(x)\rangle_a \ ,$$

where $\langle \ \rangle_a$ indicates the expectation value with the probability law $P_a(x)$. It is easy to show that

$$\langle \ln P_a(x)\rangle_a = -\ln\sqrt{2\pi} - \ln\sigma_a - \frac{1}{2} \ ,$$

$$\langle \ln P_b(x)\rangle_a = -\ln\sqrt{2\pi} - \ln\sigma_b - \frac{\sigma_a^2 + (m_a - m_b)^2}{2\sigma_b^2} \ ,$$

and consequently,

$$K_{\mathrm{u}}(P_a\|P_b) = \frac{1}{2}\frac{(m_a - m_b)^2}{\sigma_b^2} + \frac{1}{2}\left(\ln\frac{\sigma_b^2}{\sigma_a^2} + \frac{\sigma_a^2}{\sigma_b^2} - 1\right) \ .$$

(2) **Gamma Probability Laws of Different Orders.** The probability laws are

$$P_a(x) = \frac{x^{L_a-1}}{\Gamma(L_a)\mu_a^{L_a}} \exp\left(-\frac{x}{\mu_a}\right) \ , \quad P_b(x) = \frac{x^{L_b-1}}{\Gamma(L_b)\mu_b^{L_b}} \exp\left(-\frac{x}{\mu_b}\right) \ .$$

The Kullback measure is thus

$$K_{\mathrm{u}}(P_a\|P_b) = (L_a - L_b)f(L_a) + L_b\ln\frac{\mu_b}{\mu_a} + L_a\left(\frac{\mu_b}{\mu_a} - 1\right) + \ln\frac{\Gamma(L_b)}{\Gamma(L_a)}$$

where $f(L) = \langle \ln x_L\rangle$ and x_L is a Gamma-distributed random variable with mean and order L.

(3) **Poisson Laws.** The discrete probability laws are

$$P_a(n) = \mathrm{e}^{-\lambda_a}\frac{\lambda_a^n}{n!} \ , \quad P_b(n) = \mathrm{e}^{-\lambda_b}\frac{\lambda_b^n}{n!} \ .$$

To calculate the Kullback measure, we consider a discrete sum:

$$\begin{aligned}
K_{\mathrm{u}}(P_a\|P_b) &= \sum_{n=0}^{+\infty} P_a(n)\ln\frac{P_a(n)}{P_b(n)} \\
&= \langle \ln P_a(n) - \ln P_b(n)\rangle_a \\
&= \lambda_a\left(\frac{\lambda_b}{\lambda_a} - \ln\frac{\lambda_b}{\lambda_a} - 1\right) \ .
\end{aligned}$$

Here again, we can check that this measure is always positive.

(4) **Geometric Probability Laws.** Consider the discrete geometric laws

$$P_a(n) = (1 - \alpha_a)\alpha_a^n , \quad P_b(n) = (1 - \alpha_b)\alpha_b^n .$$

We have

$$K_{\mathrm{u}}(P_a\|P_b) = \langle \ln P_a(n) - \ln P_b(n)\rangle_a$$
$$= \ln \frac{1 - \alpha_a}{1 - \alpha_b} + \langle n\rangle_a \ln \frac{\alpha_a}{\alpha_b} .$$

It can then be shown that

$$\langle n\rangle_a = \frac{\alpha_a}{(1 - \alpha_a)} ,$$

$$K_{\mathrm{u}}(P_a\|P_b) = \ln \frac{1 - \alpha_a}{1 - \alpha_b} + \frac{\alpha_a}{1 - \alpha_a} \ln \frac{\alpha_a}{\alpha_b} .$$

Solution to Exercise 5.5

(1) We seek the probability law $P_s(n)$ which minimizes $K_{\mathrm{u}}(P_s\|P_b)$ for a given value of $K_{\mathrm{u}}(P_s\|P_a)$. The problem can thus be formulated as minimizing

$$K_{\mathrm{u}}(P_s\|P_b)$$

with the constraints

$$K_{\mathrm{u}}(P_s\|P_a) = A \quad \text{and} \quad \sum_{j=1}^{+\infty} P_s(n) = 1 .$$

The appropriate Lagrange function is

$$\Psi = K_{\mathrm{u}}(P_s\|P_b) - \mu K_{\mathrm{u}}(P_s\|P_a) - \lambda \sum_{j=1}^{+\infty} P_s(n) ,$$

where μ and λ are the Lagrange parameters. We then have

$$\Psi = \sum_{n=1}^{+\infty} P_s(n) \ln \frac{P_s(n)}{P_b(n)} - \mu \sum_{n=1}^{+\infty} P_s(n) \ln \frac{P_s(n)}{P_a(n)} - \lambda \sum_{j=1}^{+\infty} P_s(n) ,$$

and the law P_s is obtained from

$$\frac{\partial \Psi}{\partial P_s(n)} = 0 = P_s(n)\frac{1}{P_s(n)} + \ln \frac{P_s(n)}{P_b(n)} - \mu P_s(n)\frac{1}{P_s(n)} - \mu \ln \frac{P_s(n)}{P_a(n)} - \lambda .$$

We then find

$$\ln P_s(n) = \frac{\mu + \lambda - 1}{1 - \mu} + \ln \left[P_b(n)^{1/(1-\mu)} P_a(n)^{-\mu/(1-\mu)} \right] ,$$

and hence,

$$P_s(n) = P_b^{1/(1-\mu)}(n) P_a^{-\mu/(1-\mu)}(n) \exp\left(\frac{\mu + \lambda - 1}{1 - \mu}\right) .$$

λ is the Lagrange parameter associated with the constraint $\sum_{j=1}^{+\infty} p_s(n) = 1$ and we can therefore say that

$$P_s(n) = \frac{1}{C(s)} P_b^s(n) P_a^{1-s}(n) ,$$

with

$$C(s) = \sum_{n=1}^{+\infty} P_b^s(n) P_a^{1-s}(n) ,$$

and $s = 1/(1 - \mu)$.

(2) We have

$$K_u(P_s \| P_a) = \sum_{n=1}^{+\infty} P_s(n) \left[\ln P_s(n) - \ln P_a(n) \right] ,$$

and

$$K_u(P_s \| P_b) = \sum_{n=1}^{+\infty} P_s(n) \left[\ln P_s(n) - \ln P_b(n) \right] .$$

The value of s^* which corresponds to $K_u(P_{s^*} \| P_a) = K_u(P_{s^*} \| P_b)$ thus satisfies

$$\sum_{n=1}^{+\infty} P_{s^*}(n) \ln P_a(n) = \sum_{n=1}^{+\infty} P_{s^*}(n) \ln P_b(n) .$$

The value of s which minimizes $C(s)$ satisfies $dC(s)/ds = 0$, i.e.,

$$\sum_{n=1}^{+\infty} \frac{d}{ds} \left[P_b^s(n) P_a^{1-s}(n) \right] = 0 .$$

This can be written

$$\sum_{n=1}^{+\infty} \frac{d}{ds} \exp\left[s \ln P_b(n) + (1 - s) \ln P_a(n) \right] = 0 ,$$

or

$$\sum_{n=1}^{+\infty} \left[\ln P_b(n) - \ln P_a(n) \right] \exp\left[s \ln P_b(n) + (1 - s) \ln P_a(n) \right] = 0 ,$$

and hence,

$$\sum_{n=1}^{+\infty} P_s(n) \left[\ln P_b(n) - \ln P_a(n) \right] = 0 .$$

It is thus clear that the value of s^* which minimizes $C(s)$ is the same as that for which we have $K_u(P_{s^*}\|P_a) = K_u(P_{s^*}\|P_b)$. [One can show that it is indeed a minimum by checking that $\partial^2 C/\partial s^2 \geq 0$.]

(3) We have

$$K_u(P_{s^*}\|P_a) = \sum_{n=1}^{+\infty} P_{s^*}(n)\big[\ln P_{s^*}(n) - \ln P_b(n)\big] \,,$$

and

$$P_{s^*}(n) = P_b^{s^*}(n)P_a^{1-s^*}(n)/C(s^*) \,.$$

Hence,

$$K_u(P_{s^*}\|P_a) = \sum_{n=1}^{+\infty} \frac{P_b^{s^*}(n)P_a^{1-s^*}(n)}{C(s^*)}\left[\ln\frac{P_b^{s^*}(n)P_a^{1-s^*}(n)}{C(s^*)} - \ln P_b(n)\right] \,,$$

or

$$K_u(P_{s^*}\|P_a) = \sum_{n=1}^{+\infty} \frac{P_b^{s^*}(n)P_a^{1-s^*}(n)}{C(s^*)}$$
$$\times\left[s^*\ln P_b(n) + (1 - s^*)\ln P_a(n) - \ln C(s^*) - \ln P_b(n)\right] \,,$$

whereupon,

$$K_u(P_{s^*}\|P_a) = \sum_{n=1}^{+\infty} \frac{P_b^{s^*}(n)P_a^{1-s^*}(n)}{C(s^*)}$$
$$\times\left\{(1 - s^*)\big[\ln P_a(n) - \ln P_b(n)\big] - \ln C(s^*)\right\} \,.$$

Now,

$$\sum_{n=1}^{+\infty} \frac{P_b^{s^*}(n)P_a^{1-s^*}(n)}{C(s^*)}\big[\ln P_a(n) - \ln P_b(n)\big] = 0 \,,$$

so that

$$K_u(P_{s^*}\|P_a) = \sum_{n=1}^{+\infty} \frac{P_b^{s^*}(n)P_a^{1-s^*}(n)}{C(s^*)}\big[-\ln C(s^*)\big] = -\ln C(s^*) \,.$$

$\Upsilon(P_a\|P_b) = -\ln C(s^*)$ is the Chernov measure for $P_a(n)$ and $P_b(n)$, where

$$\Upsilon(P_a\|P_b) = -\ln\left[\sum_{n=1}^{+\infty} P_b^{s^*}(n)P_a^{1-s^*}(n)\right] \,.$$

(4) In the case of probability density functions, we have

$$\Upsilon(P_a\|P_b) = -\ln\left[\int_{-\infty}^{+\infty} P_b^{s^*}(x)P_a^{1-s^*}(x)\mathrm{d}x\right] \,.$$

Solution to Exercise 5.6

(1) We have

$$
\frac{d}{ds}\ln C(s) = \frac{d}{ds}\ln\left[\sum_{n=1}^{+\infty} P_b^s(n)P_a^{1-s}(n)\right]
$$

$$
= \frac{1}{C(s)}\frac{d}{ds}\sum_{n=1}^{+\infty}\exp\left[s\ln P_b(n) + (1-s)\ln P_a(n)\right]
$$

$$
= \frac{1}{C(s)}\sum_{n=1}^{+\infty}\left[\ln P_b(n) - \ln P_a(n)\right]
$$

$$
\times \exp\left[s\ln P_b(n) + (1-s)\ln P_a(n)\right].
$$

Hence,

$$
\frac{d}{ds}\ln C(s) = \frac{1}{C(s)}\sum_{n=1}^{+\infty} P_b^s(n)P_a^{1-s}(n)\ln\frac{P_b(n)}{P_a(n)}.
$$

Since $C(0) = 1$, we have

$$
\left.\frac{d}{ds}\ln C(s)\right|_{s=0} = \sum_{n=1}^{+\infty} P_a(n)\ln\frac{P_b(n)}{P_a(n)}.
$$

Now,

$$
K_u(P_a\|P_b) = \sum_{n=1}^{+\infty} P_a(n)\ln\frac{P_a(n)}{P_b(n)}.
$$

We therefore deduce that

$$
\left.\frac{d}{ds}\ln C(s)\right|_{s=0} = -K_u(P_a\|P_b).
$$

Noting also that $C(1) = 1$, we have

$$
\left.\frac{d}{ds}\ln C(s)\right|_{s=1} = \sum_{n=1}^{+\infty} P_b(n)\ln\frac{P_b(n)}{P_a(n)},
$$

and hence,

$$
\left.\frac{d}{ds}\ln C(s)\right|_{s=1} = K_u(P_b\|P_a).
$$

(2) We set

$$
\ln C(s) \simeq A + Bs + Cs^2.
$$

Now $\ln C(0) = 0$, so $A = 0$. Moreover, $\ln C(1) = 0$, whereupon $B + C = 0$, and

$$
\ln C(s) \simeq Bs - Bs^2 = Bs(1 - s).
$$

We see that the minimum is obtained in this approximation when $B - 2Bs = 0$, i.e., $s = 1/2$. We then obtain

$$\mathcal{B}(P_a \| P_b) = - \ln \left[\sum_{n=1}^{+\infty} \sqrt{P_b(n) P_a(n)} \right].$$

9.5 Chapter Six. Statistical Physics

Solution to Exercise 6.1

(1) $U = qE_1 + pE_2$.
(2) $\sigma_U^2 = pq(E_1 - E_2)^2$.
(3) Since $q = 1 - p$, we find $p = q = 1/2$, which corresponds to the case where the states have the same probability of being occupied.
(4) In thermodynamic units, we have $S = -k_B p \ln p - k_B(1 - p) \ln(1 - p)$.
(5) $\partial S / \partial p = 0 \Longrightarrow p = 1/2$. The entropy is thus maximal when the energy fluctuations are maximal.
(6) $T \to 0 \Longrightarrow \beta \to +\infty$ and hence, $p \to 0$. We deduce that $\sigma_U^2 \to 0$.
(7) $T \to +\infty \Longrightarrow \beta \to 0$ and hence, $q \to 1/2$ and $p \to 1/2$. We deduce that $\sigma_U^2 \to (E_1 - E_2)^2/4$. The variance of the energy fluctuations is maximal.
(8) p is an increasing function of the temperature T and $p = 1/2$ when $T = +\infty$. The entropy is an increasing function of p when p lies in the interval $[0, 1/2]$. It is therefore an increasing function of T.
(9) We have

$$P_\beta(X) = \frac{1}{Z_\beta} \exp \left[-\beta \sum_n E(n) \right],$$

where

$$Z_\beta = \sum_{E(1)=E_1, E_2} \sum_{E(2)=E_1, E_2} \cdots \sum_{E(N)=E_1, E_2} \exp \left[-\beta \sum_n E(n) \right].$$

When the temperature tends to zero, each particle tends to be in the state with the lowest energy. Only the state with $E(n) = E_1$, $\forall n = 1, \ldots, N$ then has nonzero probability.

Solution to Exercise 6.2

(1) We have

$$Z_\beta = \sum_{E(1)=\epsilon_1, \epsilon_2} \sum_{E(2)=\epsilon_1, \epsilon_2} \cdots \sum_{E(N)=\epsilon_1, \epsilon_2} \exp \left[-\beta \sum_n E(n) \right]$$

$$= \sum_{x_1=-h, h} \sum_{x_2=-h, h} \cdots \sum_{x_N=-h, h} \exp \left[-\beta \sum_n (E_0 + \alpha x_n) \right].$$

If we set

$$z_\beta = \exp\left[-\beta(E_0 + \alpha h)\right] + \exp\left[-\beta(E_0 - \alpha h)\right],$$

then $Z_\beta = z_\beta^N$. We have

$$z_\beta = 2\exp(-\beta E_0)\cosh(\alpha\beta h),$$

where cosh is the hyperbolic cosine. It follows that

$$Z_\beta = \exp(-\beta N E_0) 2^N \cosh(\alpha\beta h)^N.$$

(2) We have

$$U_\beta = -\frac{\partial}{\partial\beta}\ln Z_\beta.$$

But

$$\ln Z_\beta = -\beta N E_0 + N\ln 2 + N\ln\cosh(\alpha\beta h),$$

and hence,

$$U_\beta = N E_0 - N\alpha h\frac{e^{\alpha\beta h} - e^{-\alpha\beta h}}{e^{\alpha\beta h} + e^{-\alpha\beta h}}$$
$$= N\left[E_0 - \alpha h\tanh(\alpha\beta h)\right],$$

where tanh is the hyperbolic tangent.

(3) When $\beta h \to -\infty$, $\tanh(\alpha\beta h) \to -1$, and so $U_\beta \simeq N(E_0 + \alpha h)$. Likewise, when $\beta h \to \infty$, $\tanh(\alpha\beta h) \to 1$, and so $U_\beta \simeq N(E_0 - \alpha h)$. We can thus say that, when $\beta|h| \to +\infty$,

$$U_\beta = N(E_0 - \alpha|h|),$$

where $|h|$ is the absolute value h. We thus find that $U_\beta < N E_0$.

(4) We have $\beta h = h/kT$. If $T \to 0$, it follows that the particles will be located for the main part in the state with lowest energy, viz., $E_0 - \alpha|h|$ and the total energy is of the order of $N(E_0 - \alpha|h|)$. There is an analogous situation if, at finite temperature, the modulus $|h|$ of the field tends to $+\infty$.

Solution to Exercise 6.3

(1) We have

$$P_\ell = \frac{\exp\left[-\beta(E_0 - hm_\ell)\right]}{\sum_{\ell=-n}^{n}\exp\left[-\beta(E_0 - hm_\ell)\right]},$$

and hence,

$$\langle m\rangle = \sum_{\ell=-n}^{n} m_\ell \frac{\exp\left[-\beta(E_0 - hm_\ell)\right]}{\sum_{\ell=-n}^{n}\exp\left[-\beta(E_0 - hm_\ell)\right]}.$$

On the other hand,

$$\frac{\partial}{\partial h} \ln \left\{ \sum_{\ell=-n}^{n} \exp\left[-\beta(E_0 - hm_\ell) \right] \right\}$$

$$= \frac{\sum_{\ell=-n}^{n} \beta m_\ell \exp\left[-\beta(E_0 - hm_\ell) \right]}{\sum_{\ell=-n}^{n} \exp\left[-\beta(E_0 - hm_\ell) \right]} ,$$

and

$$\ln Z_{\beta,h} = N \ln \left\{ \sum_{\ell=-n}^{n} \exp\left[-\beta(E_0 - hm_\ell) \right] \right\} .$$

It does indeed follow that

$$\langle m \rangle = \frac{1}{N\beta} \frac{\partial \ln Z_{\beta,h}}{\partial h} .$$

(2) Likewise, we show that

$$\frac{1}{N} \frac{\partial^2 \ln Z_{\beta,h}}{\partial h^2} = \beta^2 \left[\sum_{\ell=-n}^{n} m_\ell^2 P_\ell - \left(\sum_{\ell=-n}^{n} m_\ell P_\ell \right)^2 \right] ,$$

or

$$\frac{1}{N\beta^2} \frac{\partial^2 \ln Z_{\beta,h}}{\partial h^2} = \left[\langle m^2 \rangle - \langle m \rangle^2 \right] .$$

Solution to Exercise 6.4

(1) We have $Z_\beta = z_\beta^N$, where

$$z_\beta = \sum_{n=0}^{+\infty} \exp(-\beta mgpn) = \frac{1}{1 - \exp(-\beta mgp)} ,$$

which implies that

$$\ln Z_\beta = -N \ln \left[1 - \exp(-\beta mgp) \right] .$$

The free energy is $F_\beta = -k_B T \ln Z_\beta$, whereby we obtain

$$F_\beta = N k_B T \ln \left[1 - \exp(-\beta mgp) \right] .$$

The entropy is given by $S_\beta = -\partial F_\beta / \partial T$, and a short calculation yields

$$S_\beta = -N k_B \ln \left[1 - \exp(-\beta mgp) \right] + \frac{mgpN}{T} \frac{\exp(-\beta mgp)}{1 - \exp(-\beta mgp)} .$$

(2) When $T \to 0$, $\exp(-\beta mgp) \to 0$ and $(1/T)\exp(-\beta mgp) \to 0$. Therefore, $S_\beta \to 0$, because

$$S_\beta \simeq \frac{mgpN}{T}\exp(-\beta mgp) \,.$$

(3) When $T \to +\infty$, $\beta \to 0$, $\exp(-\beta mgp) \simeq 1 - \beta mgp$ and hence,

$$S_\beta \simeq -Nk_{\mathrm{B}}\ln(\beta mgp) + \frac{mgpN}{\beta mgpT}(1 - \beta mgp)$$

$$\simeq Nk_{\mathrm{B}}\left[\ln(k_{\mathrm{B}}T) + 1 - \ln(mgp)\right] \,.$$

(4) When $p \to 0$, $S_\beta \to +\infty$. This is understandable because, when $p \to 0$, the number of degrees of freedom diverges as we tend toward a continuum of states.

Solution to Exercise 6.5

The energy of the j th particle can be written in the form

$$E(j) = x_j E_B + (1 - x_j)E_A \,,$$

where x_j describes the state of the j th site, x_j is equal to 0 if the adsorbed particle is of type A and 1 if it is of type B. The Gibbs distribution leads to

$$P_\beta(x_1, x_2, \ldots, x_N) = \frac{1}{Z_\beta}\exp\left\{-\beta\sum_{j=1}^{N}[x_j E_B + (1 - x_j)E_A]\right\} \,,$$

where

$$Z_\beta = \sum_{x_1=0,1}\sum_{x_2=0,1}\cdots\sum_{x_N=0,1}\exp\left\{-\beta\sum_{j=1}^{N}[x_j E_B + (1 - x_j)E_A]\right\} \,.$$

We deduce that the probability of observing the j th site in state x_j is

$$P_{j,\beta}(x_j) = \frac{1}{z_\beta}\exp\left\{-\beta[x_j E_B + (1 - x_j)E_A]\right\} \,,$$

where

$$z_\beta = \sum_{x_j=0,1}\exp\left\{-\beta[x_j E_B + (1 - x_j)E_A]\right\} \,.$$

The probability of finding a particle in state A is thus $P_A = \exp(-\beta E_A)/z_\beta$, whilst the probability of finding one in state B is $P_B = \exp(-\beta E_B)/z_\beta$. If N is the total number of particles, we must therefore have

$$C_A(T) \propto \frac{1}{z_\beta}\exp(-\beta E_A) \,,$$

and

$$C_B(T) \propto \frac{1}{z_\beta} \exp(-\beta E_B) \, .$$

We thus find that

$$\ln \frac{C_A(T)}{C_B(T)} = \beta(E_B - E_A) \, ,$$

where $\beta = 1/k_B T$.

Solution to Exercise 6.6

(1) We have

$$z_\beta = \sum_{n=0}^{+\infty} e^{-\beta n E_0} = \sum_{n=0}^{+\infty} a^n \, ,$$

where $a = e^{-\beta E_0}$. We deduce that $\ln z_\beta = -\ln(1-a)$, and since $Z_\beta = z_\beta^N$,

$$\ln Z_\beta = -N \ln \left(1 - e^{-\beta E_0}\right) \, .$$

(2) $F_\beta = -k_B T \ln Z_\beta$, and therefore

$$F_\beta = k_B T N \ln \left(1 - e^{-\beta E_0}\right) \, .$$

Since $U_\beta = -\partial \ln Z_\beta / \partial \beta$, it follows that

$$U_\beta = N E_0 \frac{e^{-\beta E_0}}{1 - e^{-\beta E_0}} \, .$$

In thermodynamic units $S_\beta = -\partial F_\beta / \partial T$, so that

$$S_\beta = -k_B N \ln \left(1 - e^{-\beta E_0}\right) + k_B N \beta E_0 \frac{e^{-\beta E_0}}{1 - e^{-\beta E_0}} \, .$$

(3) When $T \to +\infty$, we have $\beta E_0 \to 0$ and hence, $e^{-\beta E_0} \simeq 1 - \beta E_0$. We can then write

$$S_\beta \simeq -k_B N \ln(\beta E_0) \simeq k_B N \ln \frac{k_B T}{E_0} \, .$$

When $T \to 0$, we have $\beta E_0 \to +\infty$ and hence, $e^{-\beta E_0} \to 0$. It follows that

$$S_\beta = k_B N \beta E_0 e^{-\beta E_0} \, .$$

We retrieve this result by considering that only states $n = 0$ and $n = 1$ are occupied, with probabilities

$$P(0) = \frac{1}{1 + e^{-\beta E_0}} \quad \text{and} \quad P(1) = \frac{e^{-\beta E_0}}{1 + e^{-\beta E_0}} \, .$$

Solution to Exercise 6.7

We have

$$U_\beta = \frac{\sum_X H(X) \exp\left[-\beta H(X)\right]}{\sum_X \exp\left[-\beta H(X)\right]} \,,$$

whereupon

$$\frac{\partial U_\beta}{\partial \beta} = \frac{1}{\left[\sum_X \exp\left[-\beta H(X)\right]\right]^2}$$

$$\times \left\{\left[\sum_X \exp\left[-\beta H(X)\right]\right]\left[-\sum_X H(X)^2 \exp\left[-\beta H(X)\right]\right]\right.$$

$$\left. + \left[\sum_X H(X) \exp\left[-\beta H(X)\right]\right]^2\right\} \,,$$

or

$$\frac{\partial U_\beta}{\partial \beta} = -\langle H(X)^2 \rangle + \langle H(X) \rangle^2 \,.$$

We conclude that $a = -1$, i.e., $\sigma_U^2 = -\partial U_\beta/\partial\beta$.

Solution to Exercise 6.8

(1) The fluctuation–dissipation theorem implies that

$$\Gamma_{MM}(t) = k_B T \sigma(t) \quad \text{for} \quad t \geq 0 \,,$$

where $\sigma(t)$ is the response function. Therefore,

$$\sigma(t) = \begin{cases} \dfrac{A}{k_B T} \exp\left(-\dfrac{t^2}{2\tau^2}\right) & \text{if} \quad t \geq 0 \,, \\ 0 & \text{if} \quad t < 0 \,. \end{cases}$$

Since $\sigma(0) = \chi_0$, where χ_0 is the static susceptibility, we deduce that $A/k_B T = \chi_0$.

(2) The impulse response is $\chi(t) = -d\sigma(t)/dt$, and therefore

$$\chi(t) = \begin{cases} \chi_0 \dfrac{t}{\tau^2} \exp\left(-\dfrac{t^2}{2\tau^2}\right) & \text{if} \quad t \geq 0 \,, \\ 0 & \text{if} \quad t < 0 \,. \end{cases}$$

Solution to Exercise 6.9

(1) We saw in Section 6.7 that the spectral density of charges $\hat{S}_{QQ}(\nu)$ at the terminals of an RC circuit is given by

$$\hat{S}_{QQ}(\nu) = 2k_{\mathrm{B}}T\frac{RC^2}{1+(2\pi RC\nu)^2} \ .$$

(2) We have

$$\lim_{R\to 0} \hat{S}_{QQ}(\nu) = 0 \ .$$

(3) We saw in Section 6.4 that the total power of the fluctuations in an extensive quantity $M(X)$ is given by

$$\langle [M(X)]^2\rangle - [\langle M(X)\rangle]^2 = \frac{\chi_\beta}{\beta} \ ,$$

where χ_β is the static susceptibility. Since we have $Q = CV$ in the static case, it follows that

$$\langle [Q(X)]^2\rangle - [\langle Q(X)\rangle]^2 = k_{\mathrm{B}}TC \ .$$

(4) We observe that the total power of the fluctuations on the capacitor plates is independent of the value of the resistance. However, we must have

$$\langle [Q(X)]^2\rangle - [\langle Q(X)\rangle]^2 = \int_{-\infty}^{+\infty} \hat{S}_{QQ}(\nu)\mathrm{d}\nu \ ,$$

and therefore,

$$\int_{-\infty}^{+\infty} \hat{S}_{QQ}(\nu)\mathrm{d}\nu = k_{\mathrm{B}}TC \ , \quad \forall R \ .$$

This result implies that

$$\lim_{R\to 0}\int_{-\infty}^{+\infty} \hat{S}_{QQ}(\nu)\mathrm{d}\nu = k_{\mathrm{B}}TC \ ,$$

and hence,

$$\lim_{R\to 0}\int_{-\infty}^{+\infty} \hat{S}_{QQ}(\nu)\mathrm{d}\nu \neq \int_{-\infty}^{+\infty} \lim_{R\to 0}\hat{S}_{QQ}(\nu)\mathrm{d}\nu \ .$$

Indeed, the function

$$\hat{S}_{QQ}(\nu) = 2k_{\mathrm{B}}T\frac{RC^2}{1+(2\pi RC\nu)^2}$$

does not satisfy the mathematical conditions which would allow us to exchange the order of the integral and the limit.

9.6 Chapter Seven. Statistical Estimation

Solution to Exercise 7.1

If the domain of definition of X_λ does not depend on θ, then the variance of the statistic $T(\chi_\lambda)$ satisfies the Cramer–Rao inequality

$$\sigma_T^2(\theta) \geqslant \frac{\left| -\dfrac{\partial}{\partial \theta} h(\theta) \right|^2}{\displaystyle\int \frac{\partial^2 \ln L(\chi|\theta)}{\partial \theta^2} L(\chi|\theta) \mathrm{d}\chi},$$

where $h(\theta) = \langle T(\chi_\lambda) \rangle_\theta$. Therefore, when the estimator is unbiased, we have

$$\sigma_T^2(\theta) \geqslant -\frac{1}{\displaystyle\int \frac{\partial^2 \ln L(\chi|\theta)}{\partial \theta^2} L(\chi|\theta) \mathrm{d}\chi}.$$

We see that if $|\partial h(\theta)/\partial \theta| < 1$, in the case of a biased estimator, the Cramer–Rao bound may actually be less than the Cramer–Rao bound of an unbiased estimator. A trivial example of a biased estimator for which the bound is zero is provided by the choice of statistic $T(\chi_\lambda) = 0$.

Solution to Exercise 7.2

(1) We have $b + 2a = 1$ and hence $a = (1 - b)/2$.
(2) If $x = -1$ or $x = 1$, i.e., whenever $x^2 = 1$, we have $p(x) = a$. When $x = 0$ and thus when $1 - x^2 = 1$, $p(x) = b$. It follows that $\ln p(x) = x^2 \ln a + (1 - x^2) \ln b$, or

$$p(x) = \exp\left(\ln b + x^2 \ln \frac{1 - b}{2b} \right).$$

(3) Considering a sample $\chi = \{x_1, x_2, \ldots, x_P\}$, the log-likelihood can be written

$$\ell(\chi) = \sum_{i=1}^{P} \ln p(x_i) = P \ln b + T_2(\chi) \ln \frac{1 - b}{2b},$$

where $T_2(\chi) = \sum_{i=1}^{P} x_i^2$. It is clear that $p(x)$ belongs to the exponential family and that $T_2(\chi)$ is its sufficient statistic. The maximum likelihood estimator is obtained from

$$\frac{\partial}{\partial b} \ell(\chi) = 0 = \frac{P}{b} + T_2(\chi) \left(-\frac{1}{1 - b} - \frac{1}{b} \right).$$

This leads to

$$\hat{b}_{\mathrm{ML}}(\chi) = 1 - T_2(\chi)/P.$$

This estimator is unbiased because $\langle T_2(\chi) \rangle = 2Pa$ and hence $\langle \hat{b}_{\text{ML}}(\chi) \rangle = b$. We thus have an unbiased estimator which depends only on the sufficient statistic for a probability law in the exponential family. It therefore attains the minimal variance. Note that it must be efficient because it is proportional to $T_2(\chi)$, which is the statistic that can be efficiently estimated, i.e., its variance is equal to the Cramer–Rao bound.

Solution to Exercise 7.3

(1) For concreteness, consider a sample $\chi = \{x_1, x_2, \ldots, x_P\}$. For $P_A(x)$, the log-likelihood can be written

$$\ell_A(\chi) = \sum_{i=1}^{P} \ln P_A(x_i) = -P \ln(2a) - \frac{1}{a} T(\chi) \,,$$

where $T(\chi) = \sum_{i=1}^{P} |x_i|$. It is clear that $P_A(x)$ belongs to the exponential family and that $T(\chi)$ is its sufficient statistic.

For $P_B(x)$, the log-likelihood can be written

$$\ell_B(\chi) = \sum_{i=1}^{P} \ln P_B(x_i) = -P \ln 2 - \sum_{i=1}^{P} |x_i - a| \,.$$

We see that $P_B(x)$ does not belong to the exponential family.

(2) For $P_A(x)$, the maximum likelihood estimator of a is obtained with

$$\frac{\partial}{\partial a} \ell_A(\chi) = 0 = -\frac{P}{a} + \frac{1}{a^2} T(\chi) \,,$$

and hence,

$$\hat{a}_{\text{ML}}(\chi) = \frac{1}{P} T(\chi) \,.$$

This estimator is unbiased and depends only on the sufficient statistic $T(\chi)$ of a probability density function belonging to the exponential family. It thus attains the minimal variance. Moreover, since the estimator is proportional to the sufficient statistic which can be efficiently estimated, it is itself efficient.

Solution to Exercise 7.4

(1) For concreteness, consider a sample $\chi = \{x_1, x_2, \ldots, x_P\}$. For $P_X(x)$, the log-likelihood can be written

$$\ell(\chi) = \sum_{i=1}^{P} \ln P_X(x_i) = (n-1)T_1(\chi) + (p-1)T_2(\chi) - P \ln B(n,p) \,,$$

where $T_1(\chi) = \sum_{i=1}^{P} \ln x_i$ and $T_2(\chi) = \sum_{i=1}^{P} \ln(1 - x_i)$. It is clear that $P_X(x)$ belongs to the exponential family and that $T_1(\chi)$ and $T_2(\chi)$ are the sufficient statistics for n and p, respectively.

(2) We have

$$\frac{\partial}{\partial n}\ell(\chi) = 0 = T_1(\chi) - P\frac{\partial}{\partial n}\ln B(n,p) \,,$$

and hence,

$$\frac{\partial}{\partial n}\ln B(n,p) = T_1(\chi)/P \,.$$

Likewise for p, we find that

$$\frac{\partial}{\partial p}\ln B(n,p) = T_2(\chi)/P \,.$$

We do not obtain explicit expressions for n and p.

(3) The change of variable $y = x/(1-x)$ corresponds to a bijective transformation and therefore $P_Y(y)\mathrm{d}y = P_X(x)\mathrm{d}x$. Moreover,

$$x = \frac{y}{1+y} \quad \text{and} \quad \frac{\mathrm{d}x}{\mathrm{d}y} = \frac{1}{(1+y)^2} \,,$$

and hence,

$$P_Y(y) = \frac{1}{B(n,p)}\frac{y^{n-1}}{(1+y)^{n+p}} \,.$$

(4) Consider now the sample $\chi' = \{y_1, y_2, \ldots, y_P\}$. We have

$$\ell(\chi') = (n-1)T_3(\chi') - (n+p)T_4(\chi') - P\ln B(n,p) \,,$$

where $T_3(\chi') = \sum_{i=1}^{P}\ln y_i$ and $T_4(\chi') = \sum_{i=1}^{P}\ln(1+y_i)$. It is clear that $P_Y(y)$ belongs to the exponential family and that $T_3(\chi') - T_4(\chi')$ and $T_4(\chi')$ are the sufficient statistics for n and p, respectively. Since

$$\frac{\partial}{\partial n}\ell(\chi') = 0 = T_3(\chi') - T_4(\chi') - P\frac{\partial}{\partial n}\ln B(n,p) \,,$$

we have

$$\frac{\partial}{\partial n}\ln B(n,p) = [T_3(\chi') - T_4(\chi')]/P \,.$$

Likewise for p, we obtain

$$\frac{\partial}{\partial p}\ln B(n,p) = -T_4(\chi')/P \,.$$

We do not obtain explicit expressions for n and p.

Solution to Exercise 7.5

(1) We must have

$$P_X(x) = \begin{cases} 1/\theta & \text{if } x \in [0,\theta] \,, \\ 0 & \text{otherwise} \,. \end{cases}$$

(2) For concreteness, consider a sample $\chi = \{x_1, x_2, \ldots, x_P\}$. In this case,

$$\hat{\theta}_{MM}(\chi) = \frac{2}{P} \sum_{j=1}^{P} x_j \ .$$

This is an unbiased estimator of θ because $\langle \hat{\theta}_{MM}(\chi) \rangle = \theta$.

(3) The likelihood is

$$L(\chi|\theta) = \begin{cases} 1/\theta^n & \text{if } x_j \in [0, \theta], \quad \forall j = 1, \ldots, P \ , \\ 0 & \text{otherwise} \ . \end{cases}$$

$L(\chi|\theta)$ is therefore maximal if $x_j \in [0, \theta], \forall j = 1, \ldots, P$ and if θ is minimal, which implies that

$$\hat{\theta}_{ML}(\chi) = \sup_j x_j \ ,$$

where the notation means that we must choose the largest value of the x_j for θ.

(4) The uniform distribution is not in the exponential family and, in contrast to the situation where the probability law does belong to this family, we cannot assert that this estimator attains the minimal variance.

(5) We now have

$$P_X(x) = \begin{cases} 1/(2\theta) & \text{if } x \in [-\theta, \theta] \ , \\ 0 & \text{otherwise} \ . \end{cases}$$

(6) Consider once again a sample $\chi = \{x_1, x_2, \ldots, x_P\}$. In this case, the estimator

$$\hat{\theta}_{MM}(\chi) = \frac{1}{P} \sum_{j=1}^{P} x_j$$

is no longer an unbiased estimator of θ because $\langle \hat{\theta}_{MM}(\chi) \rangle = 0$. We can choose the estimator in the sense of the second order moment. We then have $\langle x^2 \rangle = \theta^2/3$.

$$\hat{\theta}^2_{MM'}(\chi) = \frac{3}{P} \sum_{j=1}^{P} x_j^2 \ ,$$

whence

$$\hat{\theta}_{MM'}(\chi) = \sqrt{\frac{3}{P} \sum_{j=1}^{P} x_j^2} \ .$$

Another possible choice would be

$$\hat{\theta}_{MM''}(\chi) = \frac{2}{P} \sum_{j=1}^{P} |x_j| \ ,$$

where $|x_j|$ is the absolute value of x_j. The choice between $\hat{\theta}_{MM'}(\chi)$ and $\hat{\theta}_{MM''}(\chi)$ can be made by comparing the bias and variance of each estimator.

(7) The likelihood is

$$L(\chi|\theta) = \begin{cases} 1/(2\theta)^n & \text{if } x_j \in [-\theta, \theta], \quad \forall j = 1, \ldots, P, \\ 0 & \text{otherwise}. \end{cases}$$

$L(\chi|\theta)$ is therefore maximal if $x_j \in [-\theta, \theta], \forall j = 1, \ldots, P$ and if θ is minimal, which implies that

$$\hat{\theta}_{\text{ML}}(\chi) = \sup_j |x_j|,$$

where the notation means that we must choose the largest value of the $|x_j|$ for θ.

Solution to Exercise 7.6

(1) The probability density function of X_λ is

$$P_X(x) = A(\sigma_0, c) \exp\left[-\frac{1}{2\sigma_0^2}(x - \theta)^2 - c(x - \theta)^4\right].$$

For concreteness, consider a sample $\chi = \{x_1, x_2, \ldots, x_P\}$. The estimator of the empirical mean is

$$\hat{\theta}_{MM}(\chi) = \frac{1}{P} \sum_{j=1}^{P} x_j.$$

It is an unbiased estimator of θ because the probability density function is symmetric with respect to θ. The Cramer–Rao bound can therefore be written

$$\text{CRB} = \frac{1}{I_F},$$

where

$$I_F = -P \left\langle \frac{\partial^2}{\partial \theta^2} \ln P_X(x) \right\rangle.$$

We have

$$\frac{\partial^2}{\partial \theta^2} \ln P_X(x) = -\frac{1}{\sigma_0^2} - 12c(x - \theta)^2,$$

and therefore,

$$I_F = P \left[\frac{1}{\sigma_0^2} + 12c\langle(x - \theta)^2\rangle\right].$$

We set $\sigma^2 = \langle(x - \theta)^2\rangle$ so that

$$I_F = P \left(\frac{1}{\sigma_0^2} + 12c\sigma^2\right),$$

and the Cramer–Rao bound is then

$$\text{CRB} = \frac{\sigma_0^2}{P(1 + 12c\sigma^2\sigma_0^2)}.$$

(2) When $c = 0$, we have $\sigma^2 = \sigma_0^2$. When $c > 0$, we then have CRB $< \sigma_0^2/P$. We can interpret this result by observing that, as c increases, the probability density of X_λ concentrates around θ, and this leads to a lower CRB.

9.7 Chapter Eight. Examples of Estimation in Physics

Solution to Exercise 8.1

(1) We must have $\sum_{n=0}^{+\infty} P(n) = 1$. $P(n) \geq 0$ implies that $a \geq 0$. $\sum_{n=0}^{+\infty} P(n) = 1$ then implies $a > 0$. However, $\sum_{n=0}^{+\infty} P(n) < \infty$ implies that $a < 1$. We deduce that

$$0 < a < 1 .$$

Further, in this case, $\sum_{n=0}^{+\infty} a^n = 1/(1-a)$ and thus $A = 1 - a$.

(2) We can write

$$P(n) = \exp\left[n \ln a + \ln(1-a)\right] ,$$

which shows that $P(n)$ belongs to the exponential family. Considering a sample $\chi = \{n_1, n_2, \ldots, n_P\}$, the log-likelihood can be written

$$\ell(\chi) = \sum_{j=1}^{P} n_j \ln a + P \ln(1-a) ,$$

and the maximum likelihood estimator is then obtained from

$$\frac{\partial}{\partial a} \ell(\chi) = 0 = \sum_{j=1}^{P} \frac{n_j}{a} - \frac{P}{1-a} ,$$

or

$$\left[\frac{a}{1-a}\right]_{\mathrm{ML}} (\chi) = \frac{1}{P} \sum_{j=1}^{P} n_j .$$

$T(\chi) = \sum_{j=1}^{P} n_j$ is the sufficient statistic for the probability law. Only statistics proportional to this one can be efficiently estimated. It is easy to show that

$$\left\langle \frac{1}{P} \sum_{j=1}^{P} n_j \right\rangle = \frac{a}{1-a} .$$

We thus find that $f(a) = a/(1-a)$ is a function of a that can be estimated without bias and efficiently.

Solution to Exercise 8.2

Let x_i^ℓ be the gray level of the ith pixel in the region under consideration and in the image labeled by ℓ. We have

$$P_\theta(x_i^\ell) = \frac{\alpha^\alpha (x_i^\ell)^{\alpha-1}}{\theta^\alpha \Gamma(\alpha)} \exp\left(-\frac{\alpha x_i^\ell}{\theta}\right) .$$

The log-likelihood is thus

$$L = \sum_{\ell=1}^{M} \sum_{i=1}^{P} \left(-\frac{\alpha x_i^\ell}{\theta}\right) + (\alpha-1) \sum_{\ell=1}^{M} \sum_{i=1}^{P} \ln x_i^\ell + MP\left[\alpha \ln \alpha - \alpha \ln \theta - \ln \Gamma(\alpha)\right].$$

The maximum likelihood estimator for θ is therefore

$$\hat{\theta}_{\mathrm{ML}} = \frac{1}{MP} \sum_{\ell=1}^{M} \sum_{i=1}^{P} x_i^\ell .$$

We saw in Section 7.10.2 that this estimator is efficient. Moreover, it can be written

$$\hat{\theta}_{\mathrm{ML}} = \frac{1}{P} \sum_{i=1}^{P} y_i , \quad \text{with} \quad y_i = \frac{1}{M} \sum_{\ell=1}^{M} x_i^\ell ,$$

which shows that it is enough to know the mean of the M images.

Solution to Exercise 8.3

Let x_i^ℓ be the gray level of the ith pixel in the region under consideration and in the image labeled by ℓ. We have

$$P_\theta(x_i^\ell) = \frac{\alpha (x_i^\ell)^{\alpha-1}}{\theta^\alpha} \exp\left[-\left(\frac{x_i^\ell}{\theta}\right)^\alpha\right] .$$

The log-likelihood is thus

$$L = -\sum_{\ell=1}^{M} \sum_{i=1}^{P} \left(\frac{x_i^\ell}{\theta}\right)^\alpha + (\alpha-1) \sum_{\ell=1}^{M} \sum_{i=1}^{P} \ln x_i^\ell + MP\left(\ln \alpha - \alpha \ln \theta\right) .$$

The maximum likelihood estimator of θ is therefore

$$\hat{\theta}_{\mathrm{ML}} = \left[\frac{1}{MP} \sum_{\ell=1}^{M} \sum_{i=1}^{P} (x_i^\ell)^\alpha\right]^{1/\alpha} .$$

We see that, rather than averaging the gray levels of the M images, we must average the gray levels to the power of α.

Solution to Exercise 8.4

Considering a sample $\chi = \{x_1, x_2, \ldots, x_P\}$ as usual, the log-likelihood can be written

$$\ell(\chi) = -\sum_{j=1}^{P} \frac{(\ln x_j - m)^2}{2\sigma^2} - P \ln(x_j \sigma \sqrt{2\pi}) .$$

This result shows that $P_X(x)$ belongs to the exponential family and that $T(\chi) = \sum_{j=1}^{P} \ln x_j$ is a sufficient statistic for m. The maximum likelihood estimator of m is obtained from

$$\frac{\partial}{\partial m} \ell(\chi) = 0 = -\sum_{j=1}^{P} \frac{m - \ln x_j}{\sigma^2} ,$$

and hence,

$$\hat{m}_{\mathrm{ML}}(\chi) = \frac{1}{P} \sum_{j=1}^{P} \ln x_j .$$

We thus have

$$\langle \hat{m}_{\mathrm{ML}}(\chi) \rangle = \langle \ln x_j \rangle$$
$$= \int_0^{+\infty} \frac{\ln x}{x \sigma \sqrt{2\pi}} \exp\left[-\frac{(\ln x - m)^2}{2\sigma^2}\right] \mathrm{d}x .$$

Setting $y = \ln x$, we have $\mathrm{d}y = \mathrm{d}x/x$ and hence,

$$\langle \hat{m}_{\mathrm{ML}}(\chi) \rangle = \int_0^{+\infty} \frac{y}{\sigma \sqrt{2\pi}} \exp\left[-\frac{(y - m)^2}{2\sigma^2}\right] \mathrm{d}y ,$$

or $\langle \hat{m}_{\mathrm{ML}}(\chi) \rangle = m$. It follows that $\hat{m}_{\mathrm{ML}}(\chi)$ is an unbiased estimator of m which depends only on the sufficient statistic of a random variable distributed according to a density function in the exponential family. $\hat{m}_{\mathrm{ML}}(\chi)$ thus attains the minimal variance. Moreover, since this estimator is proportional to the sufficient statistic, it must be efficient.

Solution to Exercise 8.5

The probability density function of the measurement x_i is

$$P_{a,\varphi}(x_i)) = \frac{1}{\sqrt{2\pi}\sigma} \exp\left\{-\frac{[x_i - a\sin(\omega t_i + \varphi)]^2}{2\sigma^2}\right\} ,$$

where $\omega = 2\pi/T$ and $t_i = iT/N$. With $\chi = \{x_1, x_2, \ldots, x_N\}$, the log-likelihood is

$$\ell(\chi) = -\frac{1}{2\sigma^2} \sum_{i=1}^{N} \left[x_i - a \sin(\omega t_i + \varphi) \right]^2 - N \ln(\sqrt{2\pi}\sigma) .$$

We thus have

$$\frac{\partial}{\partial a}\ell(\chi) = -\frac{1}{\sigma^2} \sum_{i=1}^{N} \sin(\omega t_i + \varphi)\left[a \sin(\omega t_i + \varphi) - x_i \right] ,$$

$$\frac{\partial^2}{\partial a^2}\ell(\chi) = -\frac{1}{\sigma^2} \sum_{i=1}^{N} \sin^2(\omega t_i + \varphi) ,$$

$$\frac{\partial}{\partial \varphi}\ell(\chi) = -\frac{1}{\sigma^2} \sum_{i=1}^{N} a \cos(\omega t_i + \varphi)\left[a \sin(\omega t_i + \varphi) - x_i \right] ,$$

$$\frac{\partial^2}{\partial \varphi^2}\ell(\chi) = \frac{1}{\sigma^2} \sum_{i=1}^{N} a \sin(\omega t_i + \varphi)\left[a \sin(\omega t_i + \varphi) - x_i \right]$$
$$-\frac{1}{\sigma^2} \sum_{i=1}^{N} a^2 \cos^2(\omega t_i + \varphi) ,$$

$$\frac{\partial}{\partial \varphi \partial a}\ell(\chi) = -\frac{1}{\sigma^2} \sum_{i=1}^{N} \cos(\omega t_i + \varphi)\left[a \sin(\omega t_i + \varphi) - x_i \right]$$
$$-\frac{1}{\sigma^2} \sum_{i=1}^{N} a \cos(\omega t_i + \varphi) \sin(\omega t_i + \varphi) .$$

We have $\langle x_i \rangle = a \sin(\omega t_i + \varphi)$, so we deduce that

$$\left\langle \frac{\partial^2}{\partial a^2}\ell(\chi) \right\rangle = -\frac{1}{\sigma^2} \sum_{i=1}^{N} \sin^2(\omega t_i + \varphi) ,$$

$$\left\langle \frac{\partial^2}{\partial \varphi^2}\ell(\chi) \right\rangle = -\frac{1}{\sigma^2} \sum_{i=1}^{N} a^2 \cos^2(\omega t_i + \varphi) ,$$

$$\left\langle \frac{\partial}{\partial \varphi \partial a}\ell(\chi) \right\rangle = -\frac{1}{\sigma^2} \sum_{i=1}^{N} a \cos(\omega t_i + \varphi) \sin(\omega t_i + \varphi) .$$

However, we have

$$\sum_{i=1}^{N} \sin^2(\omega t_i + \varphi) = \sum_{i=1}^{N} \sin^2(2\pi i/N + \varphi) = \frac{N}{2} ,$$

$$\sum_{i=1}^{N} \cos^2(\omega t_i + \varphi) = \sum_{i=1}^{N} \cos^2(2\pi i/N + \varphi) = \frac{N}{2} ,$$

and

$$\sum_{i=1}^{N} \cos(\omega t_i + \varphi) \sin(\omega t_i + \varphi) = \sum_{i=1}^{N} \cos(2\pi i/N + \varphi) \sin(2\pi i/N + \varphi) = 0 \, ,$$

which leads to

$$\left\langle \frac{\partial^2}{\partial a^2} \ell(\chi) \right\rangle = -\frac{N}{2\sigma^2} \, , \quad \left\langle \frac{\partial^2}{\partial \varphi^2} \ell(\chi) \right\rangle = -\frac{a^2 N}{2\sigma^2} \, , \quad \left\langle \frac{\partial}{\partial \varphi \partial a} \ell(\chi) \right\rangle = 0 \, .$$

The Fisher information matrix is therefore diagonal:

$$\overline{\overline{J}} = \begin{pmatrix} \dfrac{N}{2\sigma^2} & 0 \\ 0 & \dfrac{a^2 N}{2\sigma^2} \end{pmatrix} \, .$$

Its inverse is then

$$\overline{\overline{J}}^{-1} = \begin{pmatrix} \dfrac{2\sigma^2}{N} & 0 \\ 0 & \dfrac{2\sigma^2}{a^2 N} \end{pmatrix} \, .$$

In the vector case, the Cramer–Rao bound is given by

$$\boldsymbol{u}^{\mathrm{T}} \overline{\overline{C}} \boldsymbol{u} \geq \boldsymbol{u}^{\mathrm{T}} \overline{\overline{J}}^{-1} \boldsymbol{u} \, ,$$

for all $\boldsymbol{u}$, where

$$\overline{\overline{C}} = \begin{pmatrix} \langle [\delta \hat{a}(\chi)]^2 \rangle & \langle \delta \hat{a}(\chi) \delta \hat{\varphi}(\chi) \rangle \\ \langle \delta \hat{a}(\chi) \delta \hat{\varphi}(\chi) \rangle & \langle [\delta \hat{\varphi}(\chi)]^2 \rangle \end{pmatrix} \, ,$$

with $\delta \hat{a}(\chi) = \hat{a}(\chi) - a$ and $\delta \hat{\varphi}(\chi) = \hat{\varphi}(\chi) - \varphi$, because we have assumed that the estimators are unbiased and hence, $\langle \hat{a}(\chi) \rangle = a$ and $\langle \hat{\varphi}(\chi) \rangle = \varphi$. We deduce that

$$\langle [\delta \hat{a}(\chi)]^2 \rangle \geq \frac{2\sigma^2}{N} \quad \text{and} \quad \langle [\delta \hat{\varphi}(\chi)]^2 \rangle \geq \frac{2\sigma^2}{a^2 N} \, .$$

Solution to Exercise 8.6

(1) We have

$$P_X(x) = \frac{1}{I_X} \exp\left(-\frac{x}{I_X} \right) \quad \text{and} \quad P_Y(y) = \frac{1}{I_Y} \exp\left(-\frac{y}{I_Y} \right) \, .$$

We set $Z_i = X_i/Y_i$. In an analogous way to Exercise 2.10, we have

$$P_Z(z) = \int_0^{+\infty} \int_0^{+\infty} \delta\left(z - \frac{x}{y} \right) P_X(x) P_Y(y) \mathrm{d}x \mathrm{d}y \, .$$

We set $\nu = x/y$, and then

$$P_Z(z) = \int_0^{+\infty} \int_0^{+\infty} y\delta(z - \nu)P_X(y\nu)P_Y(y)\mathrm{d}\nu\mathrm{d}y$$

$$= \int_0^{+\infty} yP_X(yz)P_Y(y)\mathrm{d}y \ .$$

Using the expressions for $P_X(x)$ and $P_Y(y)$, we obtain

$$P_Z(z) = \frac{1}{I_X I_Y} \int_0^{+\infty} y \exp\left(-\frac{yz}{I_X} - \frac{y}{I_Y}\right)\mathrm{d}y$$

$$= \frac{I_X I_Y}{(I_Y z + I_X)^2} \ .$$

We have $\rho = (z - 1)(z + 1)$, and it is therefore a bijective function of z. We can thus write

$$P_\rho(\rho)\mathrm{d}\rho = P_Z(z)\mathrm{d}z \ .$$

This implies that

$$P_\rho(\rho) = \frac{2I_X I_Y}{\left[I_X + I_Y + \rho(I_Y - I_X)\right]^2} \ ,$$

which can be written more simply as a function of u:

$$P_\rho(\rho) = \frac{1 - u^2}{2(1 - u\rho)^2} \ .$$

The log-likelihood of a P-sample $\chi = \{\rho_1, \rho_2, \ldots, \rho_P\}$ is

$$\ell(\chi) = P\ln\frac{1 - u^2}{2} - 2\sum_{i=1}^{P} \ln(1 - u\rho_i) \ .$$

We thus see that this probability density function does not belong to the exponential family when u is the parameter of the probability law.

(2) We have $\beta = \ln z$, which is a bijective transformation, so that we can write

$$P_\beta(\beta)\mathrm{d}\beta = P_Z(z)\mathrm{d}z \ .$$

We deduce that

$$P_\beta(\beta) = \frac{I_X I_Y \exp\beta}{(I_Y \exp\beta + I_X)^2}$$

$$= \left[\sqrt{\frac{I_Y}{I_X}}\exp(\beta/2) + \sqrt{\frac{I_X}{I_Y}}\exp(-\beta/2)\right]^{-2} \ ,$$

which can be written in terms of $\gamma = \ln(I_X/I_Y)$:

$$P_\beta(\beta) = \Phi(\beta - \gamma) \quad\text{with}\quad \Phi(\eta) = \left[\exp\left(\frac{\eta}{2}\right) + \exp\left(-\frac{\eta}{2}\right)\right]^{-2} \ .$$

The log-likelihood of a P-sample $\chi = \{\beta_1, \beta_2, \ldots, \beta_P\}$ is

$$\ell(\chi) = -2 \sum_{i=1}^{P} \ln \left[\exp\left(\frac{\beta_i - \gamma}{2} \right) + \exp\left(-\frac{\beta_i - \gamma}{2} \right) \right] .$$

We thus find that this probability density function does not belong to the exponential family when γ is the parameter of the probability law.

(3) Note that the probability density function for β is symmetric about γ with a finite first moment. The estimator $\hat{\gamma}(\chi) = (1/P) \sum_{i=1}^{P} \beta_i$ of the first order moment is therefore unbiased.

(4) The log-likelihood is

$$\ell(\chi) = -2 \sum_{i=1}^{P} \ln \left[\varphi(\beta_i - \gamma) \right] ,$$

where $\varphi(x) = \exp(x/2) + \exp(-x/2)$. We thus have

$$\frac{\partial}{\partial \gamma} \ell(\chi) = -2 \sum_{i=1}^{P} \frac{\varphi'(\beta_i - \gamma)}{\varphi(\beta_i - \gamma)} ,$$

where $\varphi'(x) = \mathrm{d}\varphi(x)/\mathrm{d}x = (1/2) \left[\exp(x/2) - \exp(-x/2) \right]$. Likewise,

$$\frac{\partial^2}{\partial \gamma^2} \ell(\chi) = -2 \sum_{i=1}^{P} \frac{\varphi(\beta_i - \gamma) \varphi''(\beta_i - \gamma) - \left[\varphi'(\beta_i - \gamma) \right]^2}{\left[\varphi(\beta_i - \gamma) \right]^2} ,$$

where $\varphi''(x) = \mathrm{d}^2\varphi(x)/\mathrm{d}x^2 = (1/4) \left[\exp(x/2) + \exp(-x/2) \right]$. We then obtain

$$\frac{\partial^2}{\partial \gamma^2} \ell(\chi) = -\frac{1}{2} \sum_{i=1}^{P} \frac{\left[e^{(\beta_i - \gamma)/2} + e^{-(\beta_i - \gamma)/2} \right]^2 - \left[e^{(\beta_i - \gamma)/2} - e^{-(\beta_i - \gamma)/2} \right]^2}{\left[e^{(\beta_i - \gamma)/2} + e^{-(\beta_i - \gamma)/2} \right]^2}$$

$$= -2 \sum_{i=1}^{P} \frac{1}{\left[\exp\left(\frac{\beta_i - \gamma}{2} \right) + \exp\left(-\frac{\beta_i - \gamma}{2} \right) \right]^2} .$$

Taking the expectation value,

$$\left\langle \frac{\partial^2}{\partial \gamma^2} \ell(\chi) \right\rangle = -2P \int_{-\infty}^{+\infty} \frac{1}{\left[\exp\left(\frac{\beta - \gamma}{2} \right) + \exp\left(-\frac{\beta - \gamma}{2} \right) \right]^4} \mathrm{d}\beta$$

$$= -2P \int_{-\infty}^{+\infty} \frac{1}{\left[\exp\left(\frac{x}{2} \right) + \exp\left(-\frac{x}{2} \right) \right]^4} \mathrm{d}x ,$$

whereupon

$$\mathrm{CRB} = \frac{1}{2PI} .$$

Solution to Exercise 8.7

(1) The noise is additive Gaussian and we can write

$$P(x_i) = \frac{1}{\sqrt{2\pi}\sigma} \exp\left[-\frac{1}{2\sigma^2}(x_i - i\theta)^2\right] \ .$$

The log-likelihood of the sample χ is therefore

$$\ell(\chi) = -\frac{1}{2\sigma^2} \sum_{i=1}^{P}(x_i - i\theta)^2 - P\ln\left(\sqrt{2\pi}\sigma\right) \ .$$

The derivative is then

$$\frac{\partial}{\partial\theta}\ell(\chi) = \frac{1}{\sigma^2} \sum_{i=1}^{P}(ix_i - i^2\theta) \ .$$

The maximum likelihood estimator is obtained by setting the derivative equal to zero, whereupon

$$\hat{\theta}_{\mathrm{ML}}(\chi) = \frac{\sum_{i=1}^{P} ix_i}{\sum_{i=1}^{P} i^2} \ .$$

We have

$$\langle\hat{\theta}_{\mathrm{ML}}(\chi)\rangle = \frac{\sum_{i=1}^{P} i\langle x_i\rangle}{\sum_{i=1}^{P} i^2} \ ,$$

and since $\langle x_i\rangle = i\theta$, we obtain

$$\langle\hat{\theta}_{\mathrm{ML}}(\chi)\rangle = \theta \ ,$$

which shows that the maximum likelihood estimator is unbiased. The second derivative of the log-likelihood is

$$\frac{\partial^2}{\partial\theta^2}\ell(\chi) = -\frac{1}{\sigma^2}\sum_{i=1}^{P} i^2 \ ,$$

so that the Fisher information is $I_F = (1/\sigma^2)\sum_{i=1}^{P} i^2$, and the Cramer–Rao bound for estimation of θ is

$$\sigma^2(\theta) \geq \frac{\sigma^2}{\sum_{i=1}^{P} i^2} \ .$$

(2) The noise is Poisson noise and we can write

$$P(x_i) = \exp(-i\theta)\frac{(i\theta)^{x_i}}{x_i!} \ .$$

The log-likelihood of the sample χ is therefore

$$\ell(\chi) = \sum_{i=1}^{P} x_i \ln i + \ln \theta \sum_{i=1}^{P} x_i - \theta \sum_{i=1}^{P} i - P \ln(x_i!) .$$

The derivative is

$$\frac{\partial}{\partial \theta} \ell(\chi) = \frac{1}{\theta} \sum_{i=1}^{P} x_i - \sum_{i=1}^{P} i .$$

The maximum likelihood estimator is obtained by setting the derivative equal to zero, whereupon

$$\hat{\theta}_{\mathrm{ML}}(\chi) = \frac{\sum_{i=1}^{P} x_i}{\sum_{i=1}^{P} i} .$$

We have

$$\langle \hat{\theta}_{\mathrm{ML}}(\chi) \rangle = \frac{\sum_{i=1}^{P} \langle x_i \rangle}{\sum_{i=1}^{P} i} ,$$

and since $\langle x_i \rangle = i\theta$, we obtain

$$\langle \hat{\theta}_{\mathrm{ML}}(\chi) \rangle = \theta ,$$

which indicates that the maximum likelihood estimator is unbiased. The second derivative of the log-likelihood is

$$\frac{\partial^2}{\partial \theta^2} \ell(\chi) = -\frac{1}{\theta^2} \sum_{i=1}^{P} x_i ,$$

which gives the Fisher information $I_F = \left(\sum_{i=1}^{P} i \right)/\theta$, whilst the Cramer–Rao bound for estimation of θ is

$$\sigma^2(\theta) \geq \frac{\theta}{\sum_{i=1}^{P} i} .$$

(3) The noise is now Gamma noise and we can write

$$P(x_i) = \frac{x_i^{\alpha-1}}{\Gamma(\alpha)(i\theta)^\alpha} \exp\left(-\frac{x_i}{i\theta} \right) .$$

The log-likelihood of the sample χ is thus

$$\ell(\chi) = -\frac{1}{\theta} \sum_{i=1}^{P} \frac{x_i}{i} + (\alpha - 1) \sum_{i=1}^{P} \ln x_i - P\alpha \ln \theta - \alpha \sum_{i=1}^{P} \ln i - P \ln \Gamma(\alpha) .$$

The derivative is then

$$\frac{\partial}{\partial \theta} \ell(\chi) = \frac{1}{\theta^2} \sum_{i=1}^{P} \frac{x_i}{i} - \frac{P\alpha}{\theta} \,.$$

The maximum likelihood estimator is obtained by setting the derivative equal to zero, whereupon

$$\hat{\theta}_{\mathrm{ML}}(\chi) = \frac{1}{\alpha P} \sum_{i=1}^{P} \frac{x_i}{i} \,.$$

We have

$$\langle \hat{\theta}_{\mathrm{ML}}(\chi) \rangle = \frac{1}{\alpha P} \sum_{i=1}^{P} \frac{\langle x_i \rangle}{i} \,,$$

and since $\langle x_i \rangle = i\alpha\theta$, we obtain

$$\langle \hat{\theta}_{\mathrm{ML}}(\chi) \rangle = \theta \,,$$

which shows that the maximum likelihood estimator is unbiased. The second derivative of the log-likelihood is

$$\frac{\partial^2}{\partial \theta^2} \ell(\chi) = -\frac{2}{\theta^3} \sum_{i=1}^{P} \frac{x_i}{i} + \frac{P\alpha}{\theta^2} \,,$$

and hence,

$$\left\langle \frac{\partial^2}{\partial \theta^2} \ell(\chi) \right\rangle = -\frac{2P\alpha}{\theta^2} + \frac{P\alpha}{\theta^2} = -\frac{P\alpha}{\theta^2} \,.$$

The Fisher information is $I_F = \alpha P/\theta^2$ and the Cramer–Rao bound for estimation of θ is therefore

$$\sigma^2(\theta) \geq \frac{\theta^2}{\alpha P} \,.$$

(4) The various maximum likelihood estimators can be written as linear combinations of the measurements. We consider the three cases below.

Gaussian Case:

$$\hat{\theta}_{\mathrm{G}}(\chi) = \sum_{i=1}^{P} a_i^{\mathrm{G}} x_i \,, \quad \text{where} \quad a_i^{\mathrm{G}} = \frac{i}{\sum_{i=1}^{P} i^2} \,.$$

Poisson Case:

$$\hat{\theta}_{\mathrm{P}}(\chi) = \sum_{i=1}^{P} a_i^{\mathrm{P}} x_i \,, \quad \text{where} \quad a_i^{\mathrm{P}} = \frac{1}{\sum_{i=1}^{P} i} \,.$$

Gamma Case:

$$\hat{\theta}_\Gamma(\chi) = \sum_{i=1}^{P} a_i^\Gamma x_i \,, \quad \text{where} \quad a_i^\Gamma = \frac{1}{\alpha P i} \,.$$

The variance of the estimators can therefore be written

$$\langle \delta\theta(\chi)^2 \rangle = \sum_{i=1}^{P} a_i^2 \langle \delta x_i^2 \rangle \,,$$

where $\delta\theta(\chi) = \theta(\chi) - \langle\theta(\chi)\rangle$ and $\delta x_i = x_i - \langle x_i \rangle$. We consider the three cases in turn.

Gaussian Case:

$\langle \delta x_i^2 \rangle_G = \sigma^2$ and hence,

$$\langle \delta\hat{\theta}_G(\chi)^2 \rangle_G = \frac{1}{\left(\sum_{i=1}^{P} i^2\right)^2} \sum_{i=1}^{P} i^2 \sigma^2 \,,$$

where $\langle\,\rangle_G$ indicates that averages are taking under the assumption that the variables are Gaussian. We then obtain

$$\langle \delta\hat{\theta}_G(\chi)^2 \rangle_G = \frac{\sigma^2}{\sum_{i=1}^{P} i^2} \,.$$

The variance of the estimator is equal to the Cramer–Rao bound and the estimator is therefore efficient.

Poisson Case:

$\langle \delta x_i^2 \rangle_P = i\theta$ and hence,

$$\langle \delta\hat{\theta}_P(\chi)^2 \rangle_P = \frac{1}{\left(\sum_{i=1}^{P} i\right)^2} \sum_{i=1}^{P} i\theta \,,$$

where $\langle\,\rangle_P$ indicates that averages are taking under the assumption that we have Poisson variables. We thus obtain

$$\langle \delta\hat{\theta}_P(\chi)^2 \rangle_P = \frac{\theta}{\sum_{i=1}^{P} i} \,.$$

The variance of the estimator is equal to the Cramer–Rao bound and the estimator is therefore efficient.

Gamma Case:

$\langle \delta x_i^2 \rangle_\Gamma = \alpha i^2 \theta^2$ and hence,

$$\langle \delta \hat{\theta}_\Gamma(\chi)^2 \rangle_\Gamma = \frac{1}{\alpha^2 P^2} \sum_{i=1}^{P} \alpha \theta^2 \ ,$$

where $\langle \ \rangle_\Gamma$ indicates that averages are taking under the assumption that we have Gamma variables. We thus obtain

$$\langle \delta \hat{\theta}_\Gamma(\chi)^2 \rangle_\Gamma = \frac{\theta^2}{\alpha P} \ .$$

The variance of the estimator is equal to the Cramer–Rao bound and the estimator is therefore efficient.

(5) It is easy to check that $\hat{\theta}_G(\chi)$ remains unbiased if the data are distributed according to a Poisson law. We have $\langle \delta x_i^2 \rangle_P = i\theta$, and the variance of the estimator is therefore

$$\langle \delta \hat{\theta}_G(\chi)^2 \rangle_P = \frac{1}{\left(\sum_{i=1}^{P} i^2 \right)^2} \sum_{i=1}^{P} i^2 i\theta = \theta \frac{\sum_{i=1}^{P} i^3}{\left(\sum_{i=1}^{P} i^2 \right)^2} \ .$$

We thus have

$$\frac{\langle \delta \hat{\theta}_G(\chi)^2 \rangle_P}{\langle \delta \hat{\theta}_P(\chi)^2 \rangle_P} = \frac{\left(\sum_{i=1}^{P} i^3 \right) \left(\sum_{i=1}^{P} i \right)}{\left(\sum_{i=1}^{P} i^2 \right)^2} \ .$$

It is easy to show that

$$\frac{\left(\sum_{i=1}^{P} i^3 \right) \left(\sum_{i=1}^{P} i \right)}{\left(\sum_{i=1}^{P} i^2 \right)^2} \geq 1 \ ,$$

and we thus deduce that the least squares estimator is not efficient, unlike the maximum likelihood estimator, because its variance is greater.

(6) We have

$$\langle \hat{\theta}_G(\chi) \rangle_\Gamma = \frac{1}{\sum_{i=1}^{P} i^2} \sum_{i=1}^{P} \langle i x_i \rangle_\Gamma \ .$$

In the Gamma case, $\langle x_i \rangle_\Gamma = \alpha i \theta$ and hence, $\langle \hat{\theta}_G(\chi) \rangle_\Gamma = \alpha \theta$. It is better to consider the unbiased estimator

$$\hat{\theta}_G(\chi) = \frac{1}{\alpha \sum_{i=1}^{P} i^2} \sum_{i=1}^{P} i x_i \ .$$

We have $\langle \delta x_i^2 \rangle_\Gamma = \alpha i^2 \theta^2$ and hence,

$$\langle\delta\hat{\theta}_G(\chi)^2\rangle_\Gamma = \frac{1}{\alpha^2\left(\sum_{i=1}^{P} i^2\right)^2}\sum_{i=1}^{P} i^4\alpha\theta^2 = \frac{\theta^2}{\alpha}\frac{\sum_{i=1}^{P} i^4}{\left(\sum_{i=1}^{P} i^2\right)^2}.$$

We thus have

$$\frac{\langle\delta\hat{\theta}_G(\chi)^2\rangle_\Gamma}{\langle\delta\hat{\theta}_\Gamma(\chi)^2\rangle_\Gamma} = \frac{P\sum_{i=1}^{P} i^4}{\left(\sum_{i=1}^{P} i^2\right)^2}.$$

It is easy to show that

$$\frac{P\sum_{i=1}^{P} i^4}{\left(\sum_{i=1}^{P} i^2\right)^2} \geq 1,$$

and we may thus deduce that the least squares estimator is not efficient, unlike the maximum likelihood estimator, because its variance is greater.

References

1. N. Boccara: *Les Principes de la Thermodynamique* (Presses Universitaires de France, Paris, 1968)
2. N. Boccara: *Functional Analysis: An Introduction for Physicists* (Academic Press, Boston, 1990)
3. C. Brosseau: *Fundamentals of Polarized Light: A Statistical Optics Approach* (Wiley & Sons, New York, 1998)
4. H.B. Callen: *Thermodynamics and an Introduction to Thermostatistics*, 2nd edn. (John Wiley, New York, 1985)
5. T.M. Cover, J.A. Thomas: *Elements of Information Theory* (John Wiley, New York, 1991)
6. W. Feller: *An Introduction to Probability Theory and its Applications* (John Wiley, New York, 1960)
7. T. Ferguson: *Probability and Mathematical Statistics* (Academic Press, New York and London, 1967)
8. B.R. Frieden: *Probability, Statistical Optics and Data Testing* (Springer-Verlag, Berlin, Heidelberg, New York, 2001)
9. C.W. Gardiner: *Handbook of Stochastic Methods* (Springer-Verlag, New York, 1983)
10. J.W. Goodman: *Introduction to Fourier Optics* (McGraw-Hill, New York, 1968)
11. J.W. Goodman: *Statistical Optics* (John Wiley, New York, 1985)
12. S. Huard: *Polarization of Light* (John Wiley, New York, 1997)
13. C. Kittel, H. Kroemer: *Thermal Physics* (John Wiley, New York, 1969)
14. E.L. Lehmann: *Testing Statistical Hypotheses* (John Wiley, New York, 1970)
15. L. Landau, E. Lifchitz, L.P. Pitaevskii: *Statistical Physics*, Course of Theoretical Physics Vol. 5, 3rd edn. (Butterworth–Heinemann, 1999)
16. L. Mandel, E. Wolf: *Optical Coherence and Quantum Optics* (Cambridge University Press, Cambridge, 1995)
17. A. Papoulis: *Probability, Random Variables and Stochastic Processes* (McGraw-Hill, New York, 1965)
18. P. Réfrégier: *Théorie du Signal: Signal, information, fluctuations* (Masson, Paris, 1993)
19. C.E. Shannon: 'A Mathematical Theory of Communication', Bell Syst. Tech. J. **27**, 379–423 and 623–656 (1948)
20. H.L. Van Trees: *Detection, Estimation and Modulation Theory* (John Wiley, New York, 1968)

Index

absolute temperature, 141
algorithmic randomness, 118
asymptotically umbiaised estimator, 171

Bayes' relation, 14
bias, 170

canonical parameter, 180
Cauchy–Schwartz inequality, 197
centered correlation function, 30
central limit theorem, 76
central moment, 11
Chapman–Kolmogorov equation, 93
characteristic function, 74
coherency matrix, 58
coherent stochastic field, 28
conditional probability density function, 15
conjugate quantities, 46
correlation coefficient, 17
correlation function, 30
    centered, 30
correlation length, 145
covariance, 16
covariance function, normalised, 28
covariance matrix, 20, 56
Cramer–Rao bound, 178
cyclostationarity, 129
    weak, 35

diffusion equation, 95
distribution function, 9, 20

efficient estimator, 179
entropy, 111
    of continuous random variable, 123
ergodicity in statistical physics, 32
ergodicity, weak, 30
estimation, 167
estimator, variance of, 172
expectation value, 11
exponential family, 180
extensive quantity, 46

Fisher information, 178
    matrix, 182
fluctuation–dissipation theorem, 152
    in Fourier space, 152
free energy, 141

Gibbs canonical distribution, 137
Green function, 53
    diffusion equation, 95

Hessian, 120
homogeneous partial differential equation, 55
homogeneous random walk, 90
homogeneous stochastic field, 34

impulse response, 47
independent random variable, 16
instantaneous moments, 29
instantaneous power of fluctuations, 45
instantaneous Stokes parameters, 223
intensive quantity, 46
internal energy, 141
isotropic stochastic field, 34

joint distribution function, 15
joint probability density function, 15
joint probability law, 16

kernel of a convolution, 47
Kolmogorov complexity, 114
Kullback–Leibler measure, 131

Lagrange multiplier method, 119, 133
likelihood, 170
linear filter, 47
linear system, 46

marginal probability law, 16
maximum likelihood estimator, 175
mean value, 11
    at equilibrium, 142
median, 12
mode, 13

natural parameter, 180

optical coherence, 41
outlier, 195

partition function, 140
phase space, 32, 138
Poisson distribution, 100
power spectral density, 44
probability density function, 9, 20
    of a sum variable, 74

quantity of information, 110

random walk, 89
    homogeneous, 90
    stationary, 90
    without memory, 89

response function, 48
robust estimator, 192

sample, 169
spatial coherence, 28
speckle noise, 81
stable probability law, 81
standard deviation, 11
static susceptibility, 142
stationarity, weak, 29
stationary partial differential equation,
    54
stationary random walk, 90
stationary system, 46
statistic, 169
stochastic field, 27
    coherent, 28
    isotropic, 34
stochastic process, 26
stochastic vector, 19
Stokes parameters, 223
sufficient statistic, 184
susceptibility, 47

temperature in kelvins, 140
temporal coherence, 28
thermodynamic equilibrium, 138
time average, 29
total fluctuation theorem, 146

variance, 11
    of an estimator, 172

weak ergodicity, 30
weak law of large numbers, 12
weak stationarity, 29
white noise with bounded spectrum, 45
Wiener–Khinchine theorem, 44, 65